1 MONTH OF
FREE
READING

at

www.ForgottenBooks.com

By purchasing this book you are eligible for one month membership to ForgottenBooks.com, giving you unlimited access to our entire collection of over 1,000,000 titles via our web site and mobile apps.

To claim your free month visit:

www.forgottenbooks.com/free874987

ISBN 978-0-265-60785-5
PIBN 10874987

TRANSACTIONS

OF THE

MANCHESTER
GEOLOGICAL & MINING SOCIETY.

EDITED BY

THE HONORARY SECRETARY.

SESSION 1906-1907.

PUBLISHED AT THE ROOMS OF THE SOCIETY,
QUEEN'S CHAMBERS, 5, JOHN DALTON STREET, MANCHESTER.

PRINTED BY ANDREW REID & CO., LIMITED, NEWCASTLE-UPON-TYNE.
MDCCCCVI.

CORRIGENDUM.

Trans. M. G. M. S., 1908, vol. xxx., page 307.   Fourth line of Mr. A. Ghose's paper, "is general in the series of deposits, and their mode of occurrence," should read "ore-bodies, but are also interesting on account of the occurrence."

TRANSACTIONS

MANCHESTER

GEOLOGICAL AND MINING SOCIETY.

EDITED BY

THE HONORARY SECRETARY.

VOLUME THIRTIETH.

PARTS I.—XXI.

FOR SESSIONS 1906-1907, AND 1907-1908.

PUBLISHED AT THE ROOMS OF THE SOCIETY,
5, JOHN DALTON STREET, MANCHESTER.

MDCCCCVIII.

CONTENTS OF VOLUME XXX.

APPENDICES.

LIST OF PLATES.

TRANSACTIONS

OF THE

MANCHESTER

GEOLOGICAL AND MINING SOCIETY.

ANNUAL GENERAL MEETING,
HELD IN THE ROOMS OF THE SOCIETY, QUEEN'S CHAMBERS,
5, JOHN DALTON STREET, MANCHESTER,
OCTOBER 9TH, 1906.

MR. HENRY BRAMALL, RETIRING PRESIDENT, IN THE CHAIR.

The following gentlemen were elected, having been previously nominated:—

MEMBERS—

Mr. JOHN HENRY CHILCOTE BROOKING, Mechanical and Electrical Engineer, 85, Northumberland Road, Old Trafford, Manchester.

Mr. CLEMENT FLETCHER, Mining Engineer, The Hindles, Atherton, near Manchester.

Mr. ALBERT EDWARD MILLWARD, Mining Engineer, Manchester Road, Accrington.

ASSOCIATE MEMBER—

Mr. MARCEL DUBOIS, 6, Rue Gounod, Paris, XVII., France.

STUDENT—

Mr. WALTER PEARCE, Mining Student, 1, Green Lane, Heaton Moor, Stockport.

The HONORARY SECRETARY (Mr. Sydney A. Smith) read the Annual Report of the Council as follows:—

ANNUAL REPORT OF THE COUNCIL, 1905-1906.

In presenting the sixty-eighth Annual Report (the second since the federation of the society with The Institution of Mining Engineers) the Council have pleasure in congratulating the members upon another successful session.

The Honorary Treasurer's statement of accounts shows that the financial position of the society is thoroughly satisfactory, although, during the year, heavy expenditure has been incurred in providing the rooms with additional furniture, etc. The increased amount of subscriptions of federated members is a noteworthy item.

During the year the elections to membership have been as follows:—1 honorary member, 10 federated members, 2 federated associate members, 3 federated student members, and a total increase of 16.

Nine members, non-federated, have been transferred to the federated members' and 1 to the associates' list; and 1 federated member has been transferred to the non-federated members' list. The withdrawals by death, resignations and other causes have been 1 honorary member, 5 federated members and 10 non-federated members.

The following table shows the changes in the list of members for the year, from which it will be noted that the number of federated members has increased to 211 as compared with 187 on last year's list, an increase of 24 during the year.

The classification of the membership for the year 1905-1906 is shown in the following table:—

Classification	Non-federated Members.	Federated Members.	Totals.
Honorary Members	12	—	12
Members, inclusive of Life Members	63	192	255
Associate Members	—	5	5
Associates	3		3
Students	—	11	11
Totals	75	211	286

While your Council are happy in having to record a smaller number of deaths during the past year than on many former occasions, they are deeply sensible of the great loss which the Society has sustained by the death of Mr. C. E. De Rance,

F.G.S., an honorary member, elected in 1877, and one who in the past made valuable contributions to the work of the Society. The Council also regret to record the death of Mr. W. M. Campbell, of St. Helens, who joined the Society in 1887; and the death of Mr. David Mills, of Farnworth, a member of the Society since 1892.

In recognition of his long and valuable services as Honorary Secretary, Mr. William Saint was nominated by the Council to become an honorary member of the Society, and was duly elected at the November meeting.

Mr. Henry Bramall, in his Presidential Address, referred to the heavy burden of local taxation of collieries and the export-tax on coal, and, dealing with the question of wages as a factor in the cost of production, submitted that the high standard of wages now maintained rendered it imperative that such costly labour should be utilized to the best advantage by the adoption of any improvements (mechanical or otherwise) tending to reduce the amount of labour, or to make it more efficient, and consequently more economical. Mr. Bramall indicated a number of ways, in which he thought economies might be effected.

In addition to the annual meeting, eight ordinary meetings have been held in the Society's rooms, and one excursion meeting has also been held during the year. The average attendance has been very good.

During the session, important papers on geological subjects were read by Mr. Joseph Dickinson, Mr. John Gerrard and Mr. William Watts; mine engineering has been the subject of papers by Mr. William Watts, Mr. Alfred J. Tonge and Mr. James Ashworth; and the application of electricity in collieries has been dealt with by Mr. Gerald H. J. Hooghwinkel and Mr. P. Barrett Coulston.

The following is a complete list of papers and short communications brought before the Society during the year 1905-1906, and published together with the discussions thereon in its *Transactions*, and also in the *Transactions* of The Institution of Mining Engineers:—

"The Elba and Clydach Vale Colliery Explosions." By Mr. James Ashworth.
"Presidential Address." By Mr. Henry Bramall, M.Inst.C.E.
"The Use of Electricity in Collieries." By Mr. P. Barrett Coulston, M.I.E.E
"The Origin of Fossil Life." By Mr. Joseph Dickinson, F.G.S.

"Fossils at Bradford Colliery, near Manchester." By Mr. John Gerrard, H.M. Inspector of Mines.

"Marine Fossils in the Banks of the River Tame." By Mr. John Gerrard, H.M. Inspector of Mines.

"The Generation of Electricity by the Waste Gases of Modern Coke-ovens." By Mr. Gerald H. J. Hooghwinkel, M.I.E.E.

"Underground Fans as Main Ventilators." By Mr. Alfred J. Tonge.

"Alternative Schemes of Pumping and Supplying Water by Gravitation for the Use of Collieries." By Mr. William Watts, Assoc. M.Inst.C. E., F.G.S.

"Geological Notes on Sinking Langsett and Underbank Concrete-trenches in the Little Don Valley." By Mr. William Watts, Assoc. M.Inst.C.E., F.G.S.

"Report of Delegate to Conference of Delegates of Corresponding Societies of the British Association for the Advancement of Science, London, 1905." By Mr. William Watts, Assoc. M.Inst.C E., F.G.S.

Mr. John Gerrard exhibited specimens of fossil shells from the marine bed (at a depth of 2,076 feet) at Bradford colliery, and also specimens from the marine bed in the banks of the river Tame at Dukinfield.

A joint excursion of the members of this Society and of the North Staffordshire Institute of Mining and Mechanical Engineers, was made, on July 30th, to the No. 4 Atherton pit of the Hulton colliery; and, under the leadership of Mr. Alfred J. Tonge, 73 members and friends inspected the underground fans described in Mr. Tonge's paper, "Underground Fans as Main Ventilators," and the electrical equipment and other plant as described in the account of the excursion issued in the *Transactions.* The Society is indebted to the Hulton Colliery Company, Limited, for the excellent arrangements made for the convenience of the party on that occasion.

The following papers, printed in the *Transactions* of the Institution of Mining Engineers, have been discussed at the Society's meetings:—

"The Mickley Conveyor." By Mr. J. W. Batey.*

"The Conveyor-system of filling at the Coal-face, as Practised in Great Britain and America." By Messrs. W. C. Blackett and R. G. Ware.†

The Council desire to call attention to one of the privileges of membership of The Institution of Mining Engineers, namely, that each federated member is entitled to attend and take part in the proceedings at the general meetings, not only of The Institution of Mining Engineers, but also of those of any of

* *Trans. Inst. M. E.*, 1905, vol. xxix., page 268.

† *Ibid.*, vol. xxix., page 449.

the seven federated Institutes, comprizing the Manchester Geological and Mining Society; the Midland Counties Institution of Engineers; the Midland Institute of Mining, Civil and Mechanical Engineers; the North of England Institute of Mining and Mechanical Engineers; the North Staffordshire Institute of Mining and Mechanical Engineers; the South Staffordshire and Warwickshire Institute of Mining Engineers; and the Mining Institute of Scotland.

Last year, it was the privilege of this society that one of its members, Sir Lees Knowles, Bart., was elected as President of The Institution of Mining Engineers, for the year 1905-1906. Sir Lees Knowles has performed the duties of this office with dignity, and to the great satisfaction of the members of the Institution. His important address at the London meeting of the Institution, in June last, was highly appreciated.

On the Council of The Institution of Mining Engineers for the year 1906-1907, this society is represented by Mr. Charles Pilkington (Vice-President), Mr. Henry Bramall, Mr. John Gerrard, Col. George H. Hollingworth, Mr. William Saint and Mr. Sydney A. Smith (Honorary Secretary).

The Court of the Victoria University recently constituted an Advisory Committee on matters relating to the department of mining, and this Society was invited to nominate two of its members. Mr. Charles Pilkington and Mr. Jesse Wallwork have been appointed by the Council to represent the Society upon the Committee.

In February last, Mr. W. Wrench Lee resigned his post with the Society, and Mr. Joseph W. Hudson was appointed by the Council as Assistant Secretary and Librarian.

Considerable improvements and additions to the library have been made during the year, several new book-cases and book-shelves have been provided, and a new catalogue of the books, maps and periodicals is being compiled, and will shortly be issued to the members. Reference to the valuable publications of the United States Geological Survey has also been greatly facilitated by the use of the card-index, and it is hoped that members will avail themselves of these increased facilities by making free use of the library.

The thanks of the Council are tendered to the authors of the various papers and other communications, for their valuable contributions to the work of the Society, and, in conclusion, the

Council desire to impress upon every member the desirability
of taking a still greater interest in this work, by regularly
attending the meetings, by introducing new members, and by
bringing forward any matter which would be of interest to the
Society and to The Institution of Mining Engineers.

The Statement of Accounts was presented on behalf of the
Honorary Treasurer (Col. George H. Hollingworth) as annexed.

The CHAIRMAN (Mr. H. Bramall) moved the adoption of the
Council's Report and Balance Sheet.

Mr. ROBERT WINSTANLEY seconded the resolution, which was
unanimously approved.

ELECTION OF OFFICERS, 1906-1907.

The following officers were unanimously elected for the
ensuing year:—

PRESIDENT:
Mr. CHARLES PILKINGTON, J.P.

VICE-PRESIDENTS:

Mr. JOHN ASHWORTH, C.E.	Mr. ALFRED J. TONGE.
Mr. GEORGE B. HARRISON, H.M.I.M.	Mr. GEORGE H. WINSTANLEY, F.G.S.

HONORARY TREASURER: Mr. GEORGE H. HOLLINGWORTH, F.G.S.

HONORARY SECRETARY: Mr. SYDNEY A. SMITH, Assoc. M.Inst.C.E.

COUNCILLORS:

Mr. H. STANLEY ATHERTON.	Mr. P. C. POPE.
Mr. E. O. BOLTON.	Mr. JOHN ROBINSON.
Mr. C F. BOUCHIER.	Mr. W. H. SUTCLIFFE, F.G.S.
Mr. VINCENT BRAMALL.	Mr. JESSE WALLWORK.
Mr. W. OLLERENSHAW.	Mr. PERCY LEE WOOD.
Mr. WILLIAM PICKSTONE.	Mr. T. H. WORDSWORTH.

HONORARY AUDITORS:

Mr. J. BARNES, F.G.S.	Mr. GEORGE H. WINSTANLEY, F.G.S.

Mr. JOHN GERRARD (H.M. Inspector of Mines) said that he
proposed Mr. Bramall's election a year ago, and then said that
Mr. Bramall would be the right man in the right place. He

was sure that the members, during the past year, had found those words to be true; no one could have better filled the office of President than Mr. Bramall had done, and it was only fitting that they should express their appreciation. He therefore proposed that they express to Mr. Henry Bramall their sincere and hearty thanks for the admirable manner in which he had performed his duty as President of the Society.

Prof. W. BOYD DAWKINS, in seconding the proposal, said that he was expressing the feelings of every member when he said that they were extremely grateful to Mr. Bramall, their retiring President, for the way in which he had conducted the business of the Society during the past year.

The resolution was carried with acclamation.

The RETIRING PRESIDENT (Mr. Henry Bramall), in acknowledging the resolution, said that he was very much obliged to the members for their expression of appreciation. When he accepted the office twelve months ago, he feared that he was undertaking something to which he would scarcely be able to give sufficient attention. However, he had done his best, and had received such excellent support from the Honorary Secretary and other officers of the Society that it would have been a very great failing on his part if he had not responded and done what he could. He would have been glad if there had been any opportunity by which he could have furthered the interests of the Society more than he had done. He (Mr. Bramall) had great pleasure in moving that the best thanks of the Society be accorded to Mr. Sydney A. Smith for his services as Honorary Secretary during the past year. Without Mr. Sydney A. Smith's assistance, the President would have given but poor satisfaction; and he thought, really, that greater thanks were due to Mr. Smith, who did everything within his power to promote the interests of the Society.

Mr. GEO. B. HARRISON (H.M. Inspector of Mines), in seconding the motion, said that he had seen, perhaps, more of the Honorary Secretary's work than the President.

The resolution was passed with cheers.

Dr.   THE TREASURER IN ACCOUNT WITH THE MANCHESTER GEOLOGICAL AND MINING SOCIETY, FOR THE YEAR ENDING SEPTEMBER 30TH, 1906.   Cr.

Dr.

	£	s.	d.		£	s.	d.
Sept. 30th, 1905.							
To Balance in bank				... ...	131	19	4
" " in Secretary's hands				... ...	2	6	4
Sept. 30th, 1906.							
To Members' subscriptions:—	£	s.	d.				
Arrears	...	...		38	3	0	
Current:							
173 Mrs ... Mbers	360	4	0				
5 date Mbers	10	10	0				
3 Associates	3	15	0				
11 Students ...	13	15	0				
37 Non-federated Mrs	36	19	0				
2 Subscribers	2	0	0				
				427	3	0	
In Advance				6	6	0	
					471	12	0
To Dividends:—							
Birkenhead Railway ...				22	16	0	
Lancashire and Yorkshire Railway				20	17	10	
					43	13	10
To Bank-interest, less commission	...				2	16	9
To Hire of rooms ...	...				12	11	0
To Sales of *Transactions* ...	...				3	18	0
					£668	17	3

Cr.

	£	s.	d.
Sept. 30th, 1906.			
By Rent, wages, and uses of rooms	160	4	5
" Printing and stationery	22	9	2
" Postages, etc. ...	22	17	8
" Reporters ...	9	9	0
" Library ...	3	18	1
" New furniture ...	34	16	6
" Sundry uses	16	17	9¼
" The Institution of Mining Engineers	275	12	7
" *Transactions*, 1904-1905	4	17	6
" Expenses of meetings ...	24	5	0
" Balance in bank	89	13	4
" " in Secretary's hands...	3	16	2¼
	£668	17	3

LIBRARY FUND.

BALANCE SHEET, SEPTEMBER 30TH 1906

Dr.

LIABILITIES.	£	s.	d.
Outstanding accounts, say	22	10	11
Balance in favour of the Society ...	1,984	12	4
	£2,007	**3**	**3**

Cr.

	£	s.	d.
Sept. 30th, 1905.			
,, Balance	1	13	9

ASSETS.	£	s.	d.
£600 Birkenhead Railway 4 per cent. Consolidated Preference Guaranteed Stock, at £118	708	0	0
£733 Lancashire and Yorkshire Railway 3 per cent. Consolidated Preference Stock, at £86½	634	0	0
Library and furniture	500	0	0
Cash in bank	89	13	4
,, in Secretary's hands	3	16	2
Arrears of subscription, £98, say ...	70	0	0
Balance of Library Fund ...	1	13	9
	£2,007	**3**	**3**

The Investments of the Society consist of £600 Birkenhead Railway 4 per cent. Consolidated Preference Guaranteed Stock, and £733 Lancashire and Yorkshire Railway 3 per cent. Consolidated Preference Stock. The certificates for these are deposited at Messrs. Williams Deacons Bank, Limited, St. Anne's Street, Manchester, and were inspected by us on October 3rd, 1906.

Audited and found correct, October 9th, 1906.

JON. BARNES,
G. H. WINSTANLEY, } *Honorary Auditors.*

GEO. H. HOLLINGWORTH,
Honorary Treasurer.

The complete list of the Council for the ensuing year is as follows :—

President:
Mr. CHARLES PILKINGTON, J.P.

Vice-Presidents:
Mr. JOHN ASHWORTH, C.E.

Mr. GEO. B. HARRISON, H.M.I.M. | Mr. ALFRED J. TONGE,

Mr. GEORGE H. WINSTANLEY, F.G.S.

Vice-Presidents (*ex officio*):
Mr. JOSEPH DICKINSON, F.G.S.

PROF. W. BOYD DAWKINS, M.A., D.Sc., F.R.S., F.G.S.

Mr. R. CLIFFORD SMITH, F.G.S.

THE RIGHT HON. THE EARL OF CRAWFORD AND BALCARRES.

LORD SHUTTLEWORTH OF GAWTHORPE.

Mr. EDWARD PILKINGTON, J.P.

Mr. HENRY HALL, I.S.O., H.M.I.M.	Mr. JOHN S. BURROWS, F.G.S.
Mr. W. SAINT, H.M.I.M.	Mr. MARK STIRRUP, F.G.S.
Mr. W. WATTS, F.G.S.	Mr. JOHN RIDYARD, F.G.S.
Mr. ROBERT WINSTANLEY, C.E.	Mr. W. S. BARRETT, J.P.

Mr. G. C. GREENWELL, M. Inst. C.E., F.G.S.

Mr. JONATHAN BARNES, F.G.S.	Mr. G. H. HOLLINGWORTH, F.G.S.
Mr. JOHN GERRARD, H.M.I.M.	Mr. HENRY BRAMALL, M.Inst.C.E.

Honorary Treasurer:
Mr. GEO. H. HOLLINGWORTH, F.G.S.

Honorary Secretary:
Mr. SYDNEY A. SMITH, Assoc.M.Inst.C.E.

Other Members of the Council:

Mr. H. STANLEY ATHERTON.	Mr. P. C. POPE.
Mr. C. F. BOUCHIER.	Mr. JOHN ROBINSON.
Mr. E. O. BOLTON.	Mr. W. H. SUTCLIFFE, F.G.S.
Mr. VINCENT BRAMALL.	Mr. JESSE WALLWORK.
Mr. W. OLLERENSHAW.	Mr. PERCY LEE WOOD.
Mr. WILLIAM PICKSTONE.	Mr. T. H. WORDSWORTH.

Honorary Auditors:
Mr. J. BARNES, F.G.S. | Mr. GEO. H. WINSTANLEY, F.G.S

Trustees:
Mr. JOSEPH DICKINSON, F.G.S.

Sir LEES KNOWLES, BART., M.A., LL.M.

New Honorary Member:

SAINT, WILLIAM, H.M.I.M.

New Members (Federated):

ALLOTT, HENRY NEW-
  MARCH
EAMES, CECIL W.
HART-DAVIES, CAPT.
  HENRY VAUGHAN

HOPKINSON, AUSTIN
HOUGHTON, HENRY
KAY, JOSEPH

KNEEBONE, C. MAITLAND
PICKUP, WILLIAM
THOMPSON, FRED J.
YOUNG, WILLIAM

New Associate Members (Federated):

MELLOR, EDWARD THOMAS | ROGERS, JAMES TAYLOR

New Students (Federated):

CROSS, CHARLES OLIVER | SPENCER, JOHN | ENTWISLE, GEORGE

Ordinary Members transferred to Members Federated:

BOLTON, E. O.
BOLTON, H. H.
KNOWLES, JOHN

LIVESEY, JOHN
ORRELL, G. HAROLD
TAYLOR, J. T.

WALL, HENRY
WALLWORK, JESSE
WATTS, WILLIAM

Ordinary Member transferred to Associate (Federated):

DICKINSON, ARCHIBALD

Member (Federated) transferred to Ordinary Member:

WALKER, T. A.

The following have ceased to be Members:
(Members Federated):

ORRELL, G. H.   . | ROGERS, WILLIAM

Ordinary Members:

ALLEN, R.
BURY, A.
CAMPBELL, H. H.

ELCE, A.
KRAUSS, J. S.
SQUIRE, J. B.
  WIDDOWSON, J. H.

TATHAM, ROGER
TURNER, ROBERT
STOCKS, F. W.

Hon. Member Deceased:

DE RANCE, C. E.

Members Deceased:

CAMPBELL, W. M. | MILLS, DAVID

FOSSIL-SHELLS FROM CHORLEY.

Mr. John Gerrard (H.M. Inspector of Mines) exhibited shells obtained from the Mountain mine measures at Chorley. One series from a marine bed comprized *Pterinopecten (Aviculo-pecten)*, *Dimorphoceras* (Goniatites), *Posidoniella*, etc. Another series from a fresh-water bed comprized *Carbonicola robusta*, *Carbonicola acuta* and *Carbonicola aquilina*. A seam of coal had been worked for some years at Chorley, and called the Mountain mine; and recently a shaft had been sunk from this seam, about 270 feet in depth, and a lower seam of coal had been found. The marine shells were found about 63 feet below the upper seam and about 1 foot below a coal-seam, 10 inches thick. The freshwater shells were found about 5 feet above the lower seam. The problem to be decided was whether the Upper seam corresponds to the Upper Mountain mine and whether the Lower seam corresponds to the Lower Mountain mine. The position of the marine bed was different from that of the marine bed in the Billinge section; or in the Rochdale, Bacup, Burnley and Accrington sections; or in the Halifax and Yorkshire sections. He (Mr. Gerrard) hoped that it might be possible to bring before the society a section of the sinking. He was indebted to Mr. James Cunliffe, the manager of the colliery, who had very kindly enabled him to obtain the specimens.

Mr. Joseph Dickinson thought that the shells were similar to those which guided Mr. P. W. Pickup in cutting through the fault at Rishton colliery.

Mr. H. Stanley Atherton said that at Chorley colliery, about 150 feet above the top workable mine, there was a self-faced flag-bed, which was very similar to that found at Spring Vale, near Darwen, This discovery was the first means of its identification in the section of the Burnley coal-field.

———

COAL IN KENT.

Prof. W. Boyd Dawkins exhibited some specimens of coal recently found at Waldershare, in Kent. Three seams were found:—The first seam, 20 inches thick, at a depth of 1,818 feet 7 inches; the second seam, 40 inches thick, at 1,881 feet 4 inches;

and the third seam, 54 inches thick, at 1,908 feet 9 inches. The seams rested upon fire-clay floors and had hard bind roofs. These coal-seams, discovered in a new locality, proved the truth of the observations which he had addressed to the Society since the year 1886. He hoped, at some future time, to give the Society the results of his enquiry into the range of the coal-fields of Somerset and South Wales, eastward into Kent.

DISCUSSION OF MR. A. J. TONGE'S PAPER ON " UNDERGROUND FANS AS MAIN VENTILATORS."*

Mr. H. W. G. HALBAUM (Birtley) wrote that Mr. Tonge's system of placing the fans belowground certainly had two strong points in its favour:—(1) The convenience afforded by the open upcast shaft, and (2) the avoidance of the usual leakage occurring at the top of that shaft. Those advantages might, or might not, according to circumstances, outweigh some rather serious counts on the other side of the argument. Mr. Tonge's assertion, however, that the system entailed " no loss whatever from leakages "† seemed to be more enthusiastic than sound, and the author, in fact, rebutted his own argument by saying that " even with underground fans, there may be sufficient leakages in each individual seam to justify the erection of fans working in series with one another.‡

Despite the advantages admitted for the system of placing the fans belowground, it was not at all clear that, on the whole, the system was superior to the ordinary practice of placing the plant entirely aboveground. One serious objection to Mr. Tonge's method was that it made the upcast shaft the seat of the higher pressure, as compared with the downcast column; and that, in fact, appeared to be an evil inseparably associated with the system. The first result of that difference of shaft-pressures was to establish a minus water-gauge against the entire ventilation of the pit. The second result was that the inevitable leakage through the separation-doors (Plate V.§) was not, as under the surface-fan system, a comparatively harmless leakage of fresh air from the downcast shaft to the upcast shaft, but a most

* *Trans. Inst. M. E.*, 1906, vol. xxxi., page 207 ; and vol. xxxii., page 143.
† *Ibid.*, vol. xxxi., page 219.   ‡ *Ibid.*, vol. xxxi., page 212.
§ *Ibid.*, vol. xxxi., page 218.

objectionable and even dangerous leakage of foul air and gases
from the upcast shaft into the great trunk intake-currents of the
mine or mines.   In the event of an accident to any one set of
separation-doors, the result would inevitably be the wholesale
fouling of the entire ventilation.   Members would readily per-
ceive the real force of that objection by reflecting that, under
Mr. Tonge's method, the pressure of any given layer of air in
the upcast shaft must necessarily exceed that of the outer
atmosphere, whilst the pressure of any given layer in the
downcast shaft must necessarily be less than that of the outer
atmosphere.   In each case, the difference of pressure obtaining
between the given layer and the atmosphere would be equal to
the pressure due to friction in the shaft overhead.   In the upcast
shaft, that difference was plus, while in the downcast shaft, the
difference was minus, so that the motive column of leakage from
the upcast shaft to the downcast shaft in any given horizontal
plane was equal to the motive column expended on friction in
both shafts above the given plane.

One could not pass over the possibility of accidents to separation-doors, it was a contingency that should not be left out of
account, and the possibility was by no means a remote one. Consider the case, say, of the A seam.   It followed from the character
of the differential shaft-pressures that, under Mr. Tonge's system,
the separation-doors would require to be hung so as to open
against the superior pressure in the upcast shaft.   A slight coal-
dust explosion on the main intake-airway of the A seam would,
just like other similar blasts, travel against the wind and back
to the downcast shaft.   Its momentum would carry it across
that shaft, and it would collide with the separation-doors, throwing them open with more or less violence, and very possibly
damaging them to such an extent as to render them useless.   The
blast of such a minor explosion, or even the shock of a heavy
shot blown out on the intake-airway might do so much, and yet
leave the fan at the other side of the upcast shaft uninjured.
In such a case, the still-revolving fan would exhaust the noxious
fumes from the intake air-way by way of the workings, and
passing them through itself would expel them, for the greater
part, through the frames of the broken doors, into the downcast
shaft; and it would do this simply because, under Mr. Tonge's
system, the downcast shaft was a region of much lower pressure

than the upcast.   Under the ordinary surface-fan arrangement,
an accident to the separation-doors merely suspended the ventilating current in the working-places; but Mr. Tonge's method,
under similar circumstances, positively transformed the current
into a death-dealing engine of destruction.   That consideration
appeared to furnish a fatal objection to the use of underground
fans as main ventilators for fiery or dusty mines.   Under such
a system, the provisions of the Coal-mines Regulation Act would
require to be reversed, and matters would have to be so arranged
that any explosive blast capable of injuring the separation-doors
should instantly and automatically put out of action the entire
series of fans at work underground.   Otherwise, pending the
getting to work of the stand-bye plant at the surface, the whole
of the men in the mine might be suffocated by the fan-driven
fumes propelled through all the workings of the mine.

Mr. Tonge was also positive about the superior economy of
his system.   His claims in that respect appeared to be rather
extravagant; and, apart from the economy of power effected at
the top of the upcast shaft, high efficiency was not evinced by
any of the data supplied in the paper.   It should not be forgotten that the field for economizing was very much less than
the field covered by the entire ventilation.   Mr. Tonge's system
could not save a single jot of the power required for shaft-friction and final velocity, neither could it effect any reduction of
pressure in the seam of maximum drag.   Shaft-friction often
absorbed 50 per cent. or even more of the total pressure, and
altogether some 60 or 70 per cent. of the ventilation-cost was due
to a kind of preferential stock upon which constant uniform
dividends had to be paid, irrespective of the question as to
whether the ventilating plant were placed on the surface or
underground.   The utmost that could be attempted was to save
a percentage of that which was itself a mere percentage of the
total cost.   From that point of view, such illustrations as that
of the five mines requiring from 1 to 5 inches of water-gauge
each,* where the one system costs 66 per cent. more than the
other, seemed to be singularly far-fetched and unpractical.
Neither Mr. Tonge nor any other engineer could save 60 per
cent. of a sum, of which 60 per cent. was already paid away, and
of which a further percentage was required for actual necessaries.

* _Trans. Inst. M. E._, 1906, vol. xxxi., pages 211 and 212.

Furthermore, if one referred to Table I.* it appeared that the
practical economical results were as shadowy as the theoretical
illustrations. It was there recorded that 25 horsepower in the
air were obtained from 69 brake-horsepower of the motors.
Those practical results accruing from Mr. Tonge's method
should be compared with the over-all efficiencies lately obtained
by motor-driven fans of the Waddle and of the Capell types
working under the ordinary system.

He (Mr. Halbaum) could not help feeling sceptical with
regard to the claims advanced, on the score of largely increased
economy, for Mr. Tonge's system of underground fans. Because
(1), as previously stated, that system of ventilation began, owing
to its unhappy distribution of shaft-pressures, by setting up a
water-gauge against its own work. (2) Air was a material that
required to be handled very gently, and Mr. D. Murgue had laid it
down, as the first principle of fan-design, that the machine should
receive the air without shock. The air at a regulator, again, was
practically killed by shock. Hence, it was notorious that shock
was, in the case of all gaseous fluids, a merciless destroyer of
pressure. Instead of a moderately sized fan at the surface, Mr.
Tonge had installed three small fans underground; the diameters varied from 30 to 45 inches, and the revolutions from 400
to 600 feet per minute. Under present conditions the fan on
the A mine was 30 inches in diameter and the water-gauge at
580 revolutions per minute was $\frac{7}{8}$ inch. The useful water-gauge
was therefore, about one-third of the theoretical water-gauge
due to that speed of the periphery: most of the rest being
destroyed by shock, as at an ordinary regulator. The acceleration which it was attempted to impart to the air was unreason-
able, and was inseparable from disastrous shock. Mr. Tonge
claimed that he had done away with regulators. The truth was
that he had simply called the real regulator by another name:
his regulator was a more wasteful machine than the ordinary one,
for it created by power, and then destroyed by power, a greater
surplus of pressure than the difference required as between the
seam of minimum drag and that of maximum drag. It was such
gratuitous wastes of power that largely accounted for the fact
that 69 brake-horsepower were required at the motors to generate
25 horsepower in the air as shown in Table I.*

* *Trans. Inst. M. E.*, 1906, vol. xxxi., page 209.

Mr. Tonge further spoke of " the convenience of being able to regulate the supply of air in one mine, as in the case of the underground fans, without affecting any other ; " * and that discovered at once the profound fallacy of which Mr. Tonge had become enamoured.   The fact was that, at Hulton colliery, the ventilating plant consisted of three units, each one of which continually reacted against the other two.   It was the case at all collieries that the ventilation of one seam reacted against the ventilation of all the others, but it was surprising to find that Mr. Tonge imagined that his particular system enabled him to evade those reactions.   He (Mr. Halbaum) was inclined to think that the system at Hulton colliery accentuated the severity of such reactions, although he did not intend to argue that point.

But with respect to the fact of the mutual resistance of the several fans, or in other words, the mutual reaction of the several ventilations on each other, Mr. Tonge could easily prove it in the following manner:—Let him take any suitable opportunity when the pit was off, and let any two of the three fans be stopped for a couple of hours ; and during this time let the third fan be run at its usual speed.   He would then find that the water-gauge of the running fan and the volume of air passing through the fan would both be altered, the one being diminished and the other augmented.   He would also find a current of air passing through each of the standing fans, but passing in a direction contrary to that taken under the normal conditions, when all three fans were at work.   For example, if the fan in the A mine acted alone whilst those in the B and C mines stood still, the flow of air produced by the A fan would take the course indicated by the arrows in Fig. 1 (Plate XI.).   If the B fan acted alone, whilst the A and C fans stood still, the situation would be that delineated in Fig. 2 (Plate XI.); and if the C fan were to run alone, whilst the A and B fans stood still, the air-currents would flow as shown in Fig. 3 (Plate XI.).   In each figure, the full-line arrows showed the air-currents of greater pressure. Thus each fan continually endeavoured to reverse the ventilation produced by the other two, and each was obliged to do work in simply bringing the air to rest, as it were, before it could propel its own current through its own mine in the proper direction. An increase in the speed of any one fan would increase the water-

* *Trans. Inst. M. E.*, 1906, vol. xxxi., page 217.

gauge and volume of air produced by that fan; and the other
two fans would have to increase their speeds and their already
inflated water-gauges simply to maintain their volumes at the
same values as those obtained before the speed of the first fan
was augmented.

It was thus quite clear that, whether a surface fan or a
number of underground fans were employed, the ventilation of
each seam would react against the ventilation of the others
Hence, to speak of the convenience of regulating the supply of
air in one seam without affecting any other was to speak of a
myth and an impossibility.

Mr. Tonge stated that a surface fan was obliged to run " at a
speed, suitable for the one mine of the three which has the
heaviest drag,"* and claimed that the employment of underground
fans avoided or evaded that condition. It would, however, be noted
that all of Mr. Tonge's fans were of similar make, namely, the
Sirocco make; and it might, therefore, be reasonably inferred
that all would have approximately the same manometrical
efficiency.   It might from that be again inferred that the fan on
the A mine, having to produce only ⅜ inch of water-gauge, would
not need to run at so great a tangential velocity as the fan on
the C mine, where the required water-gauge was 1⅜ inches;
and still less would it need to have such a tangential speed as
that of the fan on the B mine, where the water-gauge required
was 1⅝ inches.   The facts of Mr. Tonge's practice, however, as
compared with this deduction from his theory, proved that the
three fans ran at practically the same tangential velocity; in
other words, they all encountered the same resistance, namely
the resistance of the mine of maximum drag.   The tangential
velocities of the three fans (calculated from the results recorded
in Table I.† were as follows:—A seam fan, 75·9 feet per second;
B seam fan, 76·6 feet per second; C seam fan, 75·0 feet per
second; the mean velocity of the three fans, 75·8 feet per second;
and the maximum and minimum speeds differed from the mean
by 1 per cent. According to Mr. D. Murgue's tables, the theoretical
depression due to this mean speed was 2·57 inches of water-gauge.
The fan on the A mine exceeded the mean speed, and the
normal water-gauge of that mine, according to Mr. Tonge, was

* *Trans. Inst. M. E.*, 1906, vol. xxxi., page 210.

† *Ibid.*, vol. xxxi., page 209.

$\frac{7}{8}$ inch. How did Mr. Tonge account for the balance? What did he mean by the "normal" water-gauge of the mine? What did he mean by speaking of his three fans "each running at the nearest speed to the mine-requirements,"[*] when all three fans were running practically at the same speed? For the tangential speed was the only speed that correlated with water-gauge. What, again, was the use of a smaller visible water-gauge, unless it was associated with a lessened tangential velocity? And where was the economic difference between destroying the surplus water-gauge at a regulator in the mine, and destroying the same surplus at an unduly contracted orifice of passage in the fan? And finally, with regard to shock, would Mr. Tonge state what, in his opinion, was the normal or radial acceleration in feet per second per second which it was attempted to impart to the air in its passage through the A fan at 580 revolutions per minute; and whether, in his opinion, that acceleration could be carried out without the severest shock?

Mr. ALFRED J. TONGE wrote that Mr. Halbaum devoted one short paragraph to an admission of two strong points in favour of the use of underground fans, and then proceeded to devote almost the whole of the remainder of his remarks to the details of probable happenings, largely based upon improbable assumptions.

Mr. Halbaum's objections to the use of underground fans from the safety standpoint appeared to be based upon exaggerated notions as to the amount of pressure-difference existing between the upcast and downcast shafts. In the case of a surface fan this might be very considerable, for it amounted to practically the whole water-gauge of the fan; and any leakage through old mines or other mouthings might amount to a very serious loss. With underground fans, the pressure-difference was very small, and remained constant, in a portion of the water-gauge on the various mines, so long as the quantity of air passing up or down the shafts remained constant; and, further, as the water-gauge increased, due to the extension of the mine or reduction in the size of the airways, the proportion of the shaft-resistance was reduced. One observed fact might be mentioned, to show how small the shaft-resistance, or the difference of shaft-

* *Trans. Inst. M. E.*, 1906, vol. xxxi., page 210.

pressure, actually was. When the three fans described in his
(Mr. Tonge's) paper were fully at work, any roadway directly
connecting the two shafts in other seams was approximately in
a state of balance, and the air-currents alternated in direction
with changes of atmospheric temperature. The actual water-
gauge readings taken at the three fans upon the quantities of air
referred to in the paper, confirming this statement, were recorded in Table I.; and on a mine water-gauge of 5 inches the
shaft-friction would be 2, 2 and 3 per cent. respectively. Included in the so-called shaft water-gauge was also the water-
gauge due to the resistance of the air-way from the fan to the
upcast-shaft, so that, as proved by experiment, the two shafts
were almost of equal pressure, more influenced by temperature
than by frictional resistance, and for all practical purposes might
be taken as reservoirs of air.

TABLE I.—RATIOS OF MINE AND SHAFT WATER-GAUGES.

Name of Mine.	Water-gauges.		Per cent.
	Mine.	Shaft and Outlet Airways.	
	Inches.	Inches.	
A ...	$\frac{7}{8}$ ...	0·10 ...	11·4
B ...	$1\frac{5}{8}$ ...	0·10 ...	6·1
C ...	$1\frac{3}{8}$ ...	0·15 ...	10·9

Mr. Halbaum's assumption that 50 per cent. or more of the
fan water-gauge was due to shaft-resistance was thus very much
beside the question in this particular instance, and in all cases
where the conditions were suitable for the use of underground fans.
If it were possible to have so large a proportion as 50 per cent., it
must almost of necessity occur in the case of a mine having low
water-gauges, with short and proportionately large air-ways and
considerable air-currents, and with restricted area in the shafts.
The latter conditions were named in his (Mr. Tonge's) paper as
being suited to the use of surface fans. It was this abnormal
shaft-resistance that caused Mr. Halbaum to foresee such catas-
trophes by the leaving open of the separation-doors; for, where
the shaft-resistance was so small, the amount of air passing
through the open doors depended rather upon the position of the
fan relative to the upcast-shaft than upon any other cause. In no
case, in the mines in question, did the whole of the air return
back through the separation-doors when open. In one case,
the air actually passed from the downcast to the upcast, and
not *vice versa* as prophesied. Any accidental leaving open of

the separation-doors would thus be less dangerous than if a surface fan were the ventilator, for the air in the latter case would pass straight from the downcast- to the upcast-shaft, leaving the workings untouched; while, in the assumed case of an explosion knocking down the doors, and the underground fan continuing to run, fresh air would still be delivered into the workings.   He might point out, however, that any assumption of the separation-doors being blown down without damage to the other parts of the mine drew upon one's imagination very far, for a simultaneous action would take place on the fan air-lock doors and casing, which were equally open to the haulage-road, and were specially arranged so as to give way under such circumstances.   This would have the effect of short-circuiting the air, and would leave things in that mine very much as in the case of a surface fan.   Were he to follow up the point he might show how, where a surface fan was at work, if an explosion occurred in one mine and the separation-doors were blown down as described by Mr. Halbaum, the air being short-circuited in that mine, would at once bring down the water-gauge of the fan, and so would not only leave the workings of the mine practically unventilated but would reduce the air passing through other mines possibly to dangerous limits.   He did not wish to minimize any probability of danger, in view of explosion, arising from the working of surface or underground fans, for nothing but good could come of forecasting such events, if the emergency was thereby better prepared for.

He thought that Mr. Halbaum would have done better if he had assumed that many of these things had been discussed and arranged for, rather than deal with them in the somewhat hypercritical manner which he seemed to have preferred, as the arrangements for putting in underground fans were not necessarily similar to those of surface fans, but this surely went without saying.

A fan must be capable of doing the maximum duty required during the lifetime of the mine; and it usually corresponded to the highest water-gauge, and, therefore, to the highest speed at which the fan would have to run.   A fan would give its maximum efficiency for a certain quantity, speed and water-gauge, and for these only.   Any variation in any of these three quantities implied a lowered efficiency.   The fan should there-

fore be designed to give its maximum efficiency at somewhere
about the middle of the life of the mine. At any other than the
best speed, there was one particular quantity and water-gauge,
that is, one particular orifice, which gave the best efficiency for
that speed. The speed at which the fan had actually to be
run was that at which it would drive the required volume of
air through the mine. As a rule, it was probable that the mine-
orifice did not coincide with that which gave the best efficiency
at this speed, but probably corresponded to a much lower efficiency.

In the early days of the mine, therefore, not only was the
efficiency low, owing to the lower speed at which it was necessary to run; but, unless the mine-orifice happened to agree with
the most efficient orifice for that speed, the actual efficiency
would be less than the best that could be obtained at that low
speed. This accounted for the somewhat low fan-efficiency
obtained at present in this the third year of the working of the
underground fans at Hulton collieries.

The requirements of each mine had first of all to be tested,
it was found that the A mine had a lower resistance than the
B and C mines, and a note in his paper was made of the fact
that it was intended to change the motor (and consequently the
speed of the fan).*

Mr. Halbaum did not appear to grasp the point that each
mine was developing, and therefore continually requiring a
higher water-gauge. This was met by increasing the size of
the motor-pulley, and, consequently, the speed of the fan, or by
reducing the artificial resistance: the former affording a coarse,
and the latter a fine, adjustment. He did not claim to have
abolished regulators as stated; but, as he had pointed out, the
amount of pressure dropped in these resistances was small compared with what would be necessary in the case of a single
surface fan. Already in the case of the B fan, the development
of the mine had required an increase in the fan-speed, and the
pulley had been changed. Mr. Halbaum's remarks on tangential velocity savoured somewhat of hair-splitting, and were
more a matter of fan-design. Practical experience proved that
the characteristic of a fan, when working at a duty much below
that for which it was designed, differed very greatly from the
theoretical characteristic, and was different for different fans.

* *Trans. Inst. M. E.*, 1906, vol. xxxi., page 208.

Mr. Halbaum had, moreover, taken no account of the blade-angles, which considerably affected the relation between the speed and the water-gauge.

He thought that Mr. Halbaum would now be prepared to admit that, where the shaft-resistance was so low, the stopping of one or more fans did not affect the other fan or fans to any appreciable extent. It was found by experience that the air-currents through the standing fans were very small, and their direction was chiefly determined by the temperature of the two shafts, varying between day and night.

Mr. Halbaum had referred to the reaction of one fan upon another, as though it was something beyond that due to shaft-friction, whereas there was no other possible cause, and this had already been shown to be too slight to be of any moment.

In modern practice, shafts were large in area and workings extensive; consequently, Mr. Halbaum would be much nearer the mark, if he assumed shafts as approaching to reservoirs of air rather than as constituting such greatly obstructed air-ways.

Mr. W. McKay's paper on " The Boultham Well at Lincoln " was read as follows:—

THE BOULTHAM WELL AT LINCOLN.

By WILLIAM McKAY.

Introduction.—The city of Lincoln and suburbs were practically dependent for the supply of water upon the river Wi-
tham, which was contaminated by the sewerage from the farms,
hamlets and towns near its banks, right away from its source.
The City Council decided to bore for a fresh supply of pure
water, and directed the Waterworks Committee to secure ten-
ders for the boring of a deep bore-hole to supply at least
1,000,000 gallons of water per day.

The contract for boring was let, and operations were commenced in October, 1901. A bed of running sand having
been found near the surface, metal tubbing, 12 feet in inside
diameter, was constructed upon the ground in segments bolted
together in the usual way, the joints being made with sheet-
lead. The tubbing was placed in position, and pressed down
by weights, and the sand and other material was taken out of
the inside. The segments of the cast-iron tubbing were 5
feet long, 5 feet wide, and $1\frac{3}{8}$ inches thick, with stiffening ribs across the centre, and all the flanges were bracketed
between the bolt-holes. The flanges were $1\frac{3}{8}$ inches thick,
and the brackets and ribs 1 inch thick. The bolts, $1\frac{1}{2}$ inches
in diameter, were spaced 9 inches apart. This process
was continued until a depth of $27\frac{1}{2}$ feet of tubbing was put
down: about $5\frac{1}{2}$ feet of the tubbing being pressed into the
underlying clay of the Lias formation, so as to keep back the
surface-water.

Erection of Machinery.—Long pitchpine baulks were placed
across the tubbing from north to south, upon which cross baulks
were placed, serving as pillars upon which other long baulks
were placed to support the engine-bed, engine, head-gear, etc.
The machinery consisted of a high-pressure horizontal engine,
with two cylinders, each 10 inches in diameter, with com-

pound gearing, fitted with a drum for a flat-rope for winding
purposes, a vertical cylinder in which a piston was placed to
work the boring tool, a back-screw to clamp the rope (and to
give slack rope when boring operations were proceeding), a
vertical multitubular boiler to work at 100 pounds pressure,
etc. There were two pulleys: one fixed on the top of the vertical cylinder; and the other served as a guide, at the back of
the head-gear, in a position between the drum and the main
pulley.

Boring and Tubing.—Actual boring operations commenced
in March, 1902. The boring tool consisted of a long bar, about
4½ inches in diameter, with a steel block at the bottom end,
a bow and ratchet at the top end and two guards, one fixed on
the bar a little above the block, and the other fixed immediately
below the bow and the ratchet. The cutters and shells were
made secure to the block with washers and nuts. When
boring, the horizontal winding-engine and the drum were at
rest, the back screw having been screwed up, and the rope
was clamped so that it could not move from that point on the
drum. The vertical cylinder then did the actual work of boring: the piston working inside this cylinder pushed up the
pulley over which the rope was conveyed, and raised the tool
attached thereto a distance of about 3 feet. The tool was
dropped automatically, and the cutters, striking on the bottom
of the hole, cut the strata by the percussive motion. The tool had
also a rotary movement, induced by the ratchet fixed at the top of
the bar. The top guard prevented the top end of the bar from
tilting sideways, so that a very straight and perfectly round
hole could be bored. The number of strokes per minute varied
with the different qualities of strata through which the hole
was being bored. The bore-hole was cleared with a shell-pump,
which brought up the loose material.

Boring was continued as long as possible, or until such time
as the sides became troublesome, and tubes were then placed
in position down the bore-hole. A length of 400 feet was bored,
and then tubes, 30 inches in diameter, were inserted, each tube
with couplings being about 18 feet long. They were screwed
together one at a time, lowered, and clamped at the top of the
well, and this process was continued until this length of tube

was placed in position. The lowest tube was fitted with a shoe
which rested on the bottom.

After putting in this length of tubes, the size of the bore-
hole was reduced, and a new block and guards were introduced
to suit the reduced diameter of the hole. After this change
had been made, boring was continued for a further depth of
200 feet, until the sides again became troublesome, and 200
feet of additional tubes, 26 inches in diameter, were placed in
position. A further reduced hole was bored for another depth
of 100 feet, and it was lined with tubes 24 inches in diameter.

The boring was continued of reduced diameter until a depth
of about 885 feet was reached; but the sides of the hole then
gave way whilst boring was in progress, fell down on to the top
of the tool, and jammed it fast. Whilst the borer was trying to
liberate the tool, the rope broke, and the tool was lost for the
time being. It was then decided to sink the well in order to
recover the tool, and to proceed to a further depth with the
boring apparatus.

Sinking.—In 1904, the writer expressed the opinion that the
700 feet of tubes could be got out, the tool recovered, and the
sinking continued to a depth of 900 feet within twelve months,
and this work was accomplished within the time specified. After
making the top of the shaft secure, rails were laid on the baulks
so that the carriage for the hoppets might run over the mouth
of the shaft.

The sinking of the shaft was commenced in April, 1904, every
care having been taken not to disturb the tubbing, because of
the danger of letting in the surface-water. To ensure this
end, hangers (Fig. 1, Plate XII.) made of iron bars, 3½ feet
long, 2½ inches wide and ½ inch thick, and twisted at the top
end, were bolted to the bottom flange of the tubbing. On the
hangers was placed a skeleton-ring (Fig. 4, Plate XII.) made of
iron bars, 2½ inches wide and ⅞ inch thick, composed of segments
made to templet, with two holes on either end.   One end of
each segment was cranked, so that when bolted together the ends
overlapped each other. Boards, 6 feet long, 9 inches wide, and
1 inch thick, were placed at the back of the ring, and wedged tight
so as to keep the sides secure, and to prevent any subsidence
below the metal tubbing. A skeleton-ring was placed every

5 feet in depth, the length of the hangers (Fig. 2, Plate XII.), and boarded behind. Each length of boards overlapped the other by about 1 foot (Fig. 8, Plate XII.).

Sinking had not proceeded very far before a cavity was found, which had been caused by the sides having given way during the previous boring operations. This cavity was filled before proceeding further with the sinking.

When the sinking had reached a depth of 37 feet below the bottom of the tubbing, a double bricking-ring was put in, formed by placing one ring, 9 inches wide and 8 inches thick, inside another, and bolting them together with pieces of plank, 21 inches long, 9 inches wide and 3 inches thick, reaching from the front to the back of each segment (Figs. 5 and 6, Plate XII.), and the brick-work lining of the shaft was built upon it. All the bricking in this length was solid work, four courses of stretchers, and one course of headers or binders laid with mortar, composed of 1 part of Portland cement to 3 parts by measure of fine riddled Trent sand (Fig. 8, Plate XII.). When bricking, all the boards, skeleton-rings and hangers were taken out, one length at a time, so as to allow the bricking to be built solid into the sides, in order to make it doubly sure that no surface-water could get down at the back of the brick-work. The top part of this length was done in quarters. Wooden segments were placed at intervals below the metal tubbing, and built in solid, so that the tubbing was efficiently supported.

The bricking scaffold was made in three parts, namely :—A centre and two wings, with hinges, pins and strong iron bars, about 2 feet long, to serve as pudlocks; and rested on the wall about $4\frac{1}{2}$ inches when bricking. When sinking below the bricking-rings, care was taken to leave as much solid ground below them as possible.

The first set of hangers, $3\frac{1}{2}$ feet long, was bolted to the bricking-ring, and the top ends of the boards were placed close to the bottom of the bricking-ring, flush with the front, so as to enable the line to be put on whenever required for putting inside holes.

The sides of the shaft were sheared back, so as to take out sufficient ground to enable the brick-work to be built without having to take out more ground below the first length of boring.

In the sinking of this shaft, several hard beds were passed through, some of which were almost entirely composed of ammonites and other shells, which in many cases could not be drilled by ratchet-machines, and hand-drilling was adopted.

When the sinking had reached a depth of 400 feet 11 inches, the shaft was reduced in inner diameter from 12 feet to 9 feet (Fig. 9, Plate XII.). The sinking of this well was somewhat more difficult than an ordinary shaft, on account of the tubes, 30 inches in diameter, being inserted down to the 400 feet level, and these had to be taken out one at a time as they were freed. Below the depth of 400 feet, the tubes, 26 and 24 inches in diameter, were removed in the same way. This process continued until the sinking reached a depth of 700 feet, and the last of the tubes had been removed. At this point another difficulty presented itself, as the bore-hole, open for nearly 200 feet below, had to be filled up. The sinking was then continued until the lost tool was recovered at a depth of 885 feet, and further until a depth of 891 feet 7 inches was reached. At this point a bricking-ring was put in, and the length bricked up; and as this was supposed to be the last length of bricking, bearer-holes were made in the upper part of it to carry a scaffold.

Boring.—About 9 feet of sinking was done below the last ring, and the bottom was levelled. A guide-pipe, 6 feet long and 3 feet in diameter, was put down and enclosed in concrete, so as to keep it in position. Another pipe, of the same dimensions, was bolted on the top of the other, and enclosed in concrete to within 1 foot of the top of the guide-pipe. Besides keeping the guide-pipes in position, this concrete made a good well-bottom, being composed of 1 part of Portland cement mixed with 5 parts by volume of broken bricks, mixed with sand and gravel. Two steel girders were placed in the bearer-holes and made fast, two other girders were placed across the fixed girders with a wooden roller on each, and when the boring was proceeding, the loose girders were placed close to the rope, one at each side, and bolted to the fixed girders, so as to serve as a stay and to keep the rope more rigid when moving up and down. Boring had not proceeded far, on account of the marl being softened by contact with water, before the sides gave way to such an extent that the tool worked at a higher level at the end of the day than at the beginning. The bore-hole was then emptied of water and loose marl. Two skeleton-rings were inserted and boarded up, and concrete was filled in behind the boards so as to support the sides. The sides were maintained by this method, but progress was slow. After passing through two hard blue bands and a rock-bed, a little water was again tried, and the sides soon became again troublesome. The bore-hole had to be cleai ·d out by the use of buckets, and another ring inserted, boardeJ, and concreted, so as to support the sides. Boring was then resumed and continued, almost without water, until the sides gave way, and then tubes, 30 inches in diameter, were inserted. Boring was again resumed, but did·not continue long on account of the red marl not being strong enough to stand, when in water. It was then decided to abandon the boring, and to recommence sinking operations until near the New Red Sandstone.

The tubes, concrete and guide-pipe were taken out, and sinking operations recommenced in June, 1905, and continued until a depth of 1,502 feet 3 inches had been reached. During the sinking of the last 150 feet, a pilot-hole was kept in advance, so as to prevent any unforeseen inrush of water. After bricking up the last length, the pilot-hole, 3 inches in diameter, was continued to a depth of 59 feet 3 inches below the last ring, and water was tapped on March 21st, 1906, at a depth of 1,561 feet 6 inches. The last 3 inches was bored in New Red Sandstone (Table I.).

TABLE I.—DIMENSIONS AND DEPTHS OF SHAFTS AND BOREHOLE.

	Length. Ft. Ins.	Depth from Surface. Ft. Ins.
Shaft : 12 feet in diameter	403 11	403 11
,, 9 feet in diameter	1,098 4	1,502 3
Bore-hole : 33 inches in diameter ...	59 3	1,561 6

After the boring rods had been withdrawn, a lead plug was put down the hole, followed by two wooden plugs, 5 feet and 3 feet long respectively, but the water could not be stopped by this means. A guide-pipe, 6 feet long and 3 feet in diameter, was fixed in position in the shaft-bottom and surrounded by cement-concrete to within 1 foot of the top of the pipe. The concrete-bottom did not stop the water, which continued to percolate through it, and rose in the shaft at the rate of several feet per day.

Bricking.—The lower 5 or 6 feet of every length of brick-work was built in solid, so as to make each length self-support-ing, even if the bricking ring should happen to give way.

Ventilation.—The shaft was ventilated by a small fan that forced fresh air through circular air-pipes, each 12 inches in diameter and 6 feet long. Bearers and pudlocks were inserted at certain distances, and the air-pipes were clamped to every bearer so as to prevent them from falling down, if the bolt should break. The shaft served as the return airway.

Strata.—The strata sunk through comprized Liassic clays, marls and shales; Upper, Middle and Lower Rhaetic marls and shales; and Keuper marls (Table II.).

TABLE. II.—SECTION OF STRATA SUNK THROUGH IN THE SINKING AND BORING OF BOULTHAM WELL, NEAR LINCOLN.

Description of Strata.	Thickness of Strata. Ft. Ins.	Depth from Surface. Ft. Ins.
Soil	4 0	4 0
Sand and gravel	18 0	22 0
Lias	618 11	640 11
Upper Rhaetic	16 0	656 11
Middle Rhaetic	18 2	675 1
Lower Rhaetic	17 10	692 11
Keuper Marls	868 4	1,561 3
New Red Sandstone	0 3	1,561 6

The Lias formation contains many fossils of various species, such as ammonites, belemnites, gryphites, and other shells. The upper portion of the sinking is in Lias, to a depth of 640 feet 11 inches, the bottom being about 620 feet below the sea-level.

The Upper Rhaetic beds of dark red marl, 16 feet thick, lie immediately between the Lias and the Middle Rhaetic beds of dark shale, 18 feet 2 inches thick, containing a large num-ber of fossils, pyritized imprints of shells and ammonites. When sinking through these strata, many loud " groumps " were heard; and, in fact, they were constantly on the move, when exposed to air. The Lower Rhaetic beds, of strong grey marl or shale and rock-band, are 17 feet 10 inches thick.

The Keuper marls, underlying immediately the Lower Rhaetic beds, comprize red marls interbanded with gypsum beds, green and blue bands, rock-beds and bands, gypsum-nodule beds and thin layers of gypsum. In sinking through this series, the only fossil found appeared to be a detached portion of a plant-stem or branch (*Voltzia*). The sinking of this shaft was completed without any serious accident.

Water-supply.—The flow of water from the pilot-hole, after the lead plug had been put down, was at the rate of nearly 9,600 gallons per day of 24 hours. After the bottom of the shaft had been closed with cement-concrete, and the guide-pipe fixed, the water percolated through the concrete at the rate of 3,600 gallons per day of 24 hours. A boring, with a hole 33 inches in diameter, was made from the bottom of the shaft at a depth of 1,502 feet 3 inches, until on approaching the New Red Sandstone, the water broke through and lifted the tool several feet, although its weight was about 2½ tons, showing that the pressure was very great. The breaking in of the water was heard at the surface, like the rolling of thunder, and the water rose in the shaft to a height of 180 feet in 15 minutes: consequently the flow must have been at the rate of 6,868,800 gallons per day of 24 hours. The water rose rapidly up the shaft to the surface-level in less than 24 hours, and continued to run away at the surface at the rate of 8,000 gallons per hour.

The boring operations are still proceeding.

APPENDIX.

SECTION OF STRATA SUNK AND BORED THROUGH IN THE BOULTHAM WELL AT LINCOLN.

Description of Strata	Thickness of Strata. Ft. Ins.	Depth from Surface Ft. Ins	Description of Strata.	Thickness of Strata. Ft. Ins.	Depth from Surface. Ft. Ins.
Soil	4 0	4 0	Ironstone	0 4	
Sand and gravel ...	18 0	22 0	Blue marl containing fossils of the ammonite family and shells	7 0	
Lias :			Blue marl containing fossils of the ammonite family and shells	31 9	
Blue marl or clay ...	5 6		Very hard shell and ammonite bed ...	3 6	
Blue marl containing small boulders and nodules, also fossils of the ammonite family and belemnites petrified wood with particles of coal, etc., and shells of various kinds and sizes	37 0		Blue marl with shells and ammonites ...	12 0	
			Very hard shell and ammonite bed ...	0 8	
Blue marl with fossils	41 9		Blue marl with ammonites and shells ...	21 10	
Ironstone ⅔ round the shaft	0 6		Very hard ammonite and shell bed ...	3 2	
Blue marl with fossils	4 9				

SECTION OF STRATA SUNK AND BORED THROUGH IN THE BOULTHAM WELL AT LINCOLN.—*Continued.*

Description of Strata.	Thickness of Strata. Ft. Ins.		Depth from Surface. Ft. Ins.		Description of Strata.	Thickness of Strata. Ft. Ins.		Depth from Surface. Ft. Ins.	
Blue marl with fossils	54	1			Very hard shell bed with ammonites .	1	6		
Very hard shell bed ...	4	0			Dark marl ..	11	1		
Blue marl ironstone bands and shells ...	9	1			Very hard shell and ammonite bed ...	1	6		
Blue marl with fossils	1	7			Dark marl ..	5	3		
Very hard ammonite and shell bed ..	8	0			Very hard shell and ammonite bed ...	5	0		
Blue marl with ironstone bands ...	13	3			Dark marl	1	0		
Very hard ammonite and shell bed ...	6	2			Dark marl	0	6		
Blue marl with fossils	4	8			Very hard shell bed ...	0	6		
Very hard shell and ammonite bed .	7	2			Dark marl ...	0	8		
Blue marl with shells	3	6			Hard shell bed ...	0	8		
Very hard shell and ammonite bed ...	20	8			Dark marl with shells	4	5		
Blue marl with fossils	4	11			Hard shell bed ...	0	9		
Blue marl with fossils	1	0			Dark marl ...	9	6		
Very hard ammonite and shell bed ...	1	0			Hardshell bed ...	0	6		
Blue marl with ammonites and shells ...	20	0			Darkmarl	6	6		
Very hard ammonite and shell bed ...	8				Hard shell bed ...	1	6		
Blue marl with fossils	2	0¼			Dark marl with shells	12	4		
Very hard ammonite and shell bed ...	1	0			Very hard shell bed ...	4	0		
Blue marl with fossils	1	0			Dark marl with fossils	4	6		
Blue marl with fossils	0	6			Hard shell bed ...	1	0		
Very hard shell bed ...	0	6			Dark marl with shells	3	0		
Blue marl with shells	2	0			Hard shell bed ...	1	6		
Very hard shell bed ...	1	6			Dark marl with shells	10	11		
Dark blue marl with shells ...	4	0			Hard shell bed ...	0	4		
Very hard shell bed ...	1	0			Dark marl with shells	1	2		
Dark blue marl with shells ...	6	0			Hard shell bed ...	0	4		
Very hard shell bed ...	0	9			Dark marl with shells	1	6		
Dark blue marl with shells ...	9	0			Hard shell bed ...	0	10		
Very hard shell bed ...	1	0			Dark marl with shells	13	5		
Dark blue marl with shells ..	9	0			Very hard shell bed ...	2	3		
Very hard shell bed ...	1	6			Dark marl with shells	6	6		
Dark blue marl with shells ...	1	6			Hard shell bed ...	1	6		
Very hard shell bed ...	0	6			Dark marl with shells	4	1		
Dark blue marl with shells	2	0	403	11	Hard shell bed ...	1	1		
Bottom of the shaft 12 feet in diameter reduced to 9 feet in diameter from this point.					Dark marl with shells	1	0		
					Hard shell bed ...	0	3		
					Dark marl with shells	1	2		
					Hard shell bed ...	0	6		
					Dark marl with shells	0	4		
					Hard shell bed ..	0	8		
Strong dark marl with shells ...	9	7			Dark marl with shells and ammonites and iron pyrites	25	7		
Very hard shell bed ...	2	0			Dark marl with shells	3	0		
Strong dark marl with shells ...	3	10			Hard shell bed ...	0	8		
					Strong dark marl with shells and ammonites ...	37	0		
					Hard rocky stone band	0	4		
					Strong dark marl with shells and ammonites	0	8		
					Hard rocky stone band	0	9		
					Strong dark marl with shells ..	2	7		
					Hard rocky stone band	0	5		
					Strong dark marl with shells	1	10		

SECTION OF STRATA SUNK AND BORED THROUGH IN THE BOULTHAM WELL AT LINCOLN.—*Continued.*

Description of Strata.	Thickness of Strata. Ft. Ins.		Depth from Surface. Ft. Ins.		Description of Strata.	Thickness of Strata. Ft. Ins.		Depth from Surface. Ft. Ins.	
Hard rocky stone band	0	3			Rocky band, grey ...	1	6		
Strong dark marl with shells	2	10			Green and red marl ...	29	11		
All the beds contained fossil imprints of shells and ammonites.					Red marl, mixed with a little green and grey marl, interlaid with gypsum from ½ inch to 2 inches	40	5		
Strong dark marl ...	6	0			Red marl with a little green interlaid with thin layers of gypsum running in various directions	34	0		
Hard stone band ...	0	4			Red marl	9	0		
Strong dark marl ...	1	2			Very hard grey marl ...	1	7		
Hard stone band ...	0	6			Red marl	14	0		
Strong dark marl ...	0	9			Very hard grey marl ..	1	0		
Hard stone band ...	0	3			Red marl . .	4	0		
Strong dark marl ...	1	0			Very hard grey marl ...	1	0		
Flaggy rock ...	5	0			Red marl with thin layers of gypsum ...	16	7		
Soft dark marl ...	5	8	640	11	Very hard grey marl ...	1	1½		
Upper Rhaetic:					Red marl with thin layers of gypsum ...	15	10½		
Dark red marl	16	0	656	11	Very hard grey marl ..	2	0		
Middle Rhaetic:					Red marl with thin layers of gypsum ...	12	3		
Black marl or shale, one mass of fossil shells, imprints pyritised ...	5	4			Hard grey marl ...	2	0		
Black marl or shale, one mass of fossil shells, imprints pyritised ...	12	10	675	1	Red marl with thin layers of gypsum ...	17	7		
Lower Rhaetic:					Gypsum bed ...	1	0		
Strong grey marl with spar or gypsum ...	6	2			Red marl with thin layers of gypsum and gypsum nodules .	28	1		
Rock band	0	3			Very hard gypsum bed	1	0		
Strong grey marl ...	11	5	692	11	Red marl with thin layers of gypsum mixed with green marl	41	2		
Keuper Marls:					Red marl with thin layers of gypsum mixed with green marl	1	0		
Hard gypsum or conglomerate bed	1	0			Red marl with thin layers of gypsum ...	23	0		
Red marl with layers of gypsum	8	8			Very hard grey marl ..	2	0	1,057	3
Red marl with layers of gypsum	1	10			Red marl with thin layers	19	0		
Gypsum bed	1	10			Strong grey marl with thin layers of gypsum	2	6		
Red marl	0	6			Red marl with hard bands of blue and thin layers of gypsum	41	10		
Gypsum bed	1	3			Blue stone mixed with red ...	1	0		
Red marl	5	7			Red marl with grey stone bands and thin layers of gypsum ...	17	9		
Gypsum bed	1	0			Grey stone, very hard	1	9		
Red marl	2	3			Red and grey marl mixed	7	0		
Gypsum bed	1	0							
Red marl	3	10							
Gypsum bed	0	6							
Red marl	2	7							
Gypsum bed	0	9							
Red marl	2	9							
Gypsum bed	1	0							
Red marl	3	2							
Gypsum bed	1	10							
Red marl	1	10							
Gypsum bed	0	10							
Red marl	9	0							
Red marl with thin layers of gypsum ...	7	10							
Green and blue marl ...	2	5							

SECTION OF STRATA SUNK AND BORED THROUGH IN THE BOULTHAM WELL AT LINCOLN.—*Continued.*

Description of Strata.	Thickness of Strata. Ft. Ins.	Depth from Surface. Ft Ins.	Description of Strata.	Thickness of Strata. Ft. Ins.	Depth from Surface. Ft. Ins.
Flaggy rock	1 0		Red, grey and blue, grey and blue very hard, with thin layers of gypsum and gypsum nodules ...	42 0	
Grey rock containing carbonate of lime ...	5 0		Very hard grey rock ...	1 0	
Mixture of grey and red marl	1 0		Red marl	3 0	
Red marl	1 0		Very hard grey rock ...	1 0	
Grey and red marl ...	0 11		Red marl with blue bands	2 9	
Red marl	1 0		Red marl with blue bands ...	31 8	
Very hard grey rock ...	5 9		Grey rock or sandstone	2 0	
Very hard rocky marl	6 10		Red marl with blue bands ...	17 10*	1,502 3
Red marl	2 2		Pilot hole, 3 inches in diameter, and borehole 33 inches in diameter commenced from this point.		
Grey and blue marl, very hard	10 6				
Red marl	4 3		Red marl, with blue bands, layers of gypsum and gypsum nodules	33 6	
Grey and blue marl, very hard	15 6		Grey and red, grey, sandstone	3 0	
Red marl	5 0		Red marl with blue bands, layers of gypsum and gypsum nodules	22 6	1,561 3
Very hard red, grey and blue marl with thin layers of gypsum ...	47 0		*New Red Sandstone :*		
Red, grey and blue marl, interlayed with thin layers of gypsum ...	50 0		Red sandstone ...	0 3	†1,561 6
Very hard rocky marl	2 0				
Red, grey and blue, grey and blue very hard, with thin layers of gypsum	41 5				
Red, grey and blue, grey and blue very hard, with thin layers of gypsum	52 7				

* Shaft nine feet in diameter finished at this point.   † Water tapped in the pilot hole on March 31st, 1906.

The RETIRING PRESIDENT (Mr. Henry Bramall) moved a vote of thanks to Mr. McKay for his paper.

Mr. T. H. WORDSWORTH seconded the resolution, which was cordially approved.

Mr. JOSEPH DICKINSON said that springs had been found in that part of England. Some of them were chalybeate, and it would be interesting to know whether any brine was found in the exploration described by Mr. McKay.

Mr. WM. McKAY said that practically no water was found, until the pilot bore-hole reached the New Red Sandstone, and then water was encountered as described in his paper. It was somewhat hard, but it could be made into good water.

The further discussion was adjourned.

To illustrate Mr H.W.G.Halbaum's "Notes on Underground Fans as Main Ventilators." Vol.XXXII, Plate

Fig. 1.– Fan working in A Seam.

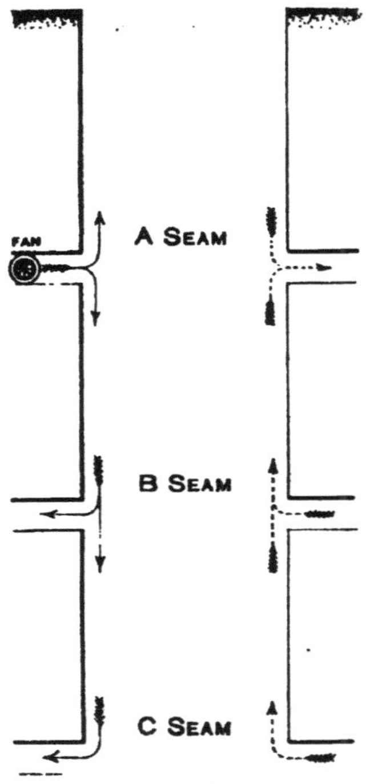

No. 3
or
Upcast Shaft.

No. 4
or
Downcast Shaft.

A Seam

B Seam

C Seam

Fig. 2.—Fan working in B Seam.

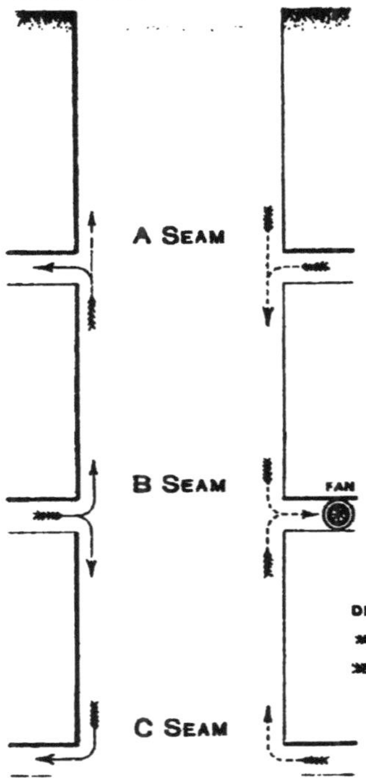

No. 3
or
Upcast Shaft.

No. 4
or
Downcast Shaft.

A Seam

B Seam

C Seam

Fig. 3.– Fan working in C Seam.

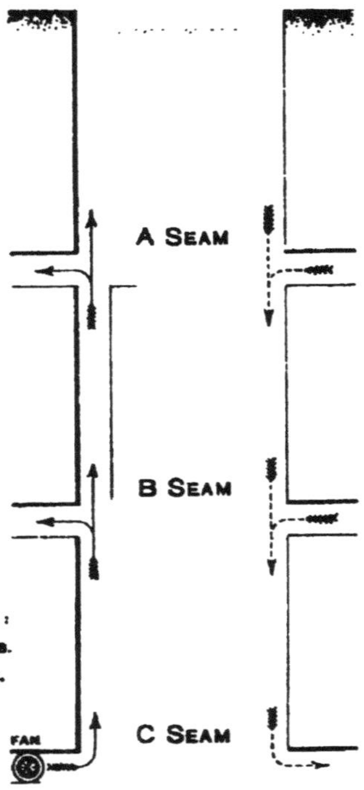

No. 3
or
Upcast Shaft.

No. 4
or
Downcast Shaft.

A Seam

B Seam

C Seam

REFERENCES.

DIRECTION OF AIR-CURRENTS:
→ HIGHER PRESSURES.
→ LOWER PRESSURES.

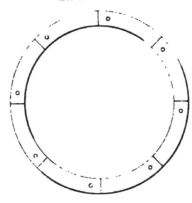

BRICKING-RING.

FIG. 9.—SECTION SHOWING REDUCTION IN DIAMETER OF SHAFT FROM 12 FEET TO 9 FEET.

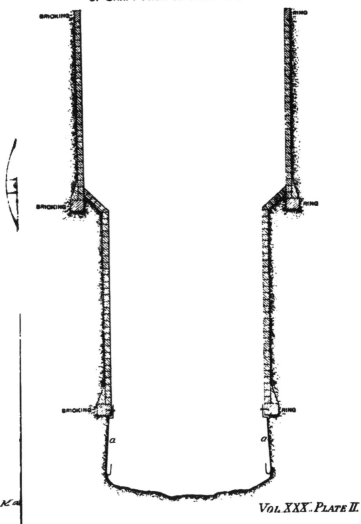

MANCHESTER GEOLOGICAL AND MINING SOCIETY.

ORDINARY MEETING,
HELD IN THE ROOMS OF THE SOCIETY, QUEEN'S CHAMBERS,
5, JOHN DALTON STREET, MANCHESTER,
NOVEMBER 13TH, 1906.

MR. CHARLES PILKINGTON, PRESIDENT, IN THE CHAIR.

The following gentlemen were elected, having been previously nominated :—

MEMBERS—

Mr. FRANCIS VERRILL BROWN, Mechanical and Electrical Engineer, 49, Deansgate, Manchester.

Mr. WILLIAM LOWBRIDGE HOBBS, Mining Engineer, 100, Bolton Road, Pendleton, Manchester.

ASSOCIATE MEMBER—

Mr. JAMES CUNLIFFE, 81, Moor Road, Chorley.

DISCUSSION OF MR. W. McKAY'S PAPER ON "THE BOULTHAM WELL AT LINCOLN."*

Mr. W. McKAY exhibited a number of fossils and minerals obtained while sinking the Boultham well.

(1) Lias formation. Ammonites: *Stephanoceras commune* and *Harpoceras bifrons* in Upper Lias, *Ægoceras planicostatum* and *annulatum*, and *Arietites Conybeari* and *Bucklandi*. Nautilus, 500 feet below the surface. Belemnites in Upper Lias. Mollusca, etc.: *Gryphœa cymbium* (variety *depressa*), *Unicardium cardioides*, *Myacites decurtata* and *donaciformis*, *Cardinia* (species), *Trigonia litterata*, *Lima gigantea*, 600 feet below the surface. *Pinna folium* in Upper Lias. Petrified and pyritized wood. (2) Middle Rhaetics or Rhaetics proper. *Avicula decussata*. (3) Keuper Marls. Gypsum, gypsum-conglomerate and gypsum-nodules. A fragment of a stem or a branch of *Voltzia*

* *Trans. Inst. M. E.*, 1906, vol. xxxii., page 245.

was found at a depth of 955 feet below the surface. The first 3 inches of the boring, below the depth of 1,561 feet 3 inches, passed through New Red Sandstone. Since then, boring operations had been continued, but he did not know the exact depth. Salt had not been found in the boring.

Mr. JOSEPH DICKINSON remarked that rock-salt overlaid the New Red Sandstone in Cheshire, and at Middlesbrough is next above the Magnesian Limestone.

The PRESIDENT (Mr. Pilkington) said that, in sinking a bore-hole near Warrington, weak brine had been found at a depth of 100 feet; and, it being found that all the deeper bore-holes in the neighbourhood had salt in them, the boring was abandoned.

Mr. W. McKAY said that, in the case of the Boultham well, it appeared that the salt had disappeared from this particular area, and that only layers of gypsum were left. In some parts of Lincolnshire, water had been found impregnated with salt; but, at Gainsborough, at a depth of 1,515 feet, splendid water had been found, and an equally good result was anticipated from the Boultham well.

The PRESIDENT (Mr. C. Pilkington) asked whether the supply of water was diminishing.

Mr. W. McKAY replied that, on the contrary, it had increased at Gainsborough, from the time that it was tapped up to the present time. He believed, however, that it was taking the supply from other districts. On this assumption, Lincoln, being on the lower level, would, no doubt, take the water from other places.

———

The PRESIDENT (Mr. Charles Pilkington) delivered the following "Presidential Address":—

PRESIDENTIAL ADDRESS.

By CHARLES PILKINGTON.

It is customary for your President to deliver an address at the opening of the session. I am sorry that I cannot give you one on some geological subject: for this Society was primarily geological, and to the geologists we owe its foundation and its establishment as a successful and useful institution. Perhaps, if geologists read mining papers and miners read geological papers, the discussions would be more lively; but scientific men are so very severe in argument, not to say vicious in their search after truth, that I dare not adopt this suggestion, and so fall back on surer ground and briefly review some of the mining problems with which we are face to face.

All professions and trades change as time goes on, owing to new inventions and methods, but mining is subject to greater changes than most other businesses, and colliery-proprietors, mining engineers and officials, are well aware that, speaking generally, they must prepare themselves to meet great and increasing difficulties, for although there is plenty of coal left in this country, most of the easily-won seams are worked out; and there remain for us, at any rate in the wellknown districts of Yorkshire and Lancashire, only those mines which, for one reason or another, could not formerly be worked to a profit.

Now, an institution like ours must devote itself to the study of the difficulties that will have to be met in altered circumstances, and it is the import of this address to suggest to the Society, and especially to its younger members, those subjects which should receive their attention, and on which the result of their work in papers would be useful to us.

Before referring to any particular subject or difficulty, I would suggest that the education of the younger members who will have to face them, should be somewhat altered. It has been the custom in Lancashire to put a "would-be mining engineer" into the plan-office, where he has had to make innu-

merable tracings, and where he has spent some four or five years
in learning to use the miner's dial and to plot accurately. Now
it is necessary that he should be taught discipline, if he has to
learn to command, and it is necessary that he should be a competent surveyor; but it is not necessary that he should spend the
best years of his apprenticeship, when his mind is most capable
of receiving impressions, in learning to becoming a past master
in the use of the dial.   It is true that the pupil is occasionally
sent to see some work carried out according to instructions; and,
if he is fortunate, he may later on have charge of a pit in connection with and under a certificated manager for a month or so; but
I think that a rather wider range should be given to him, if he is
being educated to take the position of a colliery manager, should
he prove to have grit enough to take it. He should know more of
mechanical engineering than was formerly thought necessary,
although he need not go into those technical details which are
better left to the expert.   He should be well grounded in electricity, and know something of building. The modern student
has a great advantage over the man of the past, for he has the
use of such excellent mining schools as those of Manchester and
Wigan (speaking of this district only), equipped with every
modern luxury in the way of capable teachers, good models and
diagrams.   But it is one thing to learn at a school and another
to have knowledge ingrained into one's system by familiar use,
and there are things which can be taught at a colliery far better
than anywhere else. He should for instance, amongst other
things, learn by personal experience something of the cost and
nature of the materials that he uses and of the coals sent to the
surface. Now, these were supposed to be more or less departmental secrets when I was young; and an inquisitive pupil was
regarded with some suspicion. There are certain things that
may have to be kept secret, but if you are educating a youngster,
and want him to become useful, the more that you let him know
the better. In saying this I do not suggest that the education of
the surveyor should be neglected: far from it, I would give him
the best facilities to learn his work, and better instruments to
work with than are supplied at some collieries. I would not
have him answer such an advertizement as appeared lately in
a Yorkshire paper:—" Wanted a surveyor at a large colliery,
30s. a week." But the talents required for surveying and the

talents required for management are not the same, and may not
be combined in the same individual. The very exactness of
detail required from the surveyor might develop into a niggling
habit of thought in a manager.

The first practical question to which I would call your attention
is that of coal-cutting by machinery.   There are some members
who have given to this question their best thought and work, and
have generously passed on to us the result of their labours; but
the general frame of mind in the past has been too cautious and
conservative. It is nearly thirty years since I first handled a
coal-cutter underground, and yet how few comparatively were in
use ten years ago. It is, to a certain extent, the fault of their
makers and introducers, who in the past claimed for them far
more than they could possibly achieve. Most of the early statements about cost and speed of cutting were absolutely unbusiness-
like, and naturally caused hard-headed mining men to fight shy
of trying the machines. But it is different now, for we have a fair
amount of independent figures to go upon, not only giving the work
done in a certain time, but the time required for preparing the
places for the machines, the cost of repairs and management,
the amount of breakdowns over a given period, and the capital
outlay. There are some twenty or thirty different machines
on the market worked by compressed air or electricity, so that
we have a much larger choice than we originally had. This is
very important, for one machine is suitable to work under one
set of conditions and another under other conditions; and there
are places in the same mine where it will be found best to vary
the type of machine even for longwall holing only. I may say
that, at the present time, we have nine different kinds working in
our pits. Then again, certain parts of the same mine may be
suitable for a machine, while others are best worked by the
old method.   It is of no use laying out every thin mine
for machinery, and sticking to it through thick and thin, where
the mine does not permit it. As in war, if you cannot beat an
enemy by artillery, you must bring up the infantry and let
both act together. Speaking generally, given a fairly hard mine,
say, 3 or 4 feet thick, with a good roof, free from faults, and
not too steep, the mining engineer who will not experiment with
machinery is behind his time.

But, even under the most favourable circumstances, coal-cutting requires an immense amount of personal care, forethought and management; and few elderly men can adapt themselves to the work, as questions that are novel in themselves occur with vexatious frequency. The man who has charge of a district or mine, where cutting machines are used, should be bred young to the work, if possible; and there is no doubt that the managers, whom we are educating now, will be far more successful in this department than the elder men of the present day.

These remarks apply equally to underground coal-conveyors, which have received little attention up to the present time. They are successful enough, as far as the work done is concerned, but they are usually very cumbrous and expensive to move. A light, easily-driven and quickly-moved arrangement is wanted, but it is only half-invented at the present time.

New pits will generally have to be deeper than the old ones, and the problems of cheap and safe winding are many and difficult. The immense weight of 3,000 feet of winding-rope, and the full load at starting, together constitute a great difficulty. Many attempts, more or less successful, have been made to overcome it: some of the best, so far as balancing the load is concerned, fill the pit with ropes, each difficult to examine, and the breakage of one of which might cause inextricable confusion and long stoppage. It seems to me that the best solution is the conical-grooved drum of some modern design.

The great cost of the pits and machinery will make it neces sary that one set of shafts shall serve a very large area, and consequently wind a large quantity of coal. Therefore the distance from the pit-bottom, at which work can be safely carried on, is now occupying attention; but an arbitrary limit, such as had been proposed, would, in many cases, put a colliery with deep pits in the Bankruptcy Court, and throw the colliers out of work. Safety is, of course, the important factor; and, when the limits are being approached, a very small area of workings worked at very high pressure, with improved ventilating machinery and large roads, may enable us to work with safety and health at great distances from the shaft.

To achieve the best result, ponies must be kept out of the

pit, and their places taken by hauling-engines driven by compressed air or electricity.

The use of coal-cutting machinery, at the far end, in a hot and distant district, driven by compressed air, will help to cool the atmosphere and to improve the ventilation. Here is a question for some member to work out: what is the best way of doing it? A long length of pipes direct from the surface is expensive to lay down and difficult to keep in order. Air-compressors, near the far end, driven by electricity, seem a plausible solution; but the heat developed at the compressor must be equal to the cold created at the face, so that there is no gain in temperature unless a small current of fresh air passes over the motor and air-compressor direct into the return-airway. I think that science may find in the future some means cheaper than the suggested use of liquid air to help us; but, in all these things, cheapness is an absolute necessity. The more power that we use in winning a ton of coal, the more it behoves us to economize the fuel which creates that power, and here we must turn for help to mechanical engineers. I know, by experience, that they are always ready, but let them see to it that they do not frighten us by excessive costs.

Greater depth means greater heat, and another enemy may have to be encountered. There are now mines in Great Britain hot enough to propagate the larvæ of *Ankylostoma*. Given a little dampness and the presence of one contaminated person, and it is certain that the whole of such a mine would be affected. We know the great difficulty, inconvenience and expense that have been incurred in Westphalia, and the stringent regulations that have been made to control the evil. I do not think that British miners have a sufficient sense of discipline to submit to such troublesome details as are there enforced, nor have our present managers a sufficiently methodical habit of mind to enable them to enforce constantly such regulations. Fortunately, we are not much exposed to this danger at present; but, with the interchange of men that now goes on between Great Britain, the Continent and tropical countries, it seems advisable that some simple and general precautions should be enforced. I think that the law of the land should be that no miner from abroad should be taken into a pit, where the temperature is regularly above 80° Fahr., without a medical certificate being obtained stating that he is free from this disease. Any district where a case is known to exist should be declared an " infected district " under the Infectious Diseases Act, and miners going from that to another hot district should be required to produce a doctor's certificate that they are free from the malady. It has been found that ankylostomiasis is easily diagnosed by an examination of the blood; and the medical authorities of the different countries and other areas should appoint properly qualified men to detect the disease. There has been a case lately in Scotland; last year there was a case in Manchester; it is known that the disease exists in Cornwall; and it seems to me that the less time we lose in establishing national precautions the better. There can be no doubt that the workmen will endeavour to have this disease included amongst the accidents for which compensation is paid; but the worst aspect of the misfortune would be the suffering and annoyance caused to the men themselves.

This brings me naturally to the subject of sanitation. From time to time, a disturbance is made about the dirty condition of many of our pits, designs of earth-closets and pails are discussed, and rules are drafted and printed, but they are difficult to enforce. The subject is an unsavoury one, and when the management's conscience is appeased by providing the apparatus and posting the rules on the headgear, the subject is often allowed to slide into its old channel. It is a somewhat difficult question, and the education of the workmen, as well as the proprietors, into a proper frame of mind on the subject appears to be the most effectual way of dealing with it.

I now come to the dust question: one of the most important of the day. According to a recent report,* the great Courrières disaster was mostly, if not entirely, due to a dust-explosion. I do not think any of us wanted any more proofs of the dust danger; but we do need more experiments as to the best way of removing it, and the deeper and hotter the mine, the greater will be the difficulty. In new pits it may be easy to keep screening and sorting arrangements at some distance from and, if pos-

* Report to H.M. Secretary of State for the Home Department on the Disaster which occurred at Courrières Mine, Pas de Calais, France, on March 10th, 1906, by Messrs. H. Cunynghame and W. N. Atkinson, 1906 [Cd. 3171].

sible, on the north-east (or opposite side to the prevailing wind)
of the downcast pit, so as to prevent the dust from descending.
But we are still face to face with the dust created in the mine, and
it must be remembered, where mechanical haulage is used, that
the dust of the main wagon-roads, although less in quantity, is
much more dangerous than it used to be, when ponies pulverized
the warrant, or shale, which, mingling with the coal-dust, reduced
its explosiveness.　In deeper and hotter mines, it would not be
advisable to water the roads, as this would have a tendency to
propagate *Ankylostoma*, and, by creating a moist and hot atmosphere, prove injurious to the colliers' health.　It seems to me that
in the deep mines of the future there are only two known things
that we can do, and these are:—(1) To use steel or iron tubs,
which will not allow the dust, made in transit, to fall on the
roads; and (2) to have periodical cleanings of the main wagon-
roads, perhaps using a jet of compressed air to remove the dust
from the crannies of the sides and roof.

If we could have something in the nature of a vacuum-cleaner,
using compressed air in an injector or fan to deliver the dust into
a long water-tank or wagon, wherein the water was agitated, it
might help us; but dust occurrs in such quantities that it is
difficult to deal with it.　Mr. John Gerrard, in his evidence before
the Royal Commission on Mines, refers to such an apparatus
having been coupled to a screen.　Some five or six years ago,
something of the kind was tried at a revolving screen at the
collieries with which I am connected, but it was not very
successful.

We are often forced to work deep coal-seams under water
bearing strata, through which our fathers would have hesitated
to sink a shaft.　There are many ways of overcoming this difficulty:—(1) The plan of sinking with a caisson, using compressed air to keep out the water: but this can only be relied on
for small depths and through soft strata.　(2) The freezing process is very difficult and expensive.　(3) The Kind-Chaudron,
using immense drills, followed by tubbing, driven downward
from the surface, is a very slow process and perhaps the most
expensive of all.　(4) The German "Tiefbohr" Company's system, which has not been sufficiently proved for large shafts.　And
(5) there remains the rough-and-tumble method of our fore-

fathers: no longer rough-and-tumble, now that it is worked by
skilled engineers, with the best pumps and winding tackle in the
world, but still the same in principle. This is doubtless in most
cases the best and cheapest; but it has its limits, and even when
successfully applied, as it has been at the Maypole colliery,
Wigan, and at Manton colliery, Nottingham, the cost of some of
the depths per foot must have been enormous.

One of the great difficulties at increased depths is the great
pressure on the pumps, which causes the sand to cut the clacks
and rams to pieces so quickly; and, when everything is pumping
at full stretch, it is often a very serious and difficult operation to
change a pump. The "Tomson," a modified pumping system,
is in use in Germany : as soon as the pit is started, if necessary,
or as soon as the volume of water and depth of lift warrant it, two
large cylindrical tanks, each suspended by two strong capstan-
cables, are placed on each side of the pit, and these tanks follow
the sinkers, and are lowered down, foot by foot. Into these
tanks, two large cylindrical buckets are dipped, using the
capstan-cables as guides, and by this means a large quantity of
water can be safely wound out of the shaft.   The sinking
pumps proper, driven by compressed air or electricity, pump into
these tanks; and, as the lift is small, say, 10 to 30 feet, there is
comparatively little cutting of the valve-seatings and rams, and
a comparatively small motor will lift an immense body of water.

The question of tubbing *versus* pumping is a very momentous
one, and after we have worked it out to the best of our ability
with the data to hand, it is difficult to arrive at a definite conclusion. It might help us if we had more figures giving the
decrease of work done at large colliery pumping-stations as
years go by. These figures might be collected from time to
time, and calculations worked out from them; they would certainly be of scientific interest, and they would lead to discussion
and might prove of great value.

Should tubbing be decided upon, what is the kind to be used?
The ordinary British method is to place the plain sides of the
tubbing inside the pit, the ribs being towards the strata, and
the joints made watertight with wooden wedges. The so-called
German method, but which I believe was first used in Great
Britain, has the flanges turned towards the centre of the pit, and
is put in as the pit goes down, each ring being concreted to the

strata behind, after it has been secured with bolts to the ring above. The segments of this tubbing are much larger than those ordinarily made in Great Britain, being about 5 feet square: all the joints are planed, a thin strip of lead being inserted, when they are bolted to each other. I think that tubbing smooth on the inside is better for ventilation and safer in case of anything falling in the pit, as there are no projecting flanges to receive a blow, but the other is more easily inserted and easier to repair. Should any papers be communicated on this subject, I hope that the question of the thickness of the tubbing will be discussed: for, although some formulæ are recorded in handbooks on mining, the information is untrustworthy, as little attention is paid to the depth and frequency of the flanges and ribs.

At the present time, a committee, appointed by the Lancashire and Cheshire Coal Association, has been preparing plans and collecting information preparatory to the establishment of a rescue-station. Leigh has been chosen as a central spot, and I hope that an office, storerooms for oxygen life-saving apparatus, and the various other appliances suitable for a rescue-station will soon be erected, together with a long gallery, in imitation of a roadway in a pit, with all the obstacles of broken timber and fallen roof through which a rescue-party might have to force its way after an explosion. The Committee hope to train men from every district of Lancashire and Cheshire, not only in the use and upkeep of the various life-saving appliances, but also to become so familiar with their use in a similar environment, that they may boldly and readily use them down the pit, should the necessity unfortunately arise. I hope that all the members of this Society will view this effort with sympathy, and do their best, in every way that lies in their power, to make the movement a success. The Committee entrusted with this work at once asked Mr. H. Hall and Mr. John Gerrard, H.M. inspectors of mines, to join in this work; and Mr. Gerrard especially has given a great deal of valuable time and thought to it.

I mention this in these rooms because H.M. inspectors of mines have been, and are, most useful and valued members of this Institution, whether as officers, or as geologists or miners; and also because in the person of Mr. J. Dickinson, who was Her late Majesty's chief inspector of mines, we have one of our oldest and most respected members.

And now, gentlemen, although I have not given you anything new in this address, or gone into any detail on any subject, I hope that my remarks may stimulate others to carry on carefully and diligently research in the various subjects mentioned; and that, through their efforts, this Society may not only receive benefit enuring to its own members, but be the means of forwarding the solution of many of those difficult problems that affect the safe and successful working of that mineral, upon which the prosperity of our country so largely depends.

———

Mr. JOSEPH DICKINSON, F.G.S., in moving a vote of thanks to the President for his address, said that while he agreed with the President on most of the points mentioned in his address, he would, if discussion were not forbidden, call attention to one on which there might be a difference of opinion. The formal examination of a collier for ankylostomiasis might, he thought, be left with the manager.

Prof. W. BOYD DAWKINS seconded the vote of thanks, which was carried with acclamation.

The PRESIDENT (Mr. Charles Pilkington), after acknowledging the vote of thanks, said that he did not think that a manager would, as a rule, be able to tell whether or not a man was suffering from ankylostomiasis.

———

Mr. MARK STIRRUP, F.G.S., read a paper on "The New and the Old Geology; and the New Ideas of Matter."

MANCHESTER GEOLOGICAL

ORDINARY M
HELD IN THE ROOMS OF THE SO
5, JOHN DALTON STRE
DECEMBER 4T

MR. CHARLES PILKINGTON,

.

The following gentleman was
ously nominated : —

MEMBER—
Mr. GEORGE ALFRED CHRISTOPHER, M
Company, Limited, Standish, W

CONGRATULATIONS TO M

At the opening of the proc
since we met in this room last, M
brated his eighty-eighth birthday.
him and at the same time congrat

Mr. DICKINSON briefly than
kindness.

DISCUSSION OF MR. SAM
"PRACTICAL PROBLEMS C

Mr. SAM MAVOR said that ma:
his paper were of a controversial
indicated a distinct departure fro
in certain districts; and, if he wa
from those who were continuir
condemned.

The PRESIDENT (Mr. Charles
pany were so afraid of the disad·

* *Trans. Inst. M. E.*, 1906, vol. xxxi

that they had not tried one.   They had several pick, the ordinary
disc, and one or two other machines.   They thought that, if the
bar machine had enough strength, and if the cutters were not
liable to be torn out by coming against iron nodules, it would be
a very useful tool.   He would like to hear whether it was easy to
keep the bar machine against the face in mines where the dip
was steep, say, 1 in 3½.

Mr. ALFRED J. TONGE said that he agreed with Mr. Mavor
when he said that it was a mistake for managers to place the coal-
cutter in a part of the pit where they knew that it could not succeed, and then expect it to succeed.   Managers should take the
advice of engineers who had had experience of coal-cutting by
machinery, and should work in accordance with their instructions; and he believed that then more coal-cutters would be
found at work in mines.   Advance in underground mining must
almost certainly be accompanied by, and accomplished through,
the introduction of machinery; and he thought that the coal-
cutter would bring forward the better conditions that many
managers were trying to get.   He rather disagreed with Mr.
Mavor's remarks about an increase in the number of gate-roads.
Mr. Mavor said that, if more coal was taken down a gate-road,
that was sufficient justification for making other gate-roads.   He
(Mr. Tonge) thought that such a statement should be qualified
by many other conditions; but he was sure that it was not quite
right to say that an added gate-road was justified according to
the amount of coal that went down it.   While gate-roads were
convenient for filling the coal quickly, they also meant greater
areas of exposed surface, and therefore greater liability to accident.   He rather preferred, where possible, a reduction in the
number of drawing roads and the carrying of one main road in
each district.

Mr. LEONARD R. FLETCHER said that his experience of coal-
cutting machines was, that almost every mine wanted catering
for separately.   They had tried two or three different kinds
driven by compressed air, at the collieries with which he was connected, but without much success.   The holing was rather hard,
and they did not work as quickly as they would have liked.
When the machine holed at the bottom of the seam, the coal
settled down on the machine.   The percentage of round coal was

decreased, because the holing removed the strongest part of the seam, and the overlying coal fell and broke into small pieces. A machine holing in the dirt on the top of the seam had been tried, but the stone was too hard; it quickly blunted the cutters, and the machine could not travel more than 15 or 20 feet without the cutters being changed. The seam was an exceptionally severe one for coal-cutters, and they came to the conclusion that the machines tried were not suitable for their purpose. They had, however, benefited from their experience, and they hoped at some future time to be able to adopt the use of mechanical coal-cutters on a large scale.

The PRESIDENT (Mr. Charles Pilkington) said that a young man, who had been trained to work with machines, would get better results from any coal-cutter than an older man who approached the subject with fixed ideas. The colliery manager and the workmen should both be trained to the work. He might mention that, in using a compressed-air coal-cutting machine, the noise of the exhausted air was confusing to those working near it.

Mr. SAM MAVOR, replying to the discussion, said that, although his paper was not written in special advocacy of the bar type of machine, he might be permitted to say that the President's fears as to the strength of the bar machine were unfounded, and that the low cost of repairs was freely acknowledged by those who had experience of it. Ironstone-nodules of small size were not troublesome; and, if of large size, the bar machine would, if the head-room permitted it, cut over them; or the bar could be swung out of the holing, the machine drawn past the obstruction, and the bar cut in again and the work would be continued; but, if a disc machine came into contact with an obstruction, considerable time was lost. With regard to the inclination, he had quoted a case in his paper, in which 1 in $3\frac{1}{2}$ was mentioned; that was the maximum inclination with which he had had to deal. When cutting across the dip (the face being advanced to the rise), no machine was easily kept up to the face, owing to the constant thrust upon the props due to the weight of the machine, but the bar type presented no special difficulty in this respect. With regard to the width between the gateways, he (Mr. Mavor) thought that the practice in Lancashire and in Yorkshire, in this respect, was at fault, in many cases at least. He admitted, however, that it was impossible to dogmatize in questions of this kind, as every case should be decided in accordance with local factors; but he thought that both in Lancashire and in Yorkshire, the distance between the gate-roads in many cases, might be decreased with profitable results.

The responsibility of using electrically-driven coal-cutters in gassy seams was a question that the mining engineer must decide; but he submitted that there were ways of overcoming this difficulty. An interesting example had come under his notice within the last few weeks:—In a naked-light pit, a thin seam of coal, about 19 inches thick and of excellent quality, had not been worked because of gas. The proprietors were reluctant to work it, as the use of safety-lamps might have been imposed throughout the colliery. After experience of the use of coal-cutters in other seams, the manager adopted their use in this thin seam also, and, by means of an auxiliary electrically-driven fan, a sharp air-current was sent along the face; the copious ventilation, in the relatively small working-area, dispelling all risk from gas. He did not suggest that this was a panacea for every case where gas was found, but he had little doubt that there were many cases where the adoption of this method would prove advantageous.

In cases where the present output per foot of working-face was small and the cost of maintaining the gateways was large, he felt assured that there was a wide field for economical working by coal-cutting machinery. There could be no doubt, however, unless the old systems of working were altered, that the machines would not be used to the best advantage. The noise created by a compressed-air machine was a perfectly valid objection, but an arrangement had been introduced by which the exhausted air was turned into an enclosed crank-chamber, and thus to a large extent the noise was muffled.

Mr. W. BOLTON SHAW (Hulton colliery) wrote that Mr. Mavor had reviewed the whole of his subject in so masterly and comprehensive a fashion that his paper would no doubt come to be regarded as a classic on coal-cutting. His views on the various aspects of the subject appeared to have been thought out with the utmost care and judgment, making it difficult to discuss his paper in any other way than by express-

ing agreement with his views. As one who had had some experience of the maintenance and repair of coal-cutters, he would
like to make a few remarks from an electrical engineer's standpoint. Mr. Mavor brought out the point, which, he thought,
was hardly sufficiently appreciated, that it was a mistake to over-
drive a cutter, that it paid better to have more machines and to
be satisfied with a moderate length of cut per shift by each, than
to try and make record cuts in order to increase the output per
cutter. The output might be increased, although that was
doubtful, but any reduction in the cost of cutting was more than
counterbalanced by the enhanced cost of repairs which increased
out of all proportion to the increase in the output, and also by the
greater frequency of breakdowns and the disorganization of the
whole system which they produced.

Mr. Mavor recorded in Table I.* those costs which were
affected by the output per cutter per shift, namely, machine-
labour and interest and depreciation of cutter, as about 10 per
cent. of the total cost. It was, therefore, evident that an increase
of even 50 per cent. in the cutting speed of the machine would
only affect the total cost by 3⅓ per cent. It ought to be possible
to obtain far greater real savings than this by studying and
organizing the other operations which made up the total cost,
and Mr. Mavor had gone very fully into the way in which this
could be done. There could be no doubt that the commercial
failure of many cutters was due to this point not being sufficiently recognized. He thought that makers were somewhat to
blame in pandering to the desire of purchasers by boasting of
the achievements of their machines in that respect. The electric
cutter suffered most from this kind of abuse. The compressed-
air machine stopped and refused to move, and there being no
flywheel-effect, as in the case of the electric cutter, little or no
harm was done to the working parts. In this respect, the
three-phase machine was superior to the direct-current, and
what many people formerly regarded as a vice was in reality
a virtue. This tendency of the three-phase machine to pull up
when overloaded certainly did prevent the machine from being
abused to the same extent as the direct-current—actual experience shewing that the cost of upkeep was distinctly less per ton
than with the direct-current machine.

* *Trans. Inst. M. E.*, 1906, vol. xxxi., page 388.

He (Mr. Shaw) had had an opportunity of comparing the working of two similar cutters by the same makers, a three-phase and
a direct-current, both cutting to the same depth in hard fire-clay
in the same seam. The direct-current machine was driven at
the highest possible speed, and the average cut per shift was
about 60 per cent. greater than with the three-phase machine: it
being impossible, for the reason stated above, to drive the
latter at greater speed. The wear-and-tear on the gearing,
shafts and bearings was at least four times as great in the direct-
current cutter. Shafts which were never bent or strained in the
three-phase machine continually gave trouble in the direct-current
cutter, and bearings and gearing, which only lasted weeks in
the latter, lasted as many months in the other machine.

With regard to the actual cost of picks and repairs, Mr.
Mavor's figures seemed very low. The results at Hulton colliery
for the same items, taken over a period of three years with a
number of cutters, both direct-current and three-phase, working
in seams ranging from 23 to 48 inches, show a minimum cost of
6·4d. per ton; and, during one year, the cost had been as high
as 11·4d. for a group of five machines. The accounts from which
these costs were taken had been most carefully kept and included
all wages, material and stores chargeable to the repairs and
maintenance of the coal-cutting machines.

Mr. Mavor suggested a system of periodically overhauling
coal-cutters after the manner of rolling stock; and this was no
doubt an excellent idea where the number of machines in the
same mine warranted it. A fairly successful system was to give
a fitter charge of two coal-cutters, not necessarily in the same
mine, and to make him responsible for keeping them in proper
working order. It was his duty to go down and see the cutter
actually working at least every other shift; and this, in addition
to giving him the opportunity of seeing the working parts in
motion, also enabled him to keep in touch with the drivers. He
was under the direction of the electrical foreman of the section,
who was thus kept well informed of the condition of all the
cutters. He was paid a time-wage somewhat lower than the standard wage, and received a bonus proportional to the amount of
coal cut each week in excess of a fixed minimum quantity.

The personal factor was certainly very important, one driver
seeming to have the knack of humouring the machine while

another shewed an unnecessary amount of brutality. A bad driver was dear at any price, and a good one was certainly worth a good wage. A man sometimes became a better man after receiving a better wage, simply because he was encouraged by it.

Mr. Mavor directed attention to keeping the shafts and gearing in proper adjustment and properly lubricated. He (Mr. Shaw) could heartily endorse this. He had experience of a number of cases of bearings being completely worn out in a few days' time owing to neglect of lubrication, followed by the impossibility of effective lubrication after the bearing clearances had become excessive. Gearing, too, very rapidly deteriorated after the relative position of the shaft had been allowed to alter owing to wear in the bearings. With bevil-gearing this was particularly the case, and he had recollection of two bevil-wheels, costing something like £10, completely ruined in three shifts owing to the want of proper adjustment. In his experience, bevil-gearing required a great deal of skill and intelligence to adjust; its use in coal-cutters should, he thought, be avoided as far as possible, and, where it was used, proper provision should be made for taking up the end-thrust. This latter was a point to which many makers seemed to pay insufficient attention. He thought, also, that all gearing, wherever possible, should be enclosed and run in a thick oil-bath. Mr. Mavor pointed out that the coal-cutter was a machine-tool of special design, working under exceptionally trying conditions. In his experience, the fitter who was responsible for repairing it should be highly skilled and intelligent, and should if possible have had some experience of the making of machine-tools. Mr. Mavor advised care in the use of lubricants. The ordinary cheap engine-oil, as used for colliery-purposes, was certainly useless for lubricating anything but the slowest running shafts of coal-cutters; and when melted with a large proportion of solidified oil, it was found very satisfactory for use on enclosed gearing.

Mr. Mavor stated that the alternating-current squirrel-cage motor, with the switch submerged in oil, afforded the greatest security, but that the use of oil in switches should be avoided if possible. Why should it be avoided? He had had such a switch in use on a coal-cutter for three years, he had had absolutely no trouble due to the oil, and he could state that it was the most satisfactory switch that he had used on any coal-cutter. So long as the

oil-switch was designed to have a proper head of oil above the
sparking points and the insulation of the switch and leads was of
such a nature that the oil had no deteriorating action on it, the oil-
switch was, in his opinion, the best switch for underground
alternating-current work. There was the objection that the
oil might leak or waste away, but this should not occur in any
properly-designed switch; and it was not so serious a difficulty
as that of keeping intact the lid-joint of the ordinary coal-cutter
switch.

Most people would agree with Mr. Mavor's statement that
the chief risk attending coal-cutters lay in the trailing cable,
and this, he thought, pointed to the moral that these cables
should be carefully designed and made of the best material.
Cheap low-grade cables should not be tolerated. He formerly
had had serious trouble with the trailing cables, chiefly owing to
the unsuitability of their design, but he had now used for several
years cables made to his own design, with entirely satisfactory
results. The main features of the construction of these cables
were as follows:—A thick padding surrounding each conductor
over the insulation; a copper-wire armouring, of small gauge,
serving as an earth-shield; and an outer braiding of hard and
durable waterproofed cord. These cables had a long life, were
exceedingly flexible and handy, and were almost proof against
breakdown due to falling materials and similar damage.

With regard to power-supply, Mr. Mavor pointed out the high
cost of generating when power was produced by an independent
plant put down for coal-cutting only. Where a power-company's
supply was available, there was little doubt that power for this
purpose could be bought more cheaply than it could be produced
by an independent supply; and where electric power was used
for other purposes, if the total amount was comparatively small,
and the demand was intermittent and of a highly fluctuating
character, the purchase of power would in many cases be the
cheapest way of obtaining it. Where a plant was erected for
coal-cutting only, the capital-cost should be kept as low as possible, consistent with good material and sufficient capacity.
Experience shewed that, with the low load-factor obtainable with
coal-cutting alone, interest and depreciation formed by far the
largest item in the cost of generating, and the extra economy
obtained by the use of high-class engines, condensers, econo-

mizers and other refinements, did not, by any means, balance the increased capital-charges. There were, however, few collieries where electric power could not be used for other purposes in addition to coal-cutting; and, where this was the case, a great saving could be effected by developing the electrical installation as largely as possible.

DISCUSSION OF MESSRS. W N. ATKINSON AND A. M. HENSHAW'S PAPER ON "THE COURRIÈRES EXPLOSION."*

Mr. HENSHAW said that the paper had been prepared for the North Staffordshire Institute of Mining and Mechanical Engineers, and he thought that it would be better if, before the discussion, he tried very briefly to run through a few of the main points of the paper.

Nos. 2 and 3 pits, photographs of which were on the table, gave an idea of the general appearance of the collieries, one pit being old and the other quite new. The fourteen shafts varied from 750 to 1,200 feet in depth, and up to 15 feet in diameter. They were generally fitted with compartments for ladders from top to bottom.

Air-compressors were installed at all the pits. At some of the new pits the winding-engines had compound cylinders. The mechanical engineering of the Courrières Company was really first-class, and compared well with good English collieries. The ventilation was effected by fans.

The Pas-de-Calais measures were below Cretaceous strata, the Upper Carboniferous formation being denuded. The Cretaceous contained large quantities of water, and in sinking the freezing process was generally resorted to. In the shafts the water was kept back by tubbing of oak, and sometimes of iron. There was a bed of impervious clay which prevented the water from getting down to the workings. The Coal-measures, 150 metres from the surface, were much faulted, the main faults taking a general direction east and west. The workings in the various seams consequently extended east and west, and the recovery-drifts took a direction north and south.

The usual practice of opening out was to drive north and south from each shaft, at vertical intervals of from 20 to 50

* *Trans. Inst. M. E.*, 1906, vol. xxxii., pages 439 and 340.

metres, main cross-measure drifts or bowettes. These cross-measure drifts intersected the seams at many points, and, as a result, a large number of separate districts were opened. These districts and bowettes were again connected by numerous vertical staples. It would be easily understood, therefore, that the workings were extremely complicated. The system lent itself to extension and intercommunication, and eventually to the joining-up of the different pits and seams. These communications were consequent, not only upon the natural conditions, but were necessary for the purpose of ventilation, owing to the practice of generally sinking single shafts. The extent of the underground workings in the pits concerned was estimated at about 100 miles of road laid with rails.

All the Courrières pits were practically free from fire-damp, and they were nearly all worked with open lamps; but, according to certain regulations, safety-lamps were required to be used in headings in the direction of unproved ground, for driving staples upwards and in the preparation of new stages at greater depths than those previously worked. Safety-lamps of the Wolf type were therefore used in Nos. 4 and 11 pits in the deep workings at 383 metres, and in certain advance-headings to the deep from the 331 metres level. At the time of the explosion 250 safety-lamps were in use here; in No. 3 pit a few safety-lamps were being used; and in No. 2 pit, 90: otherwise all were open lights.

During the writers' investigations, which covered practically all the workings, they found no trace of gas whatever. Some fire-damp had, however, been observed in recent years. Dust was plentiful, and no watering or other method of dealing with it was adopted.

Blasting was generally resorted to for stone-work and coal-getting; but none of the explosives were submitted to official tests as in England, although they were required to comply with certain formulæ.

The explosives used were No. 1 Favier powder and the grisounites, couche and roche. These explosives compared very closely with some of the British permitted explosives. The explosives were supplied by the company, the cost being deducted from the men's wages. The shots were fired by detonators and safety-fuses, where naked lights were in use; and by electricity where safety-lamps were used.

Five only of the pits were directly concerned in the explosion.

They were Nos. 4-11: these were together, No. 3 single, No. 2 single and No. 10 single. No. 11 pit was a winding and downcast shaft. No. 4 was an upcast. No. 3 pit was divided by a partition to a certain depth, and was both an upcast and a downcast. No. 2 was an upcast, and No. 10 was a downcast.

Mr. Henshaw next described the Cécile fire. This fire was found between March 6th and 7th at No. 3 pit. It was first discovered by smoke and "stink" in a return-airway through old workings between the levels 280 and 326 metres in the veine Cécile. It was dealt with by the erection of stoppings in the roadways leading from the area affected. Seven stoppings were put into the roads, five on the upper and return sides, and two on the lower or intake side. On the night previous to the explosion, the general manager had been down the pit, and before coming out had seen the stoppings completed and closed. At six o'clock in the morning he handed over the charge to M. Barrault, the manager of No. 3 pit, who went down to complete the work. The latter was killed by the explosion, which occurred at seven o'clock.

After the explosion, it was found that two of the stoppings near the shaft had been blown down, but instead of being blown out had been blown in. There was no trace of flame near them. The stoppings on the intake side were undisturbed, and there was no evidence of explosion, although flame had passed near by on the level 326. The indications of force shewed that the explosion had come from some other quarter, and passed by the stoppings. Personally, he was quite certain that the Cécile fire had nothing whatever to do with the explosion. This point was fully dealt with in the paper.

On March 10th, 1,665 men and boys went down the pits Nos. 2, 3, 4 and 11. The evidence of the explosion at the surface was the emission of quantities of smoke and dust. At Nos. 4 and 11 pits, there was some damage to the roof, and at No. 3 pit to the landing-floor. At Nos. 2 and 10 no evidence reached the surface, except smoke at No. 2 a few minutes after the explosion.

Rescue-parties quickly got to work. M. Bar, the general manager, Mr. Domezon, the divisional engineer, and Mr. Bousquet, pit-manager, went down No. 11 pit by the ladders. The cage was fast in the pulleys. It was repaired and got to work, and they found 100 or more men at the 331 level who had mostly come from the 383 level and were waiting for the cage.

Twenty-five other men and boys were rescued from this pit up to 11 o'clock at night.

At No. 3 pit the general manager, M. Petitjean, found that the fan was not exhausting from the mine. He tried to descend, but the cage was fast at the bottom. He detached the rope of the lower cage, and replaced the upper cage by a hoppet. On descending the shaft, he found it at 150 metres completely blocked by débris of ladders, ladder-compartments, and partitions. During the day he managed to clear some of the obstruction, but it was impossible to get through.

The engineers, M.M. Voisin and Pégheaire, went down No. 2 pit but were partly asphyxiated, the former getting his leg broken. At half-past seven this shaft became impassable, by reason of after-damp coming up. The rescue-party then proceeded with their work from No. 10 pit. The chief inspector of the district, Mr. Léon, with two assistants, arrived at eleven o'clock, and by a decree of the State which required, in accidents of this nature, that the State inspectors should take charge, they assumed full control of the operations as soon as they arrived. At six o'clock in the evening Mr. Delafond, the inspector-general of mines, arrived from Paris. He descended No. 11 pit and visited the others. In the meantime many men and boys had been brought out by Nos. 11 and 10 pits. At half-past seven the general manager in No. 3 shaft thought that he heard cries from below the obstruction referred to. A rescue-party went down by No. 10 pit, and found thirteen men at the 303 landing at No. 3, who had escaped from various parts of the mine.

On Sunday, March 11th, a consultation was held between the State engineers, the engineers and surveyors of the company, and engineers of neighbouring mines, to consider what should now be done. They believed that there were no other survivors; and, in addition to the danger from after-damp, the possible development of the Cécile fire gave cause for anxiety. The fire of course, at this time, was generally believed to have caused the explosion; and, in order to secure the safety of the explorers, the question of ventilating the wrecked workings was discussed, and it was then advised that the best course to adopt would be to reverse the ventilation. The work was started on Sunday evening and went on during Monday. But on Monday it was found that the air would not reverse. It was practically standing in No. 2, and

would not go down No. 4. Another plan was resorted to: No.
11, which had been a downcast, was closed; No 4 was retained
as an upcast; and No. 2 was converted from an upcast to a downcast, with No. 3 as the sole upcast. No. 2 pit was now the only
pit from which the work of rescue could proceed.

On March 12th the Germans arrived from Westphalia. They
did no rescue-work, although they were instrumental in
recovering bodies.

On Tuesday, March 15th, five days after the explosion,
a fire was discovered in the Joséphine workings. It was decided
to erect stoppings near No. 2 pit in the bowette 340 leading to
the veine Joséphine, and one stopping in the bowette 306 leading
to the seam Julie above the fire. These stoppings closed the only
remaining entrance to the pits affected by the explosion. On
March 17th it was decided to attack the fire, and, the stoppings
being replaced by iron doors, the work of extinguishing the fire
was commenced on the 18th by means of a range of pipes and
water under pressure. In this work the Salvage Corps from
Paris and Westphalia, with their respiratory apparatus, did good
service. On the 27th, the fire having diminished considerably,
the stopping in the bowette 306 was opened.

On March 30th the exploring party, having gone in by the
route 306, found thirteen men alive, making their way slowly
towards No. 2 pit. They had kept themselves alive by eating the
provisions of their dead comrades, and corn from the stables.
They had been working to the extreme south of No. 3 fire at the
time of the explosion. Great excitement was naturally caused by
this marvellous escape, as there appeared even then a possibility
of others living entombed in mines. During the two days following the whole of the workings of Nos. 2 and 3 pits were examined
and searched in all seams without result; and it was equally
urged that Nos. 4 and 11 pits, which had been closed since March
12th, should be opened and searched. There was, however, to
be considered the effect of ventilation on the Cécile fire at No.
3 pit, and the hurried completion of some repairs which had
been going on to the winding-engine at No. 11 pit. On April
2nd the search of Nos. 4 and 11 pits was commenced, and three
days afterwards another survivor was found alive near No. 4
pit. His name was Burthon: he had survived the explosion
twenty-five days, and when found seemed incapable of realizing
his position. On May 20th a second fire was found in the

Joséphine north-east of No. 3 pit.   It was promptly dealt with by the erection of stoppings, but eleven bodies were enclosed.

As to the cause of the explosion, he (Mr. A. M. Henshaw) remarked that the pits produced little fire-damp, and the best evidence of that was that none was found after the explosion. When the members considered the area covered by the explosion, it was impossible to believe that even a sudden outburst of fire-damp could have fouled so extensive a range of workings, and caused such an explosion as would account for all the effects observed.   The roads were dry and dusty, especially the parts traversed by flame.   The disastrous result was to be attributed to coal-dust alone.

Before the authors' visit, some of the main roads had been cleaned, without indications being recorded.   They traced the indications to a common source, and this led them to the north side of No. 3 pit at the 1,070 feet (326 metres) level.   On May 18th, one of the writers went into the road in the Marie seam from the Joséphine seam, and found leading indications; and on May 22nd, Mr. Heurteau found a blown-out shot-hole in the face of the Lecœuvre heading in the Joséphine seam.   This shot-hole, as the point of origin, was consistent with all the other indications found. The most probable explanation was that the shot in question had missed fire on the previous day; that at the time of the explosion the men were engaged in cutting out the shot; and that in doing so they struck the detonator and exploded the charge.   He (Mr. Henshaw), therefore, attributed the explosion to a blown-out shot and coal-dust.

He (Mr. Henshaw) hoped that his brief remarks would enable the members to follow the discussion with interest.   He directed particular attention to the important lesson to be drawn from the disaster, that the great extent of the explosion and the terrible loss of life were due to the presence of dry coal-dust in the roadways and workings of the mines: and in this, the most disastrous explosion ever recorded in the history of coal-mining, the dangers of coal-dust were, in his opinion, most clearly demonstrated.

Mr. HENRY HALL (H.M. Inspector of Mines) said that it was absolutely necessary that all the facts should have been brought before the members in the interests of the British mines.   The members would all agree that the enquiry had been put into most excellent keeping, in the hands of Mr. Atkinson and Mr. Henshaw. It remained for the members to enquire into the matter fully. He hoped that the result would be that some steps would be taken such as would render so appalling a disaster impossible in this country. So far as the criticism of the paper was concerned, it struck him as most singular that the whole of the information depended upon what the authors thought them-selves. When they had an enquiry in Great Britain, an endeavour, as far as possible, was made to get information from those who worked in the pit on previous days, and from any of the survivors. Information of that kind was absolutely absent in this enquiry, and that he thought was possibly the greatest drawback that could be mentioned with regard to the enquiry itself. The paper itself was most complete and reflected great credit on the authors. Speaking generally, he (Mr. Hall) thought that the conclusion of the authors of the paper was right—that the disaster was caused solely by coal-dust—but the details were open to criticism, where, for instance, they endeavoured to fix the origin of the explosion. The members all knew that this was most difficult, and it very seldom had been done with absolute certainty. He thought, however, that the authors could be excused if they had not quite satisfied the members with regard to the site of the explosion. They had stated that the cause was a blown-out shot; but they had still to prove that the shot had previously missed fire. In his experi-ence, a coal-dust explosion started, as it were, gradually, and increased in force and in volume as it travelled. He could not imagine that a coal-dust explosion could start in the desperate manner described by the authors. With regard to the general conclusion that coal-dust had caused the explosion, it behoved the members to consider what they were going to do. If the members believed that coal-dust was the absolute cause of this loss of life they would have to undertake watering on a very different scale from what they had been accustomed to. It behoved them to satisfy themselves, however, that in this respect they were taking precautions against an absolute danger. He (Mr. Hall) would recommend the coal-trade to establish an experimental station, not 150 feet long but 1,500 feet long, and see whether that model gallery could be blown up by coal-dust in the way suggested. If it could not be blown up in this way, then mine-owners ought not to be called upon to undertake the precautionary watering of their mines. He was afraid that mining engineers in France were unprepared for such an explosion as that which occurred; and it seemed to him that they had very little knowledge of how to proceed after the disaster, and many steps were taken of very doubtful utility. He was not quite certain that mining engineers in Great Britain were in a much better position. When an explosion occurred in this country, there was no one upon whom the duty devolved of saying what steps should be taken to rescue those left in the mine. He thought that some action should be taken, so that some person or some committee should be consulted before any mine was closed, while men dead or alive remained in it. To put the responsibility, however, solely on H.M. inspector of mines was almost more than any man should bear.

Mr. JOHN GERRARD (H.M. Inspector of Mines) congratulated the authors on the success which they had attained; and the fulness of the information given proved the immense pains that they must have taken. It was fitting that such a terrible catastrophe, the greatest in the annals of mining, should be thoroughly enquired into by British mining engineers, and he thought that it would be impossible to have found anyone better qualified than the authors of the paper. One could not but be struck by the extraordinary facilities rendered them : the owners of the collieries, their engineers, and the inspectors of mines must have received the authors with open arms; and for his part, he was anxious that this should be fully recognized.

The cause of the immense loss of life at Courrières opened out a wide field for discussion—the shafts, the roadways connecting the workings with the shafts, coal-dust, explosives, underground fires, discipline, mines-inspection, etc. If there was anything to be learned from that terrible disaster, it was their duty to apply the lessons to British mines with a view to the prevention of a similar loss of life in this country. Mr. A. M. Henshaw had spoken of the cause of that disaster as being a blown-out shot. He (Mr. Gerrard) was not going to differ from him, but he would like to examine the question for a moment, because on previous occasions shots had been called blown-out shots, which, to his mind, were not blown-out shots at all. Could this be directly called a blown-out shot? In point of fact, it would not have been fired if the detonator had not been exposed. It was supposed that the detonator was struck by a pick. The starting of the force in this case was quite different from that seen so many times at the start of a coal-dust explosion. The smashing of the air-pipe, which took the air from the face, was in itself interesting; and, if fire-damp had been found, the members could understand the bursting of the pipes in that extraordinary outward manner, or coal-dust in the pipe might explain it. He regretted that there was no information with regard to the amount of ventilation that passed through the mine. He asked whether the Government officials, in taking possession of the mine, acted on their own initiative absolutely or whether they consulted with the engineers.

Mr. A. M. HENSHAW replied that they consulted with the engineers.

Mr. GERRARD remarked that it was foreign to his experience. It had always been customary to work together with the engineers that assembled; and nothing was done without the approval of the engineers. The direction of operations by Government officials seemed to be open to very serious question, and was extremely undesirable. One of the most interesting points in this valuable paper was the analysis of the dust of the roads, distinguishing the dust that could feed an explosion and the dust that did not sustain one. There were many questions that he would have been glad to ask, but the large number of members present, some of whom would wish to join in the discussion, stopped him from going farther.

Mr. JOSEPH DICKINSON said that he had read Messrs. H. Cunynghame and W. N. Atkinson's report,* and Messrs. W. N. Atkinson and A. M. Henshaw's paper, but there were still many details which had not yet been brought before the members. Fire-damp was almost entirely absent, although in 1904 a miner was burnt by the ignition of fire-damp by a naked light, and some fire-damp was met with in the lowest level in Nos. 4 and 11 pits in 1903, 1904 and 1905. It seemed that all hope of any survivors being in the pit appeared to have been abandoned up to the time of the appearance of 13 men, twenty days after the

---

* *Report to H.M. Secretary of State for the Home Department on the Disaster which occurred at Courrières Mine, Pas de Calais, France, on March 10th, 1906*, by Messrs. H. Cunynghame and W. N. Atkinson, 1906 [Cd. 3171].

explosion. It was stated that the blast did not appear to have been so violent as in some British explosions, yet that there were traces of violence in some parts and notably in the intake-airways. Then in connection with the fire, which was discovered in the Cécile seam before the explosion, it was a rather strange coincidence that the stoppings were closed just a little time before the explosion took place. The mode of re-entry after the explosion did not meet with general approval. Four days after the explosion, there was a general strike of the miners throughout the Pas-de-Calais district. A new trade-union was formed, and five days afterwards the Minister of the Interior had an interview with the officials. The iron door for passage, which was introduced in each stopping, when extinguishing the first fire in the Joséphine seam gave rise to the allegation that the restoration of the mine was more the object in view than the rescue of the entombed miners; and that resulted soon afterwards in a change being made. Bitter attacks found expression against the explorers, although some of them lost their lives in their endeavours. The miners' agents on the Commission of Inquiry made a premature minority report, throwing the responsibility on the owners, alleging repeated warnings to them that the mine was dangerous. The majority report, however, followed and cleared the engineers of all blame, and testified to their exertions. A member of the Mines Commission of the Chamber of Deputies visited the mine and took evidence; and, afterwards, in a debate in the Chamber, criticized the owners, to the effect that the State engineers were insufficient in number and the functions of the miners' agents too restricted,—the result being to rely upon the declaration of the Government to ascertain the responsibilities, and, if occasion arose, to enforce all requirements of the law. Meanwhile, the *Parquet* at Béthune acquitted the State engineers of all blame in the recent operations; and it was stated that medical examination of the bodies had established the opinion that of the bodies recovered none, so far as ascertained, had survived the day of the explosion.

For the purposes of his report, Mr. W. N. Atkinson went underground eighteen times between May 4th and 18th, and between June 22nd and 29th, the first view being made 55 days after the explosion. Mr. Henshaw also seemed to have made ample views. Before the commencement of the view, therefore, exploration had effected many changes, and road-ways formerly dry and dusty had become wet with water. Steam also from the water used in extinguishing the first fire in the Joséphine seam had moistened some roadways.

After tracing the direction of the explosion, the writers came to the conclusion that it was caused through a blown-out shot in the Lecœuvre heading, that probably the shot was fired while the men were attempting to cut it out, and that in all probability it was a dust-explosion without the presence of gas. The writers, however, agreed that the actual cause might never be ascertained; and, in coming to this conclusion, they gave some good, although not completely convincing, reasons. The mines were singularly free from fire-damp, and were chiefly worked with naked lights. Partially coked dust was observed in the parts traversed by flame. The first fire in the Joséphine seam was attributed to the flame of the explosion; and the second fire was attributed to the same cause. The former fire was found in the return-airway of the Cécile seam three or four days before the explosion, the explosion following soon after the closing of the stoppings, and as the explosion did not appear to have occurred at the fire, it was supposed that gas was distilled from the fire into the Lecœuvre heading. Local opinions, deserving of notice, concurred in holding that the explosion began in the Lecœuvre heading, but that it was started by gas, and that this cause was supported by part of the air-pipes in the heading being burst outwards. The parallel gallery on the rise of the Lecœuvre heading was not working at the time of the explosion, and being much fallen, was not explored after the explosion. Of the four pits in which the loss of life occurred, two of the shafts were entirely and the third partly used as upcasts for the return-air. Whether therefore, the explosion was entirely due to the blown-out shot, helped on by dust only, or whether impure return air contributed, were points that might be properly considered without prejudice. The distinguishing features were 13 pits, all connected by roads underground. If the origin was a shot, apparently that ignition might light coal-dust. If fire-damp aided, naked lights were unsafe in dusty mines, unless watered. The explosion opened out a wide field for experiment and discussion. The experiments made by Mr. Henry Hall* with gun-

* Report of Experiments to test the Effects of Blasting with Gunpowder in Dry and Dusty Colliery Workings in the Entire Absence of Fire-damp, by Mr. Henry Hall, 1890; and Report made by desire of the Secretary of State to the Royal Commission on Explosions from Coal-dust in Mines, by Mr. Henry Hall, 1893 [C.—7.185].

powder and other explosives fired from a cannon into sprinkled
coal-dust in an old shaft showed clearly that such dust could
be ignited and some force developed, yet in these experiments
as seen by himself the force was less destructive than in ordinary
fire-damp explosions.   Similar lack of vigour was noticed at the
Courrières collieries.   He (Mr. Dickinson) had produced a spark
many times from compressed air alone, but never repeated from
the same body of air once exploded.   It would be interesting to
know the amount of ventilation at the Courrières collieries,
as it was not stated by the authors.   Also, whether among such
extensive workings any new opening was being made on the dip,
from   which   atmospheric   pressure   might   have   helped   fire-
damp to ascend into the workings.   Such sudden appearances
had occurred; or the stoppings, shutting off the fire in the
Cécile seam, might have disarranged the ventilation.   Of the
total number of safety-lamps enjoined by the regulations, 250
were used in Nos. 4 and 11 pits and the other 90 in No. 2 pit.
All these lamps were in use in the pits that exploded.   The
rescue-operations had been criticized, but he would say that such
operations required nerve and care.   As to this point it should be
noted that after-damp and fire-damp are poisonous gases unless
diluted, and therefore pits containing such might not be entered
with impunity for a longer time than a person could hold his
breath unless he was provided with some reliable breathing
apparatus.   It was satisfactory to know that the report of the
officials of the Courts of Justice on the responsibilities and points
of law was expected to contain much valuable evidence.

The PRESIDENT (Mr. Charles Pilkington) remarked that there
appeared to have been a certain amount of interference by the
State engineers in the management of the pits, with which he
certainly did not agree.   He could hardly believe that it was a
fact.   It was the last thing that a British inspector of mines
would desire.   The engineers and managers, responsible for a
pit before an explosion, were the men to take charge of it after
the accident.   H.M. inspectors of mines were, of course, always
present on such occasions, and rendered valuable assistance, and
their advice was always gladly received; but, if they took the
management out of the hands of the colliery engineers, he could
quite understand that things might go wrong.

MANCHESTER GEOLOGICAL AND MINING SOCIETY.

ORDINARY MEETING,
HELD IN THE ROOMS OF THE SOCIETY, QUEEN'S CHAMBERS,
5, JOHN DALTON STREET, MANCHESTER,
JANUARY 8TH, 1907.

MR. CHARLES PILKINGTON, PRESIDENT, IN THE CHAIR.

The following gentlemen were elected, having been previously nominated : —

MEMBERS—
Mr. JAMES FILES, Mining Engineer, 402, Bolton Road, Clifton, Manchester.
Mr. T. OLIVER CROSS, Mining Engineer, 77, King Street, Manchester.

ASSOCIATE MEMBER—
Mr. WILFRID BENJAMIN WAINEWRIGHT, Los Angeles, California, United States of America.

CONFERENCE OF DELEGATES OF CORRESPONDING SOCIETIES OF THE BRITISH ASSOCIATION FOR THE ADVANCEMENT OF SCIENCE, YORK, 1906.

Mr. WILLIAM WATTS, delegate of the Society to the meeting of the Corresponding Societies of the British Association for the Advancement of Science, held at York on August 2nd and 7th, 1906, read his report, as follows : —

I attended both meetings of the Corresponding Societies on the 2nd and 7th of August, 1906, and beg to submit the following report.

The opening meeting was presided over by Sir Edward Brabrook, C.B., who, in his opening address, dealt with the advantages local societies derived by being affiliated and associated with the British Association for the Advancement of Science. The meeting was the twenty-second of the Corresponding Societies and the first at which associated and affiliated members were present. No societies were represented but those which published original papers dealing with scientific investigations, and it should be regarded as an honour to be affiliated with the British Association, to stimulate the members to further efforts in the work in which they are engaged. Members of Corresponding Societies have opportunities and advantages of acquiring local information not available to casual visitors, and it was pointed out that the British Association cultivates the study of all sciences bearing upon the welfare of mankind. Careful investigation of the principles of science in every part of the country enlarges the bounds of human knowledge and makes one take a deeper interest in the laws of nature.

In the survey of the sectional work done by the Association, economic science was not overlooked. The method of dealing with wealth leads to public good, even in the management of a household.

The investigation of the unemployed problem is of importance, also the consideration of the decline in the birth rate, with all the moral consequences it implies. The employment of women, feeding of children at schools, and the question whether marriages should be permitted where the contracting parties, by feebleness of mind and body (or in other ways) are unfit to maintain their offspring in comfort and happiness, are economic questions affecting local societies in different ways in different districts.

Each locality has its own ancient monuments, its own connection with past history, its own mixed population, with special racial affinities, its own ancient customs, some of their roots going far down into the past; its own folklore, its own dialect, its own place names, and thus every local society has an interest in working out for itself its own anthropological history.

It is only by the help of local societies that this can be done satisfactorily.

Education was referred to by the President who suggested, by way of illustration, that boys in seaport towns should be taught seamanship, boys in business towns, book-keeping, kindred trades and so on. Children, as a rule, fall into the businesses and trades followed by their fathers, and so it has been for generations past. All cannot be book-keepers. There must be tillers of land, sowers of seed, and harvest reapers, and other allied trades contingent upon these occupations.

Noah's three sons, Shem, Ham and Japhet, set the example of separate trade industries, also the diversity of languages when they separated. Environment is a great power in teaching and training children the stern duties of life. The great object of education is to develop the thinking faculties and to teach a child to think for itself and rely more and more upon its own resources for success. The healthy surrounding of children, at home and in school, was also referred to and reference made to the good effect a sound mind in a sound body would have on the future population of the country.

One sees daily evidence of the want of careful training of children. Parents ignorant of feeding and nursing their offspring; caring more for the training and rearing of animals and plants, and in perfecting themselves in games or so-called sports, than in the careful training of their own children.

Sir Edward Brabrook also spoke of the legitimate provinces of conference in inducing and encouraging free communication with each other, and learning what has been done, so as to avoid wasteful time and duplication of work by others engaged in like pursuits. "Iron sharpens iron; so man sharpeneth the countenance of his friend."

At the conclusion of the opening address, Dr. Hugh Robert Mill introduced the subject of Meteorology.

He commended the attention of scientific societies to the intellectual advantages to be derived by members from the study of Meteorology; not merely that it may help them in predicting the weather from day to day but enable them to foretell the changes of weather by careful and accurate observation.   He spoke of the use of good instruments which always assist to make observations more valuable.

Experiments with kites were referred to and the results which await the student in the measurement of moisture in the air, also the revision of the tables by means of which the humidity is calculated.

The study of local climate in which the co-operation of local societies is invited, was alluded to, and the importance of establishing well equipped and managed observatories under careful control as suggested, and Dr. Mill promised to render any assistance he could if required. Two elements of study are much needed, viz., sunshine records and rainfall observations.

The duration of sunshine is measured by a recorder but at the present time there is no accurate map of the average annual sunshine in the British Isles, yet the relation of sunlight to health and agriculture is of great importance.

The distribution of rain and the pressure of wind are local phenomena the study of which corresponding societies could take up and by the erection of a series of stations secure continuity of observations and make it possible to obtain general averages of rainfall over a larger area, so that the limits and distribution could be foretold with some degree of certainty, and also the intensity and duration.

This information would be beneficial to farmers in preparing for and gathering of their crops.   Some parts of the country are well provided with rain gauges, but they are not uniformly distributed and Dr. Mill expressed a wish that gentlemen who had the necessary time, and could afford the expense of establishing gauges would do so; not only for the scientific value of the work done, but for the pleasure of collecting the data from day to day and thus assist the Association.

It was also urged that corresponding societies should assist in seeing that proper instruments are provided and the records duly forwarded to the Meteorological office for tabulation and comparison.

A discussion followed Dr. Mill's paper in which several delegates took part agreeing in the main with the Dr.'s suggestion that more active work was required in the weather forecasts, to make them more reliable and beneficial to everyone engaged in the various trades of the nation influenced by Meterological laws.

The second meeting was held on the 7th of August, Mr. John Hopkinson, F.L.S., F.G.S., taking the chair.

The Chairman pointed out that the second conference of the delegates was held at York twenty-five years ago, the first being held at Swansea in 1880. He impressed upon the delegates the importance of preparing annual reports to their respective societies, detailing the work done at the conferences, in order that their own members should know from year to year what subjects were brought forward for discussion and thereby inspire a deeper interest in the work of the Association generally.

The importance of taking permanent photographs in all parts of the country wherever change and decay were noticeable, was recommended in order to obtain lasting impressions of the interesting land features—whether natural or artificial—for at no other period had the destruction and

mutilation been more rife than at the present time, and never before had
so much interest been taken in the preservation, change and decay of the
surface features of the land. The ravages of time should not be overlooked,
such as the encroachment of the sea upon the land, landslips, denudation
of moorland glens, and the widening action of rivers and destruction of
forests. Indeed, everything of scientific interest relating to surface changes,
the Chairman said, is worth representation by the camera. For this purpose
camera and photographic clubs should be established and worked by country
societies in conjunction with some authoritative society, or committee, in
touch with the British Association. This is the only way permanent results
can be secured and compared, and made valuable by combination.

Several minor papers were submitted by members of the conference,
dealing, in the main, with the photo-survey movement and urging upon
societies the desirability of furthering the movement by establishing local
depôts in which the pictorial prints should be placed for reference.

The physical features of the land slowly change, not only naturally
but by the agency of man. Waste land is year by year diminishing, bogs
are drained, pastures slowly creep up the hills, forests vanish, while new
plantations spring up where they did not previously exist, and the opening
of mines and quarries alter the natural scenery. These changes should be
recorded.

In addition to the opening of canals and railways, the expansion of
large towns changes the scenery which should also be placed on record, not
forgetting the ethnological survey of the people of the different countries,
which also is changing with the trades, dress, occupation, habits and amuse-
ments, by means of freer communication and education of the people.
Every corresponding society could assist in aiding the work of the British
Association and your delegate hopes that the Manchester Geological and
Mining Society, one of the largest societies in the North of England, will
do its fair share of original scientific work, and that next year your delegate
will attend the conference with a report worthy of the name of the society.

The PRESIDENT (Mr. Charles Pilkington) complimented Mr.
Watts on his able report and the suggestions it contained. He
did not know whether they would be able to make such reports as
Mr. Watts had indicated, but he was certain it would be a very
good thing if they could. Perhaps Mr. Watts would tell them
which point he more particularly desired to have taken up.

Mr. WILLIAM WATTS said he thought the subject of topo-
graphy might be taken up by the members of the Society, and not
only should changes in the features of the land be noted, but
records of sunshine and rainfall might very well occupy attention. The report was much condensed, and was only an epitome
of what took place at the meetings.

Colonel Geo. H. Hollingworth moved a vote of thanks to Mr. Watts for his report, which was seconded by Mr. John Ashworth and carried unanimously.

The President (Mr. Charles Pilkington) said he quite agreed with the remarks about the education and preparation for future work in life. When anyone in his part of Lancashire met a little girl going to school and asked, " What are you going to be?" the answer usually was, " Please, sir, a teacher," that was all. They learnt little concerning the right way of looking after a house, or the babies, or what might be useful to them in after life. They wanted to be teachers. He believed the education authorities were trying to alter this state of things. With regard to rain gauges, one of the difficulties he anticipated was the constant attention that they would require if they were to be of any use.

Mr. William Watts, after thanking the members for their vote of thanks, said that one point that was insisted upon at the York meeting was the desirability of fixing more sunshine recorders and rain gauges. Dr. Mill of the Meteorological Department wanted more of these gauges fixing to cover a wide area of the country, and he thought local societies might very well take up the task of recording the sunshine and rainfall in their various districts. Although he had under his charge a number of rain gauges covering a wide area of the country it was not fully represented in this respect, and more were required. In some districts many miles lay between one gauge and another, leaving districts uncovered by these instruments which might be represented.

HORIZONTAL AND VERTICAL SECTIONS OF COAL-MEASURES FROM RISHTON, LANCASHIRE, TO PONTEFRACT, YORKSHIRE.

Mr. John Gerrard (H.M. Inspector of Mines) exhibited horizontal and vertical sections of the Coal-measures from Rishton, in Lancashire, to Pontefract, in Yorkshire. The section had been prepared, on the Lancashire side, by Mr. William Pickup, of

Rishton colliery, who had shewn the position of the Upper and Lower Mountain mines, in connection with the district of Rishton; by Mr. George Elce, who had shewn the position of the Arley mine, and the Upper and Lower Mountain mines at Altham; and by Mr. Edgar O. Bolton, of Burnley colliery, who had shown in a very clear manner the position of the seams in the Burnley district. The first portion was carried to the end of the Lancashire basin, then came the moorlands through the Millstone and other grits until the Yorkshire coal-field was reached. The section shewed the Halifax seams which were said to correspond to the Mountain mines of Lancashire; it was extended to Bradford, Low Moor and Cleckheaton, until it included the higher seams, such as the Middleton Main, which might or might not have some relation to the Arley mine; and finally it showed the thicker seams near Normanton, and so on to Pontefract. The section was really the outcome of a discussion which took place some six or seven years ago. A Committee was appointed of members of this Society and of the Midland Institute of Mining, Civil and Mechanical Engineers, to work together and try to correlate the seams of Lancashire with those of Yorkshire. He (Mr. Gerrard) did not say that the section carried matters very much forward in connection with the work of correlation, but it was a first step, and shewed the position of the different coalseams; and upon this section mining engineers could advance their favourite theories so far as they chose to go. The section now exhibited belonged to friends in Yorkshire, and before handing it over to them he thought that it might be shewn at this meeting. The distance covered by the section was about 60 miles. The absolute break between the two districts was clearly shown. He was very grateful to the gentlemen he had named for the work that they had done.

A hearty vote of thanks was accorded to Mr. John Gerrard for the trouble that he had taken in the matter, and for exhibiting the section.

DISCUSSION OF MR. J. T. STOBBS' PAPER ON "THE
VALUE OF FOSSIL MOLLUSCA IN COAL-MEASURE
STRATIGRAPHY."*

Mr. JOHN T. STOBBS said that zones of freshwater mollusca
had already been established over most of the British coal-
fields. These zones were first worked out in North Staffordshire; and with this key to their sequence it was comparatively an easy matter to go to the other coal-fields and see how
far the sequence held in them. The presence of these different
forms was a true indication of the coal-seams in their vicinity.
It was everybody's experience that one never found anything
unless one looked specifically for it, and that one always found
what one looked for specifically. If these forms were diligently
looked for, he believed that they would be found. In Yorkshire,
workers were taking up this matter in earnest, and he did not
doubt that in a very few years they would be able to add very
much to their store of knowledge of this question. Mining engineers required some scheme for the unification of their knowledge of the Coal-measures, so that when a man, who had been
trained in one coal-field and knew its sequence, went to another
coal-field he could, almost at once (if the knowledge were codified
and registered and recorded), pass from the sequence of the new
coal-field to that with which he was already familiar.

To some extent that need was met by classification of the
Coal-measures, and classification had been taken up by different
workers and based on different grounds. Almost everybody had
adopted the Upper, Middle and Lower Coal-measure divisions
until the last few years. That classification had the advantage of
being rough-and-ready, but a division, a threefold division, of
3,000 feet or more of Coal-measures was not, and never could be,
refined enough for the practical needs of the mining engineer,
who wanted to know with much greater particularity where he
was, in proving faults, in putting down bore-holes, and in sinking, than simply to know whether he was in the Upper, Middle or
Lower Coal-measures. Another disadvantage of that scheme was
that the Upper, Middle and Lower divisions of one district were
not correlative with the Upper, Middle and Lower divisions
respectively of other coal-fields. That was a serious disadvan-

* *Trans. M. G. M. S.*, 1905, vol. xxix., page 323; and *Trans. Inst. M. E.*,
1906, vol. xxxi., page 485.

tage; people acquainted with the characteristics of, say, the
Middle Coal-measures of one coal-field, when they went to
another coal-field were often misled.   They found that certain
organic forms were either present in, or absent from, the Middle
Coal-measures in the new district, and the result was that they
had to get their bearings afresh, *ab initio*, for the new coal-field.
The Upper, Middle and Lower terms had been adopted by palæo-
botanists; but the trouble was that they had so many transition
series.   As a matter of fact, the Coal-measures were one great
transition series so far as fossil plants were concerned, and that
introduced an amount of uncertainty into the correlation by
plant-remains that was most perplexing to the practical mining
engineer.   Then, there was a recent classification based on
colour, such as a red, or a grey, or a pale-grey, or a red-and-grey
series.   Now, no classification of the Coal-measures based on
colour could be of much service to the mining engineer.   Colour
was, geologically, one of the most uncertain things, depending,
as it did, so largely on fortuitous changes produced by weathering, and it was unsafe to base any classification of the
Coal-measures on the fact that a majority of the strata were either
grey or red.   Practical mining engineers had attempted the correlation of the coal-seams of individual fields by reference to
their physical characters, and by comparing sections of coal-
seams of a certain thickness and quality, and lying at a certain
depth above or below other seams of a certain thickness and
quality.   They had, on such data, joined up one seam with
another, and that was what he called " the arm-chair method of
correlation."   It was based on the variable data above-mentioned; and a classification that was based on such variable data
could not be reliable.   Eventually he had been driven to try
the system of correlation by the mollusca.

There was, he (Mr. Stobbs) found, considerable scepticism as
to the value of these fossil shells, and it rested with those who
believed in this system to shew that it was of practical value.
Coal-measure mollusca had been very largely neglected, and it
was only within the last few years that they had been shewn
to be general and wide in their distribution.   He thought, however, the fact that eight zones had been determined, and that
the uppermost and the lowest extended from the Northumberland and Durham coal-field to the South Wales coal-field, ought
to prove that the subject deserved systematic work and investi-

gation by mining engineers.   The complete palæontological sequence of each coal-field would only be determined by diligent and even laborious effort on the part of residents, and local workers were greatly needed.   Each shell would teach mining engineers something, if their observation were only subtle enough. Slight variations of the same species were at present observable, and those variations in time, as men got educated to look at them aright, would convey useful knowledge as to the true horizon of the strata.

Instances where the mollusca would be of special service, and examples where they had already proved so, might be cited: in coal-fields like those of Lancashire and Yorkshire, where the Coal-measures were overlain unconformably by newer rocks (Triassic or Permian), great care should be exercised when sinking through them to the underlying Coal-measures.   It was of the first importance that the mining engineer should know, when he had sunk or bored through the former, the horizon reached in the Coal-measures, and undoubtedly mollusca would be able to help him most.   His (Mr. Stobbs') confidence on this point was based on experience gained in mining operations, as he had proved the utility of mollusca in determining horizons even in borings.   In the Cheadle coal-field, a bore-hole was put through measures which had never been passed through before and were not exposed at the surface, and it was important that the parties responsible for the bore-hole should know what seams had been passed through.   The fossils found in a core, 3 inches in diameter, enabled him to identify the Four-feet seam; and a further examination of the cores shewed a new marine bed about 81 feet above the Dilhorne coal-seam of that district.   About a year later, when a shaft was being sunk in the same coal-field, he found the same marine bed, and he was able, therefore, to state that 81 feet lower down they would get the Dilhorne coal-seam; and some months later that coal was found at that exact depth.   Of course, great care was required; the shells should be authoritatively recognized, and their diagnostic value should be properly assessed.*   He (Mr. Stobbs) appealed to the younger members to assist him, as workers were needed in every coal-field. In taking up this work they would have the satisfaction of feeling that they were making discoveries and adding to scientific

* *Trans. M. G. M. S.*, 1905, vol. xxix., page 336.

knowledge, and that they were acquiring skill in recognizing
the different forms and in looking for them. He assumed that
the members would agree that the engineers of the future had
their work cut out for them, and it was the more necessary that
in this generation they should settle the question of the distribution of mollusca in the Coal-measures, so that this problem, at
any rate, would not be a future source of uncertainty and anxiety.
In the future, as a result of accurate accumulated work in different coal-fields, this question, now in portions of the sequence in
a state of undesirable uncertainty, would be so established that it
would be possible to define the position of any seam or band in
the Coal-measures with the certainty that one now, in the case
of winding-engines of given power, determined the quantity of
coal that could be raised from a certain depth in a given time.

Mr. H. STANLEY ATHERTON urged the desirability of forming
a band of workers in this field of enquiry who would render to
Lancashire something of that service which Dr. Wheelton Hind
and Mr. J. T. Stobbs had rendered to North Staffordshire.

Mr. A. RUSHTON exhibited fossils found in the roof of the
Wigan Nine-feet or Trencherbone seam, at Maypole colliery,
Abram. They formed part of a shell-bed lying on the roof,
within 3 feet of the coal-seam. He had worked the Trencherbone
seam in the Manchester district, and the Wigan Nine-feet in
collieries lying to the north of Wigan, but he had not found
these shells previously in the roof of that coal-seam. They
appeared to be of a local distribution at Maypole colliery.

Mr. WILLIAM OLLERENSHAW remarked that mining engineers
and others who had experience in coal-mining were agreed that
a more reliable method and system of correlating coal-seams was
required, as it had been repeatedly proved that the present
system was unreliable. The Two-feet or cannel coal-seam, in the
Manchester district, in the course of a few miles, altered its
physical peculiarities several times. At Ashton-under-Lyne, it
was an ordinary bituminous coal, 2 feet thick; at the Dukinfield
collieries, about 2 miles eastward, it varied from 2 feet to 6 feet
in thickness, and was a valuable cannel-seam; a little further
eastward, it again altered its physical features, and was found at
Hyde colliery, with occasional belts of cannel; and at Denton
colliery, as an ordinary seam of coal, 2 feet thick. At Denton

colliery, the strata between the Great and the Roger coal-seams varied in a very short distance from 100 feet to about 60 feet in thickness, with a corresponding alteration in the character of the strata. These examples proved the unreliability of the system of determining or correlating seams by their appearance and the thickness of the intervening strata, or by the physical features of a coal-seam. He agreed that, if mining engineers would proceed on the lines advocated by Mr. Stobbs, it would prove a much safer method, and that the present unsatisfactory correlation of coal-seams would be improved in the future.

Dr. WHEELTON HIND said that he was pleased to find that his work on the Coal-measure mollusca of twelve years ago was bearing good fruit, and delighted that Mr. Stobbs was proving the value of these fossils in practical mining. He (Dr. Hind) approached these fossils from a biological standpoint at first, and it was only later on that the accumulation of facts of distribution shewed him the value of certain species as accurate indices of horizons in the Coal-measures. He had attempted to sketch out the distribution of the genera *Carbonicola*, *Anthracomya* and *Naiadites*, as far as was then possible, in the strata of the various coal-fields of Great Britain. One zone of great importance, characterized by the presence of *Anthracomya Phillipsi*, denoted the top of the workable coal series in most of the coal-fields. At Bristol, a higher zone occurred, but this was recognized by its flora. Marine could be readily distinguished from non-marine bands, as the structure of marine shells, their ornament, large teeth, and anatomy differed markedly from those of fresh-water forms. In marine bands, Gasteropoda, or coiled shells, Cephalopoda or chambered shells, and Brachiopoda occurred with lamellibranchs, but were never found in freshwater beds where the shells belonged to the *Unio* type, just as in the rivers and lakes of to-day.

Mr. WILLIAM PICKUP said that a correction should be made in the paper.* The words " Upper Mountain " should read " Upper Foot." The marine shells referred to were not found above the Upper Mountain, they were always found above the Bullion or Upper Foot coal, a seam lower in the series than the Upper Moun-

* *Trans. M. G. M. S.*, 1905, vol. xxix., page 329, Lancashire coal-field, paragraph (d).

tain, and generally found by itself, but sometimes it combined
with the Lower Mountain mine.

Mr. JOHN GERRARD sincerely hoped that some of the younger
members would take up this work. He was quite sure that if
they once got interested in it they would go on and never regret
their perseverance. The work was not only interesting, but
exceedingly valuable in connection with mining.

Mr. WALTER BALDWIN wrote that Mr. Stobbs and Dr.
Wheelton Hind were to be congratulated upon the good work
that they had done in Staffordshire. They had shewn that
"zoning" the Coal-measures could be carried out upon a sound
basis. He regretted that, in the Lancashire and other coal-fields,
mining engineers had been slow to search for and to adopt these
useful indices. The key which Mr. Stobbs and Dr. Wheelton
Hind had discovered in Staffordshire would, he believed,
undoubtedly open the doors of other coal-fields. He regretted that
in the Lancashire coal-field, the material, at present to hand, was
scanty, and the field-workers were few; and, until both were
increased, the value of mollusca would not appeal to the practical
man. He hoped that the effect of Mr. Stobbs' paper and the
present discussion would be that all members, who were engaged
in sinking or driving drifts, would look out for fossil mollusca; let them note as nearly as possible the vertical distance
above or below the nearest seam of coal, and submit the fossils
to some recognized expert for determination. The result of such
observations, when collected, would prove most useful and of
great value to all engaged in the coal-mining industry as well as
to the general geological student.

Whilst engaged as resident engineer, a few years ago, in
sinking a deep puddle-trench for a reservoir, he (Mr. Baldwin)
encountered three thin coal-seams associated with a marine band
containing *Goniatites, Orthoceras, Pterinopecten*, etc. He was
thus able to refer the coals to the Holcombe or Brooksbottoms
seams of the late Mr. E. W. Binney and consequently recognize
their position in the Third Millstone Grit series; and by these
means, he had recognized the same seams at various places since.
At Cheesden, on the same horizon, large calcareous nodules
occurred, such as overlay the Bullion mine or Upper Foot of
Lancashire; and, in some of these, the writer believed that he

had recognized *Glyphioceras paucilobum* (Phillips). The marine horizon above the Bullion mine of Lancashire had been well established by the late Mr. E. W. Binney, Prof. Edward Hull. and others, and he (Mr. Baldwin) had used it over large areas. He had, in company with Messrs. W. H. Sutcliffe and W. A. Parker, recognized the following fauna at Shore, near Little-borough. This list was, however, incomplete:—*Posidoniella lævis* (Brown), *P. subquadrata* (Hind), and *P. sulcata* (Hind); *Pterinopecten papyraceus; Cœlonautilus* sp.; *Orthoceras cinctum* (de Koninck), *O. obtusum* (Brown), *O. Browni* and *O. subsulcatum;* *Glyphioceras reticulatum* (Phillips); *Dimorphoceras Gilbertsoni* (Phillips); and *Gastrioceras carbonarium* (L. von Buch), and *G. Listeri* (Martin). A very similar assemblage holds good for Starring, Dearnley, Dulesgate and Besom Hill, near Shaw. About 135 feet above the Arley mine at Sparth, near Rochdale,* he (Mr. Baldwin) had found *Carbonicola acuta* in large numbers associated with *C. turgida* (not common) and *C. robusta* (rare); and *Naiadites modiolaris, N. triangularis, N. carinata* and *N. elongata.* These were accompanied by that wonderful collection of Arthropoda discovered last year by Mr. W. A. Parker and exhibited and described by Dr. Henry Woodward at the meeting of the British Association for the Advancement of Science held at York. He (Mr. Baldwin) asked whether any Arthropoda had been found at or about this horizon above the Cockshead coal-seam of North Staffordshire. In his opinion, this horizon ought to be watched at many localities, as well in Yorkshire, as in Nottinghamshire, Lancashire and in North Staffordshire.

Mr. JOSEPH DICKINSON, F.G.S., wrote requesting that the name of the coal-seam with which the well-known fossil-horizon occurred in the Lower series of the Lancashire coal-field, might be corrected in the present paper.† It was erroneously called the Upper Mountain instead of the Upper Foot coal-seam: the Upper Mountain being a seam higher in the series than the Upper Foot seam. In East Lancashire, this well-marked horizon

* "Notes on the Palæontology of Sparth Bottoms, Rochdale," by Mr. W Baldwin, *Transactions of the Rochdale Literary and Scientific Society,* 1903-1905, vol. viii., page 82; and "*Bellinurus bellulus* from Sparth, Rochdale," by Mr. W. Baldwin, *Transactions of the Manchester Geological and Mining Society,* 1903, vol. xxviii., page 198.

† *Trans. M. G. M. S.,* 1905, vol. xxix., page 329, Lancashire coal-field, paragraph (d).

continued over and with the Foot coal alone. But in North Lancashire, the Foot coal dipped down to the Gannister coal, and the two together formed the Four-feet Mountain mine in North Lancashire.

Mr. J. T. Stobbs accepted the correction concerning the use of the name Upper Foot instead of Upper Mountain mine printed in his paper. Referring to Mr. Rushton's remarks, he said that he would like to know whether it had been actually demonstrated that the Wigan Nine-feet seam was the Trencher-bone seam. He appreciated Mr. Ollerenshaw's reference to the necessity of having some more reliable guide to the nature of coal-seams than the lithological features that had hitherto been so much trusted by mining engineers.

The President (Mr. Charles Pilkington) suggested that a committee of the members should be formed to carry on these investigations.

MANCHESTER GEOLOGICAL AND MINING SOCIETY.

ORDINARY MEETING,
HELD IN THE ROOMS OF THE SOCIETY, QUEEN'S CHAMBERS,
5, JOHN DALTON STREET, MANCHESTER,
FEBRUARY 12TH, 1907.

MR. CHARLES PILKINGTON, PRESIDENT, IN THE CHAIR.

The following gentlemen were elected, having been previously nominated:—

MEMBER—

Mr. HENRY BATSON BEALES, Mechanical and Electrical Engineer, 64, Cross Street, Manchester.

ASSOCIATE MEMBER—

Mr. FRIEDRICH SCHEMBER, 9 Liechtenstrasse, Vienna.

DISCUSSION OF MR. OTTO SIMONIS' PAPER ON "LIQUID AIR AND ITS USE IN RESCUE-APPARATUS."*

Mr. OTTO SIMONIS (London) demonstrated and explained the aerolith, a liquid-air rescue-apparatus. He (Mr. Otto Simonis) was not a mining engineer, but a fire engineer; and it was for the purpose of combating smoke that the apparatus was first designed. The apparatus was now used at Baron Rothschild's coal-mines in Austria, and by the London Fire Brigade. At Baron Rothschild's mines, a plant had been installed at a cost of £1,400 for the production of liquid air; about 5 gallons per hour were made, and in addition compressed oxygen was produced with the same plant. This plant proved to be not too big, as they had found employment for the liquid air in a variety of ways.

The PRESIDENT (Mr. Charles Pilkington) expressed the fear that the aerolith apparatus would easily get damaged when

* *Trans. Inst. M. E.*, 1906, vol. xxxii., page 534.

worn on the back; because, in a mine after an explosion, the
wearer would have to crawl on his hands and knees, and the
knapsack or bag would be rubbed against the roof. He suggested that the clock, placed at the top, and intended to warn
the wearer when the stock of liquid air was nearly exhausted,
might be placed, if possible, in a less exposed position.

Mr. OTTO SIMONIS replied that the bag was not an essential
part of the apparatus, as it could be removed, and the apparatus
would still work efficiently.   He, however, preferred to use the
bag, as it afforded an auxiliary reservoir for extraordinary cases.
The man who was using the apparatus would become aware in
due time of his danger, by feeling the bag and by the diminution
of the volume of air passing through his lungs.

Mr. JOHN GERRARD (H.M. Inspector of Mines) asked what
quantity of air was left in the knapsack after the clock had
given warning.

Mr. OTTO SIMONIS said that the wearer could set the clock
to give warning when the supply was half done, or when it
would be exhausted in another half-hour.   Bottles filled with
liquid air could be taken into the mine and kept there for use
in emergencies.

The PRESIDENT (Mr. Charles Pilkington) said that he had
seen many appliances designed for rescue-purposes.   He was
a member of a committee which had to deal with this particular
subject, and there were possibilities in this apparatus that, in
his view, were almost better than anything else he had seen.
He was favourably impressed with it, but he wanted to see it
tested in rough work.   The aerolith apparatus was very much
lighter than other kinds, and he hoped to see it severely
tested at one of the rescue-stations. He thanked Mr. Simonis, on
behalf of the members, for giving this demonstration of its use.

Mr. GEORGE H. WINSTANLEY, in moving a vote of thanks to
Mr. Simonis, said that he had for a long time taken a deep
interest in the question of rescue-apparatus and read a paper
ten years ago before the Society in which, curiously enough, he
strongly advocated the establishment of properly-equipped

rescue-stations throughout the coal-fields of Great Britain.* The aerolith apparatus struck him as having features which strongly recommended it as a practical appliance. The great difference between it and many other appliances was its simplicity. The difficulty hitherto had been to dispose satisfactorily of the carbon dioxide, which tended to increase in quantity in the initial volume of air that was breathed over and over again with additions of oxygen. In this apparatus no attempt was made to breathe the same air again; a great advantage would be found, too, in the easily portable character of the bottles of liquid air as against the large steel cylinders of oxygen for replenishing the apparatus in actual use. The former (the bottles of liquid air) could more easily be taken into the mine, and carried forward to some convenient place for the purpose of replenishing the knapsack when necessary; and consequently the rescue-party could carry on their work for a longer period, with less anxiety as to their supply of respirable air.

Mr. JOHN GERRARD (H.M. Inspector of Mines), in seconding the motion, said that he was not going to commit himself to anything more, than to say that, theoretically, the aerolith apparatus appeared to have advantages over the oxygen principle on which other systems depended. He preferred to wait, before pronouncing a definite judgment, until the apparatus had been tested at one of the rescue-stations; and he sincerely hoped that, before very long, they would have a station in Lancashire where they could prove whether this apparatus was practicable or not. Their President was the chairman of the committee that was working for this end, and it was very desirable that its labours should soon be successful. He thought that such an appliance as this might be useful in dealing with underground fires. He held very strongly that there was a field for such apparatus. Fortunately, they did not now have in Lancashire disastrous explosions; but they were constantly having to deal with underground fires, and they had had, in the last few years, several disastrous underground fires which had been most difficult indeed to deal with. It seemed to him that, apart from explosions, there was a field for the use of these appliances in dealing with the extinc-

* Transactions of the Manchester Geological and Mining Society, 1897, vol. xxv., page 148.

tion of underground fires.   He was sure that they were all very grateful to Mr. Simonis for coming to Manchester, and giving to the Society so full an explantion, because the people of Lancashire liked to be in the van of progress.

The motion was cordially adopted.

Mr. OTTO SIMONIS said that he did not claim that his apparatus was perfect in every detail, such as the type of mouthpiece, etc., and therefore he welcomed any suggestions.   He had always held that the rescue-appliances hitherto in use possessed drawbacks which were absolutely against them from a practical point of view.   The weight and complication were fatal disadvantages, independently of the fact that there were certain dangers in the use of caustic soda and other chemicals.

——

Mr. THOMAS H. WORDSWORTH read the following paper on " Cage-lowering Tables at New Moss Colliery ":—

CAGE-LOWERING TABLES AT NEW MOSS COLLIERY.

By T. H. WORDSWORTH.

When pits were shallow and only single-decked cages were in use, there was no need for cage-lowering tables; but, as they became deeper, the extra length of time spent in winding, coupled with the great weight of the rope compared with the load lifted, led colliery owners to adopt larger cages. As the size of the pits rendered it impossible to put more tubs on a deck, cages with two, three, four and up to six decks were used.

Where two-decked cages only are used, duplicate landings may be put in both at the surface and underground, and a double set of men may be employed; and this reduces the period of decking to a minimum. In other cases, keps are put in at the pit-bottom; and, by carefully adjusting the ropes, the tubs for each deck are changed simultaneously at the top and the bottom. This system has the disadvantage that, if any hindrance takes place at the top or bottom of the shaft, both sets of men are stopped and the length of time occupied in changing is considerably increased; besides which, the engineman needs a signal from the top and bottom for each deck, and, should he slightly overrun, the cage is likely to be damaged.

The first cage-lowering table seen by the writer was at the West Riding collieries, Normanton, although he believes that they were originally used in Lancashire. The table in use at the West Riding colliery was balanced by weights hanging in a small pit, half of these being hung on the end of the rope and the other half supported on a stage, so that they were picked up after loading the second deck of a four-decked cage. The use of these supplementary weights, whilst reducing the necessary brake-power, caused complications when winding men, and it was only possible to ride men on two of the decks.

The tables now in use at New Moss colliery have been evolved from this: they are so arranged that it is possible to lower the empty cage and table, and in this way men can ride on all the decks. The balance-weights have been taken out of the pit and put in sight. The apparatus may be divided into three parts:—(1) The table on which the cages alight; (2) the brake-arrangement for giving the hooker-on control of the tackle; and (3) the balance-weights for bringing back the table when the cage is lifted from the bottom.

The table is built of steel joists and channels (figs. 1 and 2, plate vi.). The main frame of the table, A, is formed of two main longitudinal joists, a and b, 13 feet 3 inches in length, 9 inches deep, 4 inches wide and $\frac{1}{2}$ inch thick, connected at each end by two cross joists, c, d, e and f, of the same section, trimmed in and secured by steel angle-irons and rivets, 18 inches from the end. This frame is further strengthened in the centre by steel channel-irons, g and h, $3\frac{1}{2}$ inches wide and 9 inches deep, placed back to back; and four short joists, i, j, k and l, trimmed into these and the end joists form the opening for the balance-rope. There are also four short channel-iron distance-pieces, m, n, o and p. Four battens, q, r, s and t, 12 inches wide and 3 inches deep are placed longitudinally on this table, so as to form a cushion between the steel cage and the steel table and save wear on the bottom of the cages.

The table, A, is carried by two steel-wire ropes, B and C, $1\frac{1}{2}$ inches in diameter, passing over two pulleys, D and E, 4 feet in diameter, resting on girders running across the pit-bottom. The attachment between the table and the ropes has been arranged so that, by taking out two pins, u and v or w and x, at either end of the table, it can be slung from the other rope and hung clear of the cages. In case of cleaning out the sump or of changing the weights, F and G, on the end of the conductors, H, or for any work done below the table, this is a great advantage.

The other ends of the ropes, B and C, after passing round two surge-wheels, I and J, 4 feet in diameter (similar to endless-rope haulage-wheels), and over two pulleys, K and L, 4 feet in diameter, placed about 20 feet above the floor, are attached to steel balance-boxes, M and N, built of angle-irons (3 inches wide, 3 inches deep and $\frac{1}{2}$ inch thick) and plates $\frac{1}{2}$ inch thick. The attachment is made by short sockets, a, with pins, b, passing through cross bars, c, inside the boxes. These boxes are filled with small scrap-iron, such as chain, bolts, etc., and run in angle-iron slides, *d, e, f* and *g*.

The brake-wheel, O, 6 feet in diameter and 9 inches wide on the face, together with the two surge-wheels, I and J, is keyed on to a shaft, P, 6 inches in diameter, supported by six pedestals, *a, b, c, d, e* and *f*. The brake is of the double-post pattern, with compound levers, *ab, cd, de, fg, ghi, hj, il* and *kl*, the leverage being about 400 to 1; and wooden brake-blocks, *m* and *n*, are used, as they have been found to give the best results. The weight of the levers is balanced by a weight, *q*, attached to a chain, passing over the pulley, *p*.

In working, the winding-engineman drops the cage, Q, on the table, A, which is then in the top position, and at the same time puts the top cage on the keps at the surface. The tubs are then changed in the bottom-deck, and, when the brake-lever is lifted, the weight of the cage, Q, is sufficient to overcome the balance-weights. The hooker-on is then able to stop the cage, with the brake, when the second deck comes level with the landing-plates; and, so on, until all the decks are loaded and the cage and the table are in the position shown in fig. 2 (plate vi.). In this way, by having a small amount of chase or slack on the winding-rope, the hooker-on at the bottom of the shaft is enabled to change his tubs practically as quickly as the change is made on the surface, and no signal is necessary until all the decks are changed and the cage is ready for sending to the surface.

After the cage leaves the bottom, the table is brought back to the top position by the balance-weights.

Too much chase or slack, however, should not be allowed on the winding-rope, as this is objectionable when picking up, and there is also a danger of the detaching-hook coming on the cage-top or down the cage-side, should there be any delay in changing the tubs at the pit-bottom. This difficulty may be overcome by using cage-chains, 10 to 12 feet long; and, with decks only 4 feet deep, it is then possible for the surface-man to deck the whole of the decks of the cage before the detaching-hook falls over the cage-side. On one occasion, when there was too much chase or slack on the rope, the detaching-hook became twisted in the cage-chains, and sheared the copper pin instead of lifting the cage from the bottom.

Old winding-ropes may be used for working the tables, but experience will show, in each case, whether this is the better policy. The writer prefers to use a special flexible rope, as it reduces the liability to stoppage during working hours; and one rope will last three or four years. In cases of overwinding, the table has the advantage that the shock to the bottom cage is considerably reduced. If men were in the cage there would be less liability of their being injured; and it has also been found that the life of the cages is considerably lengthened.

In case of accident or breakdown of any portion of the tackle, the table can be placed in the bottom position, as shown in fig. 2 (plate vi.) and winding continued until the end of the shift or such time as the repairs can be finished. But if the pit has been previously fully occupied in winding, the reduction of the output would vary from 10 to 20 per cent.

It is not suggested that cage-lowering tables should take the place of hydraulic decking-cages, with simultaneous changing for all decks, but there are many instances in faulted and steep districts or thin seams, where it is practically impossible to get sufficient coal to the pit-bottom to keep the winding-engine working full time and so afford a return on the capital necessary to instal hydraulic decking-plant. The writer is of opinion that, where three or four decks are used and the output does not warrant an expenditure on simultaneous changing plant with balance-cages, the cage-lowering table is the best means of dealing with three or four decked cages, especially if a spiral drum be used, as it reduces the number of signals to a minimum and leaves the top and bottom of the shaft practically independent of each other.

————

The PRESIDENT (Mr. Charles Pilkington) expressed the obligation of the members to Mr. Wordsworth for his paper and the accompanying drawings. It was, no doubt, an important apparatus that Mr. Wordsworth had described, and one that saved a great deal of time in decking. There were several appliances, very similar in character, in use at the Clifton and Kersley collieries, with which he himself was associated; but in the latter case a chain was used, instead of a rope.

Mr. GEORGE B. HARRISON (H.M. Inspector of Mines), in moving that the thanks of the members be accorded to Mr. Wordsworth, stated that he did not remember having seen an apparatus like the one described by Mr. Wordsworth before he came to Lancashire, where, at some pits, there were as many as six decks in the cages, and this apparatus afforded great advantages, particularly to the engineman, who was thereby relieved of a great strain.

Mr. G. H. WINSTANLEY, in seconding the motion, remarked that the apparatus described by Mr. Wordsworth included new details and improvements, which had considerably enhanced the value and interest of his paper. The members present had been interested, and the drawings clearly demonstrated the advantages of the apparatus.

Mr. T. H. WORDSWORTH said that, if the winding-ropes were properly adjusted, there was no trouble with them.

The motion was cordially adopted.

The Institution of Mining Engineers
Transactions. 1906 1907.

VOL. XXXIII, PLATE VI.

"To bles at New Moss Colliery."

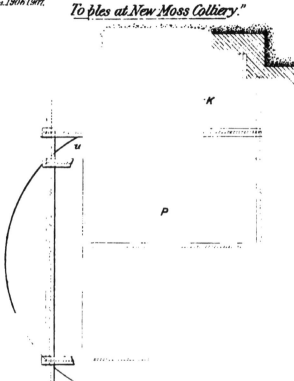

Scale, 6 Feet to 1 Inch.

MANCHESTER GEOLOGICAL AND MINING SOCIETY.

———

ORDINARY MEETING,
HELD IN THE ROOMS OF THE SOCIETY, QUEEN'S CHAMBERS,
5, JOHN DALTON STREET, MANCHESTER,
MARCH 12TH, 1907.

———

MR. CHARLES PILKINGTON, PRESIDENT, IN THE CHAIR.

———

The following gentlemen were elected, having been previously nominated :—

MEMBERS—
Mr. TOM STONE, Mining Engineer, The Collieries, Garswood, near Wigan.
Mr. PERCY HOUSTON SWANN WATSON, Mining Engineer, 11, Trafalgar Square, Ashton-under-Lyne.

ASSOCIATE—
Mr. JOHN GALLIFORD, Colliery Manager, 479, Edge Lane, Droylsden.

STUDENT—
Mr. H. HARGREAVES BOLTON, Jun., High Brake, Accrington.

———

DISCUSSION OF MR. W. E. GARFORTH'S PAPER ON "A NEW APPARATUS FOR RESCUE-WORK IN MINES."*

The PRESIDENT (Mr. Charles Pilkington) said that the committee, entrusted with the work of fitting up a station, near Tyldesley, for rescue-apparatus and for teaching men how to use it, would start operations at once, and he hoped that six months hence there would be a place connected with the collieries of the district, where mining engineers and others could see rescue-apparatus at work and be trained in its use, or to which they could send their men to be trained.

Mr. W. E. GARFORTH gave a demonstration of the working of the Weg apparatus. He would have preferred that the members had attended at Altofts to see the apparatus in use in

* *Trans. Inst. M. E.*, 1906, vol. xxxi., page 625 ; and vol. xxxiii., page 180.

the experimental gallery, under conditions which approximated
as nearly as possible to those found in a mine after an explosion—as regarded damaged roadways, contracted passages and a
noxious atmosphere.   He had been engaged on the work of the
life-saving apparatus for many years, and still considered that it
could be improved.   The supply of oxygen was sufficient for $2\frac{1}{2}$
hours, if the wearer were undergoing exertion, or enough for 6
hours when the wearer was resting.   The purifier contained about
$2\frac{1}{4}$ pounds of caustic potash and caustic soda, which absorbed the
carbon dioxide given off by the person wearing the apparatus.
The apparatus had been made entirely at the colliery, but when
it was put into the hands of a surgical-instrument maker he
apprehended that the weight, at present 30 pounds, would be
materially reduced.   The apparatus did not interfere with the
hearing of the wearer, and the goggles over the eyes could be
removed (when advisable) without interfering with the other parts
of the apparatus.   As only a small portion of the head was
covered, perspiration was as far as possible unchecked.

Mr. JOSEPH DICKINSON said that had Mr. Garforth carried
his history of rescue-appliances 13 years prior to Sir Henry
T. De la Beche and Dr. Lyon Playfair's invention, it would
have comprized one on a similar basis for which 50 guineas and
a silver medal was presented by the Society of Arts and the
Royal bounty of £100 by King George IV.*   The inventor
was Mr. John Roberts (of Upton-and-Roberts safety-lamp fame).
The apparatus consisted of a mask, with goggles and inhaling
arrangements through the interstices of a sponge immersed in
varied saturated solutions of lime-water, chloride of lime, caustic
soda, and sometimes pure water, according to the impurity likely
to be met with.   The expired air was not passed through the
inhalator.

Personally, the only breathing-appliance used by himself
(Mr. Dickinson) on many explorations in foul air was the 1846
bag of Glauber salts and lime.   The mine was on fire, following an explosion in 1851, and Mr. Goldsworthy Gurney's treatment with the fumes from burning coke, salt, and sulphur had

* "Apparatus to enable Persons to breathe in Thick Smoke, or in Air
loaded with Suffocating Vapours," by Mr. John Roberts, *Transactions of the
Society of Arts*, 1825, vol. xliii., page 25 and plate i. ; and *Report from the Select
Committee on Accidents in Mines*, 1835, page 262.

failed to extinguish the conflagration. The bags with the mixture were used for some hours, most of the explorers being sick;
and eventually the bags were discarded, and the fire was
approached in the old-fashioned way.

Rescue-apparatus were now available of better form, and
possibly some such might have saved some time; but the
saving would probably have been less than some persons might
suppose. On such occasions, when survivors might possibly
be rescued or valuable property saved, advances were made which
under ordinary circumstances would be inexcusable. Dormant
side-accumulations of foul air were then passed by, and left for
future clearance. Exploration also followed underneath smoke
and gas to close a hole in a blown-out stopping, and close touch
was kept with the advancing air-column. In front of this, among
débris, it would often be imprudent for apparatus-men to advance
far. The common tests for gas—elongated flame and blue cap
for fire-damp, dulled flame or extinction for black-damp, sensitive
creepings for after-damp and white-damp—did not delay much,
and with practice and care were reliable. The good results
occurring from the lessening number of fire-damp explosions
were it seemed so diminishing opportunities for re-entering that
special education was now suggested; consisting of the establishment of central training-stations with rescue-apparatus and a
practising-gallery, which deserved favourable consideration.

Mr. W. E. GARFORTH pointed out that the apparatus would
be considerably lightened if it were charged to last a man for
only an hour. He asked Mr. Dickinson to say, from his unequalled experience in rescue-work, if at times he had not found
it of the utmost advantage for a man to be able to get as soon
as possible into a pit when an explosion had occurred in order to
ascertain the state of affairs underground. He did not hesitate
to say that five men, who had been trained in the gallery in
the use of this apparatus, could go into the most noxious atmosphere, and give relief to men who might be suffering. He held
that if they could only save one life, it was worth all the trouble
that he had taken in perfecting the Weg apparatus. With this
apparatus a man would also be able to travel a certain distance,
in order to ascertain if any part of the mine was on fire. On the
occasion of an explosion at Altofts colliery, a fortnight was lost·

before they were able to reach a certain point where a fire was
discovered; but with this apparatus they might have discovered
the fire on the second day, instead of suffering great risks and
anxiety actually incurred.

The PRESIDENT (Mr. Charles Pilkington) said that he was in
favour of rescue-apparatus being provided, but there was a difficulty in the matter of weight.   Thirty pounds was a great
weight for a man to carry a long distance.   As to the general
utility of rescue-appliances, it had been stated that no lives
were saved through their instrumentality after the Courrières
explosion; but it must be remembered that there were no apparatus in the district at the time of the disaster.

Mr. JOHN GERRARD said that occasionally the need arose, in
exploring mines after explosions and in connection with fires,
to use all the appliances that science and experience had suggested: thus mice and birds, by reason of their greater suscepti-
bility to carbon monoxide, gave early intimation of its presence.
The statement that rescue-apparatus failed at the Courrières
mines, he emphatically declared, was most unfair, as the rescue-
apparatus had not a chance to save life.   It was perfectly true that
men came out of the pit alive after Mr. G. A. Meyer arrived at
the Courrières collieries, but everybody at that time firmly believed that there was not a living man in the pit, and everybody
acted on that assumption.   He (Mr. Gerrard) thought that Mr.
Garforth was deserving of all praise for his persistent efforts
to perfect this apparatus.   Fortunately, Lancashire had been
spared disastrous explosions of recent years; but not infrequently
trouble had arisen through fires which might much sooner
have been extinguished, if such an apparatus as that which Mr.
Garforth had designed had been available for use.   Life also
would have been saved.   In connection with fires, there was,
he was persuaded, a big field for the use of this kind of apparatus in Lancashire.   He was hoping that before long they
would have a rescue-station in Lancashire, where men could
accustom themselves to the use of these appliances.   He (Mr.
Gerrard) raised the question of what could be done in the
event of a man equipped for rescue-purposes passing through
foul air to a place where the air was tolerable and finding men

alive there.   Mr. G. A. Meyer had made a supplementary apparatus to bring out men in such cases.   He moved that the best
thanks of the members be given to Mr. Garforth.

Mr. JOHN GALLIFORD seconded the motion, which was
cordially approved.

———

Mr. JOHN GALLIFORD read a paper on " A New Reflector for
Safety-lamps."   The reflector, made of aluminium, 5 inches in
diameter, can be fitted, above or below the standards or poles
protecting the glass of a Clanny type of safety-lamp, by means
of a spring clip or a bayonet joint.

———

MANCHESTER GEOLOGICAL AND MINING SOCIETY.

ORDINARY MEETING

HELD IN THE ROOMS OF THE SOCIETY, QUEEN'S CHAMBERS,
5, JOHN DALTON STREET, MANCHESTER,
APRIL 9TH, 1907.

MR. CHARLES PILKINGTON, PRESIDENT, IN THE CHAIR.

The following gentlemen were elected, having been previously nominated : —

MEMBER –
Mr. WILLIAM OLDFIELD, Mining Engineer, West View, Minsterley, Shropshire.

ASSOCIATE—
Mr. GEORGE REYNOLDS WYNNE, Hope Cottage, Tarvin Road, Chester.

Mr. W. H. COLEMAN read the following paper on " The Cook Calorimetric Bomb " : —

THE COOK CALORIMETRIC BOMB.

By W. H. COLEMAN.

Introduction.—In bringing this new pattern of bomb calorimeter before your notice, it is necessary to preface the description with a few words on fuel-valuation.

This subject may be divided into three heads:—(1) Sampling the fuel; (2) determining its calorific value; and (3) comparing different fuels, so as to find out which will serve the purpose in hand most economically.

(1) *The Sample.*—The chief difficulty in the way of determining the calorific value of a fuel is the fact that it is not at all easy to obtain an average sample. Unless the sample is what it professes to be, that is, of the same composition as that of the whole bulk of the fuel under examination, it is useless to waste time and money in making elaborate calorimetric determinations. If the sample is to represent a seam of coal, several freshly hand-cut portions should be taken from different parts of the seam; taking care that a strip is cut from a fresh face from top to bottom of the seam and that it contains everything, whether coal or dross, that would be sent to the surface as coal, and nothing else.

If the sample represents a delivery of a particular quality of coal, then a truck or a cart-load should be selected at random from a number of freshly-filled trucks or carts. Special care should be taken that the loads are average ordinary loads, which have been filled for delivery, without any attempt to pick the quality. The whole load should be tipped, well mixed, and quartered down until a sample of about 1 hundredweight is obtained. Great care must be taken that this contains the right proportions of large, small and fine coal. The sample should then be either put through a stone-breaker, or ground roughly in a mortar-mill. It should then be again quartered, until a sample weighing 2 or 3 pounds is obtained. This must then be ground in a suitable mill to a moderately fine granular powder. It is necessary, when determining the moisture in a coal, to be sure that it is not too finely ground, and it should be bottled at once. The sampling, grinding, etc., should be quickly done, so that the coal may not lose moisture.

(2) *The Calorific Value.*—Three ways offer themselves for determining the calorific value of a fuel:—(1) An ultimate analysis of the coal may be made, and its theoretical calorific value may be obtained by calculation. This method, though accurate, takes a considerable time and requires the services of an expert

FIG. 1.—THE COOK CALORIMETRIC BOMB.

chemist. (2) The moisture, ash, volatile matter and fixed carbon may be estimated, and the calorific value calculated from one of the numerous formalæ given by different authorities. This method is simple, and can be carried out by an average intelligent laboratory-assistant; but the results, though useful in checking deliveries, the calorific value of which has been determined by the first method from a previous sample-delivery, are not sufficiently accurate to discriminate between several different qualities of coal. (3) The actual calorific value may be determined by means of a calorimeter.

In principle, all calorimeters depend upon burning a weighed quantity of coal and observing the rise in temperature imparted to a known quantity of water by the heat given out during the complete combustion of the fuel. Several forms of calorimeters have been proposed.

(a) *Lewis-Thompson Calorimeter.*—In this instrument, the weighed sample of coal is mixed with potassium chlorate or some other oxygen-carrying body and burnt in a bell-jar immersed in water contained in the calorimeter. The products of combustion bubble upwards through the water, to which they communicate their heat.

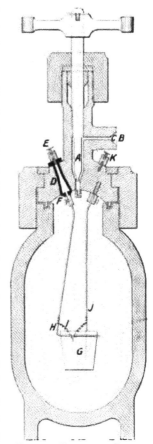

(b) *William-Thomson, Fischer and Darling Calorimeters.*—These calorimeters are all modifications of a similar principle. The coal is burnt in a crucible placed in a bell-jar immersed in water in the calorimeter. It is not mixed with any oxygen-carrying material, but a stream of oxygen is caused to pass into the bell-jar. The products of combustion escape, as in the first-mentioned apparatus. In the Fischer calorimeter, the gases pass through a special chamber.

(c) *Parr Calorimeter.*—In this apparatus, the weighed coal is mixed with sodium peroxide, which serves to supply the oxygen necessary for combustion, and the caustic soda produced absorbs and combines with the carbon dioxide given off by the burning fuel. The mix-

Fig. 2.—The Cook Bomb.
Scale, 2 Inches to 1 Inch.

ture is placed in a closed metal chamber immersed in the water of the calorimeter, and ignited.

(d) *Berthelot-Mahler Bomb.*—In the new modification of this instrument (fig. 1), the weighed sample of coal is placed in a platinum crucible, G, which is then placed on the wire supporting ring, H (fig. 2). In order to ignite the coal, a fine platinum wire, I, is connected to the supporting ring, H, and to the other wire, J, depending from the cover. The latter, carrying the crucible, is gently lowered into the bomb and screwed up tightly. Oxygen, under pressure, is then admitted to the bomb until the gauge registers a pressure of about 21 atmospheres. The valve, A, is then shut, and the oxygen-cylinder is disconnected. The bomb is then carefully lowered into the copper calorimeter vessel, into which about 2 litres of water, slightly below the temperature of the room, is poured; and the wires of the battery are connected to the bomb, at E and K, the circuit being open. This vessel stands in another double-walled copper vessel filled with water at the room-temperature, and surrounded by felt to prevent loss or gain of heat from surrounding objects.

The agitator is set to work, and the rise in temperature of the water is observed; and, when this is constant, the circuit is closed, and the coal is burnt. The temperature of the water rises, and the maximum point is noted. Then the increase in the temperature of the water in Centigrade degrees, multiplied by the weight of the water and the water-equivalent of the apparatus, gives the number of calories evolved by the combustion of the weight of coal taken. If 1 gramme had been taken, then this number represents the calorific value in calories; and if not, then the figure is divided by the weight of coal taken. The calories multiplied by 1·8 give the British thermal units per pound of coal. The evaporative power of the fuel from and at 212° Fahr. is obtained by dividing the calorific value, expressed in British thermal units per pound by the number, 966, representing the latent heat of vaporization of water in Fahrenheit units. Several forms of this calorimeter have been proposed:— The original Mahler bomb; the Bryan-Donkin bomb; and the Cook-Berthelot-Mahler calorimetric bomb shown in figs. 1 and 2.

This latter form has several improvements, and has been found very satisfactory in use. The chief improvements are the gas-valve, A, and the gas-inlet, B. In older forms of the instrument, the gas entered at the top instead of at the side, and the valve had to be slightly loosened to disconnect the copper pressure-tube. Other minor advantages are the following:—The gas-inlet valve does not easily get out of order. The stirring arrangements are very simple. The connections have coned

joints, C, and do not require washers.   The insulation, D, of
the plug, E, is protected from injury by high temperature at the
moment of explosion, by a quartz plate, F.   Care must be taken
that the apparatus is tight; and, to ensure this, the lead-washer
should be examined to see that it is not burred.

Several points require a little attention : the coal should be
dried at a temperature of about 110 Cent. (230° Fahr.); and
should be finely powdered, and then made into briquettes with
the small briquetting machine.

The water-equivalent of the apparatus is best found by determining the calorific value of pure cellulose (4,200 calories) or
naphthaline (9,692 calories).   A correction for cooling ought to
be made for very accurate experiments.*   If, however, the water-
equivalent of the apparatus is determined by burning naphthaline, and such a quantity is used as to give about the same rise of
temperature as the weight of coal taken, this correction may be
neglected if the determination is only wanted to be within ½ or
1 per cent. of the truth.

It may also be advisable to correct the results for the nitric
acid and sulphuric acid formed.   This can be done by washing
out the bomb after the experiment, and estimating the quantities
of nitric acid and sulphuric acid formed.

The room, in which calorimeter-determinations are made,
should have a temperature of about 60° Fahr., and the water should
be about 2° Cent. (4° Fahr.) below the temperature of the room.

Of course, a calorimetric determination does not tell all there
is to be known about the suitability of a coal for any particular
purpose.   The length of flame, the quality of the ash (whether
easily fusible or not), and other factors must be taken into consideration; but, given several samples of coal, all of which are
otherwise suitable for the intended use, the most economical coal
can be selected by the calorimeter.   If, from the result of the
determination, the number of calories obtained for any given
sum of money be calculated, it is easy to pick out the most
economical fuel.

* This is made by observing the rate of the rise of temperature that takes
place when the bomb is placed in the calorimeter and before the charge is fired, and
also the rate of the fall of temperature during a similar period of time after the
maximum temperature has been reached. From these observations, the loss by
radiation during the experiment can be calculated and a correction applied.

At the present day, when every little detail has to be considered in order to make a profit, there is no doubt that the coal which can be guaranteed to give the best result for the money will find the readiest sale.

Anyone who is desirous of going further into the question of valuing fuel may refer to the excellent paper by Mr. J. B. C. Kershaw.*

———

The PRESIDENT (Mr. Charles Pilkington) thought that the instrument was very valuable, and suggested that, when a coal-owner wished to ascertain the true value of his coal-seam, or of a particular portion of a seam, samples taken from all points of the seam should be duly tested in the calorimeter by a competent person.

Mr. COLEMAN said that a sample of 7 or 8 pounds of coal should be sent, broken into nuts, and then the final grinding could be done quickly.

The PRESIDENT asked what allowance was made for ash.

Mr. COLEMAN said that no allowance was made for ash in the coal, as it would have the same influence in the calorimeter as in the furnace. The coal was burnt with pure oxygen. An electric current with a pressure of 12 to 20 volts would suffice to ignite the charge.

The PRESIDENT said that the calorimetric bomb was valuable, but he thought that it was more for use in the laboratory than at a colliery.

Mr. SYDNEY A. SMITH said that the calorific value of a coal could be ascertained by this instrument, but it did not indicate the value of the coal as a fuel suitable for a particular purpose; and much depended on the rate of combustion. In a recent case, there was a dispute about a shipment of coal. The calorific value was excellent, but it could not be burnt, except at a very slow rate, in the ordinary grates of water-tube boilers, without forced draught. In ascertaining the value of coal as fuel, that

* "Fuel Analysis for Steam Users," by Mr. J. B. C. Kershaw, *The Engineer*, 1906, vol. cii., pages 314 and 337.

point must be considered, because the value of a coal for a particular purpose depended on the rate of combustion. Some coals burnt rapidly, others burned slowly, and yet all might have the same calorific value.

Mr. COLEMAN admitted that even when the calorific value of a coal had been determined, the question as to the coal's suitability for a particular purpose was not settled. But given several coals, at different prices, that were suitable as regarded their burning qualities, the calorimetric bomb would indicate the one that produced the most heat for the least money. The question of ash was of the utmost importance in some cases.

A vote of thanks was accorded to Mr. Coleman for his interesting paper.

MANCHESTER GEOLOGICAL AND MINING SOCIETY.

ORDINARY MEETING,

HELD IN THE ROOMS OF THE SOCIETY, QUEEN'S CHAMBERS,
5, JOHN DALTON STREET, MANCHESTER,
MAY 14TH, 1907.

MR. CHARLES PILKINGTON, PRESIDENT, IN THE CHAIR.

The following gentlemen were elected, having been previously nominated : —

MEMBERS—

Mr. ALFRED ACKROYD, Ellerslie, Victoria Crescent, Eccles.

Mr. HUGH FRANK TAYLOR, Mechanical Engineer, Sandy Croft, near Chester.

Mr. THOMAS WILLIAMS, Mining Engineer, 5, Westbourne Grove, Hexham.

Mr. Frederick J. Thompson read the following paper on " The Rock-salt Deposits at Preesall, Fleetwood, and the Mining Operations Therein " : —

THE ROCK-SALT DEPOSITS AT PREESALL, FLEET-WOOD, AND THE MINING OPERATIONS THEREIN.

By FREDK. J. THOMPSON.

Introduction.—This paper is divided into four parts, comprising:—(1) The history and the geology of the rock-salt deposit at Preesall; (2) the method of sinking shafts for rock-salt; (3) the methods of mining rock-salt; and (4) a description of the Preesall salt-mine worked by the United Alkali Company, Limited.

(1) *History and Geology.*—The earliest record of any underground exploration, near Fleetwood, is that of a bore-hole, A (fig. 2, plate x.), put down by the Royal Engineers in 1860 in search of water for the troops then stationed at Fleetwood, the town at that time being entirely dependent on surface-wells for its water-supply. This bore-hole was carried to a depth of 559 feet, the whole of the distance being bored in Keuper marls—with the exception of a few feet, near the surface, in the usual surface-drift; but no water was found.

In 1872, the district on the Preesall side of the river Wyre was mapped out by a syndicate, with the intention of putting down twenty bore-holes in various places in search of iron-ore, as it was thought possible that such might exist on the south side of Morecambe bay, as well as at Barrow-in-Furness and other places on the north side (fig. 1, plate x.). The site of each of these twenty bore-holes was fixed, and some of them were bored simultaneously, but the only available records are those of Nos. 2, 8, 9 and 17 (fig. 2, plate x.). Rock-salt was struck in No. 2 bore-hole in 1872, and signs of rock-salt in two of the others; and there is no doubt that the discovery of rock-salt prevented the completion of the original design of twenty bore-holes.

In 1875, No. 17 bore-hole was put down about ¾ mile to the north-east of No. 2 bore-hole in which rock-salt had been struck,

and red sandstone was found at a depth of 45 feet from the surface. The boring was continued to a total depth of 564 feet from
the surface, and indicated the existence of an enormous supply
of fresh water.   The map (fig. 1, plate x.) shows the relation
of Fleetwood on the south side of Morecambe bay to Barrow-in-
Furness on the north side.   All previous geological surveys had
indicated that the red-sandstone fault was 8 miles eastward of
the site of No. 17 bore-hole, which had struck sandstone at
Preesall.   The author, in 1906, constructed a geological section
running east-and-west through the rock-salt deposit, and the
late Mr. C. E. de Rance expressed the opinion that it was
substantially correct.   A reproduction of this section is shown
in fig. 3 (plate x.).

Nothing further was done in the development of the minerals,
as the state of the salt-trade did not warrant such, except that in
1885, a shaft around No. 2 bore-hole, which had previously been
sunk from the surface into the rock-head, was carried still lower
into the rock-salt bed until the bottom was reached.   This shaft
indicated a total thickness of 340 feet of rock-salt stratified in
several layers of very good and some of inferior rock-salt and
marl.

The general theory amongst geologists as to the origin of the
various rock-salt deposits found in this country is that, at some
remote period in the world's history, considerable quantities of
sea-water had, owing to changes in the earth's surface, been
separated from the main body of the sea; and that solar evaporation concentrated such bodies of sea-water until salt was
deposited.   This flooding of the same area, happening again and
again, would account for the deposits of marl and inferior rock-
salt, which originally would be deposited as mud and salt.   It
is not the author's intention to express any opinion as to this
theory; but it is obvious to anyone that salt could not have been
in solution in sea-water, unless it had been dissolved from the
solid mass in some previous formation.

Nothing was done in the way of serious development until
1889, when the reconstructed Fleetwood Salt Company, Limited,
took over the undertaking, and the author was engaged by them
to take charge of the development of the salt-beds.   The old
shaft previously mentioned was converted into a shaft for pumping brine, the small amount of natural brine found in it on the

rock-head being augmented by putting water down bore-holes, which were bored down to the rock-head at a short distance away from the shaft. Salt-works for the evaporation of the brine were erected on the Fleetwood side of the river Wyre and were connected to the brine-pumping station by a line of pipes, which were laid across the estuary of the Wyre (fig. 2, plate x.). It is not the author's intention to deal with the question of brine-pumping or salt-making, except to state that both operations have attained considerable proportions, especially since the erection of the Ammonia-soda works at Fleetwood by the United Alkali Company, Limited, who acquired the whole of the Fleetwood undertaking from the Fleetwood Salt Company, Limited, in 1890. It may be stated at this point that salt was required by the company as a raw material in the manufacture of various products involving the use of chlorine or sodium.

(2) *Sinking Shafts.*—In 1893, the United Alkali Company, Limited, decided to sink shafts for the purpose of mining rock-salt in a dry state, and they were commenced in that year. In sinking shafts for the purpose of mining rock-salt it is necessary to ensure the absolute exclusion of all water or moisture; as, rock-salt being extremely soluble, great havoc would be caused if water were allowed to enter any salt-mine. This must be effected, before the rock-salt bed is entered, in carrying down the shafts.

The first stratum passed through was the boulder-clay overlying the Keuper marls, and this was gone through by two timbered shafts, each 7½ feet square. A considerable amount of water was found in various places in sand-beds in the boulder-clay, besides enormous boulders, which occupied in some places the whole area of the shaft in sinking, and had to be blasted before they could be raised to the surface. The boulders were mainly of limestone, and many of them showed extraordinarily clear ice-markings of the glacial period.

The Keuper marls were reached with both shafts, which were continued until the marls were found in such a condition as to afford a safe foundation; at these points in each shaft, a brick-work foundation was built in with blue bricks and Portland cement; and from this foundation, iron tubbing was carried to the surface. This tubbing consisted of cast-iron socket-and-

spigot pipes, 6 feet in diameter. The space between the back of
these tubes and the timbering of the shaft was filled with clay-
puddle, and the absolute exclusion of the water was thus effected.
The shafts were afterwards continued downwards through the
Keuper marls to the rock-salt bed, and were stopped, for the
time being, in a good bed of rock-salt, about 450 feet from the
surface.

(3) *Mining.*—As an interval of 50 years had elapsed, prior to
this, since any rock-salt mine had been opened out, the author
had no previous personal experience to guide him as to the best
method of opening out a mine in the shortest possible time. A
rock-salt mine is different from most mines, owing to the fact
that it does not consist of a series of passages or ways; but it is
only one large excavation or cavern hewn out of the solid rock-
salt, with solid masses of rock-salt left in at various places for
the purpose of supporting the roof (fig. 4, plate x.). The usual
way of opening out such a mine had been to commence enlarging the two shafts, around their bottoms, until they met and
formed one excavation, and afterwards this operation was continued all around. This meant that for some time, owing to the
smallness of the excavation, very few men could be put into the
work, as blasting operations would render the presence of a
large number of men undesirable, because of the danger.

It was ultimately decided to employ a compressed-air tunnelling-machine, to drive headings in various directions into the
rock-salt at as rapid a rate as possible. This machine, supplied
with air from a compressor stationed on the surface, drove headings 5½ feet in diameter at the rate of 360 feet per week. A
heading was driven first from one shaft to the other in an
easterly direction and continued a considerable distance in that
direction, and afterwards a heading was driven in a westerly
direction; and, while so driving in this westerly direction, men
were put into the eastern heading to widen it out northward and
southward. Another heading was driven by the machine in a
northerly direction and other men put into the western heading,
so that in a very short time there was a considerable area of
excavation, 6 feet in height. The widening out of the headings
was done by rotary undercutting machines, worked by compressed air, and these machines have been retained until this

day, four being now at work. The bed of rock-salt that is being mined is about 40 feet thick, and it dips in a north-westerly direction at the rate of 1 in 3½.

The rock-salt is mined by making a "roofing," A, 6 feet in height at the top of the bed; and the roof thereof afterwards forms the ceiling of the mine (fig. 5, plate x.). The roofing is done by machine-undercutting, in lengths of 105 feet—the distance between each pillar, the pillars being 60 feet square. Rock-salt is much harder to mine than coal, and an ordinary coal-undercutter will not touch it. The special undercutting machines made for this purpose, undercut to a depth of 2½ feet only.

The roofing lays bare the whole of the lower part of the bed, say, 34 feet thick, and this is worked by blasting from the face as in open quarry operations. Black gunpowder, in compressed cartridges, is used for the purpose of blasting all over the mine. The rock-salt, after blasting, is slid down the face to the floor of the mine where it is filled into hoppets, B, trammed to the shaft, and wound to the surface in the usual way (fig. 5, plate x.).

Members who have had any experience of the ordinary rotary undercutter, will no doubt be aware that, in undercutting from one face to another, the two ends of the cut cannot be touched, but have to be got out by other means, as the rotary under-cutter will not start at, or finish right to the end of the cut. This necessitates getting out the two ends either in advance of the cut or after the cut has been made by the undercutter. In the case of rock-salt mines, it has been found easier to get out these ends after the long cut has been made; and, for this purpose, Ingersoll-Sergeant coal-cutting or punching machines are used. These machines are also used in other parts of the mine, where the range of cutting is so short that it will not pay to run the rotary undercutter.

It is in the direction of drilling holes for blasting that the most rapid strides have been made in advance of the old methods employed in Cheshire and elsewhere, and it has been the means of great economy over the original methods of drilling holes. The original method of drilling holes in rock-salt was by using a long bar of iron, steel-pointed at each end, as a jumping drill, worked by hand. By this means one man can drill a hole, 1½ inches in diameter and 5 feet deep, in about 45 minutes. The author first introduced into salt-mines worm-drills, operated by ratchets, worked by hand, the Elliott drill being the most suitable for rock-salt mining owing to the fact that the pressure on the feed can be regulated by a friction-brake. This is especially necessary in rock-salt owing to its varying nature, several degrees of hardness being gone through in a single hole. Later, compressed-air motors were applied to the end of the feed-screw of these drills, the result being eminently satisfactory, and, at the present time, practically the whole of the drilling in the mine is done by means of drills driven by compressed-air motors. It is now quite easy for one man to put in a hole 1½ inches in diameter and 5 feet deep in 5 minutes, by means of one of these motor-driven drills.

(4) *Description of Mine and Plant.*—The total thickness of the rock-salt bed at the Preesall mines is approximately 380 feet; but, as stated previously, although homogeneous, it contains strata of varying quality and only a portion of it is mined. There are two levels in the Preesall salt-mine: the floor of the upper mine being about 470 feet from the surface, and that of the lower mine 900 feet. In opening these mines, a thickness of about 40 feet was mined in the upper level; but, some three years ago, the shafts were carried down to the bottom of the salt-bed, and it was found that a thickness of about 22 feet of rock-salt at the bottom was of very superior quality. At the present time, mining operations are confined to thicknesses of 22 feet in the upper mine and of 22 feet in the lower mine.

The same winding-engine is used for both mines, a larger drum and a longer rope being used for the bottom mine than those used for the upper mine, the difference in the weight of the load being made up by a balance-weight on the rope of the upper mine. Cages are not used in the shafts. There are two guide-ropes, EE, in each shaft, and guides, D, are attached thereto (fig. 5, plate x.). The rock-salt is loaded in the mine into large wooden hoppets, locally called " rock-salt tubs." These hoppets, placed on the frame of a bogie, are not attached to it in any way, and when fully loaded they contain about 1½ tons of rock-salt. They are trucked to the shaft-bottom by the men who load them. The empty tub goes down the shaft and is placed on a bogie waiting to receive it, the three hanging chains, attached to it, are unhooked and the bogie and the empty tub are pushed away; the bogie and the loaded tub are moved into the place of the empty tub, and the three chains are hooked and the loaded tub hoisted to the surface.

There are two systems of signalling, one being electric and the other by an ordinary signal-line, pulled by hand, which actuates a hammer and gong on the surface. Similar return signals are also employed, the one system of signalling being simply a standby for the other in case of failure.

The dip of the strata necessitates the use of several compressed-air haulage-engines in the mine. Air for driving the haulage-engines, the undercutting machines and the drills, is supplied from an air-compressor, fixed on the surface, capable of dealing with 2,250 cubic feet of free air per minute. The air has a pressure in the mine of about 70 pounds per square inch.

The upper and lower mines are lighted by electricity; Cooper-Hewitt mercury-vapour lamps have been found extremely effective for this purpose; while for general lighting, and in other places for local lighting, tantalum-lamps have now taken the place of ordinary carbon-filament incandescent lamps. No gases of any description are encountered in the rock-salt beds, and natural ventilation is found to be all that is required, assisted by the exhaust-air from the various engines.

The mine is capable of turning out 3,500 tons of rock-salt per week, containing an average of 96.5 per cent. of chloride of sodium. About one half of the present output is consumed at the various works of the United Alkali Company, Limited, at Widnes, St. Helens, Runcorn, Flint, Glasgow, Irvine and Bristol. This portion of the output is ground by specially designed machinery to a fine grain, and it has been the means of ousting higher-priced manufactured salt from the various chemical processes of the United Alkali Company, Limited. This grinding process is peculiar, inasmuch as it mechanically removes the marl-impurities contained in rock-salt, these impurities going from the screens in the shape of tailings. The grinding machinery is capable of reducing 60 tons per hour to the grain of common white salt.

The mine is situated about 1 mile from the estuary of the river Wyre at Fleetwood, and there is a railway of standard gauge to the Company's pier on the river Wyre, where vessels up to

15

1,500 tons burden can be loaded. A large percentage of the output is shipped direct to the Continent.

In conclusion, the author expresses his thanks to the directors of the United Alkali Company, Limited, for permission to give the information respecting their operations contained in this paper.

Specimens of rock-salt of various qualities and colour, and also of the Keuper marls, overlying the rock-salt, were produced for inspection.

———

Prof. W. Boyd Dawkins moved that a vote of thanks be tendered to Mr. Thompson for his paper.

Mr. Joseph Dickinson, in seconding the motion, said that Mr. Thompson's paper related to a branch of mining in which he (Mr. Dickinson) had been much interested, having in his time been in every rock-salt mine in England and Ireland. He had been several times in the first shaft sunk at Preesall, and had seen the development as now described. The paper gave the depth of the two workings in the mine, one under the other, and the distance between the supporting pillars, leaving members to infer from the section that a sufficient thickness of rock-salt was left for roof-strength between the pillars. The height of the workings was similar to that of many old mines, but it had been greatly exceeded in recent Irish mines, where the dip was about the same as at Preesall. Under-cutting or holing by machinery had during many years been successfully used in other rock-salt mines. He (Mr. Dickinson) was much impressed with the rapid rate at which the mine had recently been extended, and the increase of brine, by putting water down the bore-holes, from a trickle to a large quantity. The quickened opening was explained by the altered mode of driving out from the shafts, aided by compressed-air drilling-machines. In addition to these modifications, the improved dressing apparatus had been mentioned.

The motion was carried unanimously.

Prof. W. Boyd Dawkins (Manchester) said that he had been very much interested in hearing Mr. Thompson's account of the occurrence of the red sandstone in the vicinity of the Preesall mine. Large quantities of water were always found in the red sandstone, where there was a fault, as in this case, with the red marl, which was practically impervious, on one side, and the red sandstone on the other. The discovery of rock-salt was extremely interesting to him, because it went to prove the existence of a salt-field under the sea, between the Isle of Man, and Preesall and Barrow in England, and Carrickfergus in Ireland. Some years ago he laid before the members an account of the discovery of rock-salt in Triassic strata in the boring at the Point of Ayr in the Isle of Man, and a thickness of over 70 feet of rock-salt was proved. The brine was now pumped and conveyed in pipes to Ramsey from the Point of Ayr, and the manufacture of salt, on a small scale, had been commenced. The rock-salt round the Irish Sea was of very great geological interest. He could not help thinking that in remote times there was a great salt-basin between the Isle of Man and the Lake country, analogous to the great salt-field of Cheshire.

Mr. JOHN GERRARD (H.M. Inspector of Mines) said that geologically it was extremely interesting to note that at Fleetwood, near Barrow, also in the Isle of Man, and again on the north-eastern coast of Ireland, rock-salt had been proved. At Fleetwood, a bore-hole seeking iron-ore; at Barrow, in the Isle of Man, and at Carrickfergus, the bore-holes were in search of coal. The Preesall mine was probably the most productive salt-mine in this country, and having seen Mr. Thompson's work at the mine he could assure the members that the intelligence which Mr. Thompson had displayed in this paper was fully carried out in the practical work of the mine. He (Mr. Gerrard) had seen in both the upper and lower mines, a band of salt, very crystalline in appearance, several inches thick; the men stated that this guided them in working the salt; and he asked whether this band ran generally throughout the proved mine. He also asked whether it was a fact, where the very pure, very white rock-salt was found, that above or below it, the salt was very impure and mixed with marl. And, lastly, in giving the time occupied in drilling the holes by machine, did the author include the time occupied in fixing and taking down the machine.

Mr. THOMPSON replied that the seam of crystalline salt was

overlaid by 6 feet of rock-salt and underlaid by 15 feet or 16 feet of good rock-salt. This seam was found in the upper as well as in the lower mine, and indicated the direction of the dip. There was nothing like it found in Cheshire. An analysis of this seam showed that it did not vary from that of the other parts of the mine, except that it was uniformly of good quality and contained no marl, and was softer than the larger bed of rock-salt. It was more like grit, although it was not in any way gritty, and the men called it the " gritty seam."

Mr. JOHN GERRARD asked whether Mr. Thompson had formed any theory as to these guiding crystalline bands and their relation to the seams around them.

Mr. THOMPSON stated that he had not formed any theory as to the relation of the crystalline bands; but they were found in both the upper and the lower mines, and could be relied upon as guides in working the rock-salt.

Mr. JOHN GERRARD said that he had been repeatedly asked by persons interested in brine and salt production why at Fleetwood so large an amount of rock-salt was mined, in addition to the pumping of a large quantity of brine. The mechanical arrangements worked admirably, and the rapid opening-out of the mine, under the conditions obtaining in this case, showed that colliery engineers had something to learn. It was an instance of the advantage of persistent effort in using machinery for opening out mines; and he thought that Mr. Thompson would bear him out in saying that the first machine tried was not altogether a success, as it had to be altered and adapted to the conditions.

Mr. H. STANLEY ATHERTON asked how it had come to pass that these salt-deposits were found so much below the present sea-level. He also asked whether in the working of rock-salt it was easier to cut in one direction than another. In other words, was it similar to working an anthracite-mine, where if the cut was made along the lines of crystallization the work was much easier, and there was an easier line of cleavage and a much better way of cutting.

The PRESIDENT (Mr. Charles Pilkington) directed attention to the supporting shaft-pillars in the Preesall mine, being on a different scale from those deemed necessary in coal-mines. He was astonished at the rapid progress of the tunnelling-machines: 360 feet a week being driven in rock-salt that was harder than coal. He wished that such a speed could be secured in coal-mines: they could not touch 360 feet a week at the mines that he had to do with. He could scarcely believe that rock-salt was harder than coal.

Mr. THOMPSON stated that the tunnelling-machine was worked continuously by three shifts of men, eight hours each. Brine and rock-salt were both worked, because both were required by the United Alkali Company, Limited. Two processes were worked, in one of which salt was required in the shape of brine. The rock-salt was mined mainly to supply the salt required at the works at Widnes and St. Helens, and a considerable quantity was required for export, because rock-salt was admitted at a lower rate of duty than manufactured white salt. Brine was raised from about ten bore-holes, each one being independent of the other, all in the neighbourhood of the Preesall mine. In his experience, a bore-hole became useless, after a certain quantity of salt had been extracted from it. The original level of the sea was not maintained because the land did not retain its level; and there must have been tremendous upheavals at times. Cutting was carried on incessantly, and the cut was made across the face.

Prof. W. BOYD DAWKINS said that salt was merely the result of the evaporation of sea-water. He would expect to find a variation in a thick bed of rock-salt like that at Fleetwood, for the very simple reason that, in various stages of the evaporation of sea-water, different minerals were deposited. It was very largely, therefore, a question of the degree of concentration of the water from which that salt had come. With regard to the present level of the sea, it was quite an accident that rock-salt should be in this relation to it in this place, but in the British Isles nearly all rock-salt was below sea-level. In the vicissitudes of time, after the Triassic rocks had been formed in the sea, they were thrown into a series of folds, and parts of these folds were worn away by denudation; and the salt had been washed out of most of them down to sea-level. Consequently, there was very little rock-salt in Britain above sea-level.

Mr. JOHN RIGBY (Winsford) said that rock-salt was found in Cheshire above sea-level.

Prof. BOYD DAWKINS asked whether it was known that rock-salt occurred above sea-level in any other part of Britain, and remarked that this was an exception to the general rule. There was a large district near Burton, whence the salt had been dissolved out, leaving the broken and disjointed red marls, and in some cases the casts of salt-crystals in the rock, as evidence that it was formerly present.

Mr. DICKINSON said that Mr. Thompson gave the geological view of rock-salt formation, and refrained from expressing his own opinion. Was it not quite possible that the sea obtained its salt from the rock-salt, rather than by the reverse process, namely, that the salt was the result of the evaporation of sea-water?

Mr. THOMPSON thought that it was the accepted theory that the earth was gradually cooling down. If they reversed the process and assumed that the earth was gradually becoming hotter, the water would become steam, and salt would be left. The same rule would apply when the earth was cooling down, and the last occurrence would be that of steam condensing into water.

Prof. BOYD DAWKINS said that the earth was originally in a heated condition, like the sun. As it gradually cooled down, it would arrive at such a stage that chlorine and sodium would combine together in the atmosphere, and a deposit of salt would be formed over the surface; and hydrogen and oxygen combining together, would form water. Thus the sea was salt from the very beginning. The salt-fields of Britain were all derived directly from the evaporation of sea-water.

FIG. 4 —PLAN OF ROCK-SALT MINE.

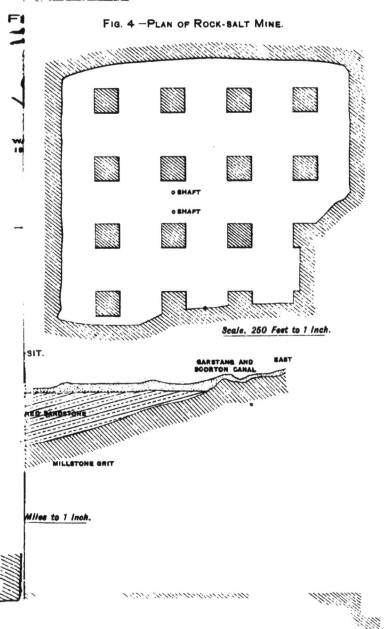

Scale, 250 Feet to 1 Inch.

Miles to 1 Inch.

MANCHESTER GEOLOGICAL AND MINING SOCIETY.

ORDINARY MEETING,
HELD IN THE ROOMS OF THE SOCIETY, QUEEN'S CHAMBERS,
5, JOHN DALTON STREET, MANCHESTER,
JUNE 11TH, 1907.

MR. CHARLES PILKINGTON, PRESIDENT, IN THE CHAIR.

The following gentlemen were elected, having been previously nominated:—

MEMBERS—

Mr. GREGORY RABY, Mining Engineer, Lota Alto, Lota, Chile.
Mr. SHERWOOD HUNTER, Mechanical Engineer, 20, Mount Street, Manchester.

DEATH OF MR. MARK STIRRUP.

Mr. JOSEPH DICKINSON said that the late Mr. Mark Stirrup was one of the oldest and most useful of their geological members. Mr. Stirrup held the post of honorary secretary of the Society for about ten years, and was elected an honorary member in the year 1904. He (Mr. Dickinson) moved that the members express their regret at Mr. Stirrup's death, and that a letter of condolence be sent to his relatives.

Prof. W. BOYD DAWKINS, in supporting the proposal, said that he experienced great difficulty in speaking about Mr. Stirrup, because of the friendship which existed between them. Mr. Stirrup was one of his oldest and best friends in Manchester, and they made many geological expeditions together, both in this country and on the Continent. He (Prof. Boyd Dawkins) felt that the Society had suffered a very great loss. Mr. Stirrup was a very high type of a man, and he did not think that any man could make him swerve in the slightest degree from what he considered to be the truth. He died in the fullness of years, and had left his mark on the science of his time.

The motion was adopted in silence: the members standing.

Mr. G. G. L. PREECE read the following paper on "Recent Improvements in the Design of Electric Cables for Collieries":—

17

RECENT IMPROVEMENTS IN THE DESIGN OF ELECTRIC CABLES FOR COLLIERIES.

By G. G. L. PREECE, M.Inst.M.E., A.M.I.E.E.

In connection with various papers communicated to the members of mining institutions, criticism had often been made that electrical engineers, when submitting the fruits of their wisdom to practical mining engineers, dealt too much in generalities, and avoided facts. The writer had this criticism in view when preparing this short paper, and considered that if he could lay before the members definite views, without qualification, as to the best types of electric cable for use under modern conditions of electrical working in collieries, he, at least, would disarm a portion of the criticism, whilst possibly his contribution might possess some practical value. He had pleasure, therefore, in submitting to the members two recent improvements in the design of electric cables which are particularly adapted for the conditions ordinarily prevailing in collieries, and which, in his opinion, are so valuable as to result in a greatly increased efficiency and length of life compared with cables hitherto designed and manufactured for colliery-purposes.

I.—The first improvement in cable design is not an entirely new type of cable, but a special feature in connection with the basis of all insulated electric cables, namely, the stranded copper conductor. The speciality is called " the patent solid strand-filling,"* and consists of the filling of the interstices between the wires composing the strand with a solid compound which will not run or become displaced by the heating of the conductor through overload. In course of manufacture the central wire, and each succeeding layer of the stranded conductor, is covered with a thin tube of plastic bituminous compound. This compound being thoroughly worked into the interstices of the strand, all capillary passages for moisture are stopped—in fact, water

* British patent, 1905, No. 4260.

cannot be forced in, even under pressure.   This type of stranded conductor is particularly recommended, in conjunction with bitumen, as the insulation or dielectric.

The advantages of such a type of cable are very obvious: there is nothing of a fibrous nature about such a cable, nor anything which can possibly take up moisture, the whole cable being waterproof.   The writer may state that water in the strand of a cable is harmful, and the majority of faults—indeed, the chief reason of deterioration in colliery-cables—are due to water getting into the strands of the conductor.   In most collieries, the water

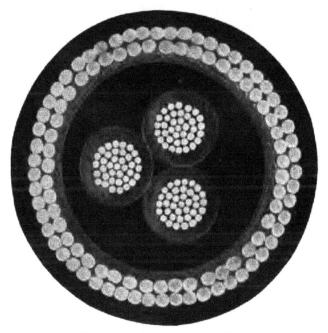

FIG. 1.—COMPLETED SOLID BITUMEN SHAFT-CABLE.

is more harmful than ordinary water.   This water enters either through the ends or through an original fault, which in many cases may be caused by mechanical damage.   It is the writer's experience that, at collieries, cable-ends are often exposed to dampness through carelessness; and owing to the difficulty of repairing a fault immediately and properly, water has often entered the cable before a repair can take place.   Neglect by the ordinary workman to inform his superiors of the occurrence of a fault or damage, and a hasty repair, made with a bit of rag or tape, come under this heading.

Once water is in a cable, actions and reactions occur and deterioration takes place, more or less rapidly, depending, more or less, on the load of the cable and the thickness of the dielectric. It is very obvious, therefore, that any cable, particularly a bitumen-insulated cable, treated so as to be absolutely proof against such deterioration by preventing the passage of water along the strands of the conductor, must be of great value and possess many advantages; and it is almost "foolproof" against the carelessness of employés. The extra cost of so treating a bitumen-insulated cable is small and trifling, compared with the increase in life given to the cable.

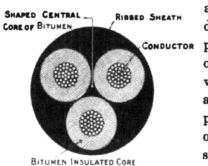

SHAPED CENTRAL CORE OF BITUMEN   RIBBED SHEATH

CONDUCTOR

BITUMEN INSULATED CORE

FIG. 2.—END SECTION OF SOLID BITUMEN SHAFT-CABLE.

II.—The second improvement in cable design which the writer wishes to describe is an entirely new type of cable, called "the patent solid three-core bitumen-cable."* The mode of manufacture is as follows:—The conductors of this cable are each separately insulated with bitumen compound, they are then laid together round a shaped central core of bitumen which

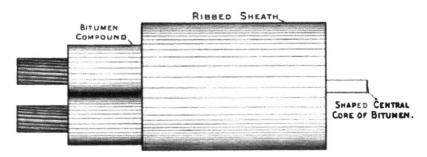

RIBBED SHEATH

BITUMEN COMPOUND

SHAPED CENTRAL CORE OF BITUMEN.

FIG. 3.—SOLID TYPE THREE-CORE BITUMEN-CABLE WITH INSULATION AND SHEATHING.

fills up the central space; and the whole is sheathed over with a solid tube of bitumen, which is forced on so as to be ribbed internally and to fit exactly the interstices between the three cores (figs. 1, 2 and 3). The cable can then be taped and protected

* British patent, 1906, No. 8238.

by armour, etc., as may be necessary. The cable being a three-core one, is used for three-phase alternating currents. Three-phase working is becoming very general in collieries; and as for electrical reasons, apart from other reasons, the three-core cable is the best, a large quantity of such cable is used.

The special construction of the solid three-core bitumen-cable obviates the introduction of fibrous material as padding or elsewhere in the body of the cables; and consequently it has many advantages over the ordinary types. The ordinary three-core bitumen-cable, for instance, is composed of three cores, each insulated with bitumen compound—some types have jute-braids or tapes directly over the stranded conductor and underneath the bitumen insulation; each of the three cores is taped, with more fibrous material, and then laid up together; and, in order to make the cable into a circular form, jute-warp or other fibrous material is used to fill the interstices between the cores and in the centre. The cable is then taped and protected in the usual manner. Cables are manufactured with an outer sheathing · of bitumen over the cores; but, in these cases, the jute-padding is still used in making the cable circular. In the solid three-core bitumen-cable, it will be noticed that the outer sheathing of bitumen, owing to its ribbed construction, itself fills up the interstices.

The advantages possessed by this cable over ordinary types may be described as follows: (1) Higher insulation and dielectric strength. The effective strength between each core and between the cores and earth in the case of the solid three-core bitumen cable is the total thickness of bitumen between the conductors, that is, twice the thickness on each conductor, and the total thickness of bitumen over each conductor and around the three cores respectively. This is twice the effective dielectric obtained in three-core cables of the ordinary construction, as, by the use of fibrous tapes on the insulated conductor and of jute padding, a potential earth is introduced immediately around each core, and if moisture creeps down, as is quite likely, such fibrous material becomes at earth potential. (2) Greater chemical protection against bad water, etc. This is obvious in view of the explanation given above that there is greater effective insulation. (3) Greater reliability in wet situations. This is also obvious as water cannot creep down the cores. This may be understood by examining a rough sample, and noting with what difficulty the outer sheathing is wrenched away from the three cores. The cable is practically homogeneous. (4) Greater mechanical protection of the cores. This is obvious owing to its construction. The cores in the ordinary type are fairly loose. (5) Minimum chance of decentralization due to overload. This is also obvious owing to its construction. The ribbed sheath holds the cores in place, gripping them tightly. (6) The solid three-core bitumen-cable combines in itself all the advantages of a perfect solid system. (7) A further advantage lies in the fact that, in consequence of the cores being gripped along their whole length by the outer sheath, and thus not slipping, there should be no hesitation in suspending this type of cable (when armoured) even for a distance of 1,200 feet. The distance in practice is limited by the weight of the cable. In fact, a cable 1,200 feet long so suspended, and weighing about 5 tons, is to be installed during the next two or three weeks in a shaft near Stoke-upon-Trent. There are many advantages obvious to colliery managers, which the writer need not point out here, in being able to suspend a cable from one point at the top of the shaft.

The stranded conductors in this type of cable, which can be and are used up to a pressure of 3,000 volts alternating, can, of course, be treated with the special solid strand-filling compound. This would then form, in the writer's opinion, absolutely the best cable possible for three-phase working in collieries and mines. The improvements being mainly connected with the bitumen-insulated type of cable, a few general remarks on this type of cable will be, therefore, a fitting conclusion to this paper.

The bitumen type of cable, as now designed and manufactured by the leading British cablemakers, is of so stable a nature, and so carefully and scientifically made, that the writer has no hesitation in recommending it as the best type of cable for use in collieries, at all pressures, at any rate up to 3,000 volts. Beyond that voltage, it may be necessary to introduce a certain amount of paper with the bitumen; but in this short paper there is no scope for the discussion of special conditions.

The bitumen-compound should be tough and resilient, and designed so as to suit varying temperatures; it must not be too soft at high temperatures, nor become brittle at low temperatures.

A bitumen-cable so designed possesses great permanency and chemical inertness, and resists very successfully the various oxidizing and other influences met with in collieries—such as bad pit-water, etc.   Some bitumen-cables are very hard, and appear very strong, mechanically, at normal temperatures; and these cables are often looked on with favour by mine managers. These physical properties are, however, obtained by the sacrifice of permanency of dielectric and chemical inertness; the hardening or stiffening usually takes the form of an addition of mineral loading-matter.   Such a course renders the dielectric liable to loss of coherence in its particles, with a great tendency to become brittle and to crack at low temperatures; and, consequently, it encourages in a much greater degree the action of the oxidizing and other influences met with in colliery-workings.

———

Mr. T. H. WORDSWORTH asked whether Mr. Preece considered it advisable to use three-core cables in shafts when the power to be transmitted was, say, 250 kilowatts at 500 volts.   In his experience, three-core cables for this amount of work became too heavy and troublesome to handle in the shaft.   He wished to know whether the filling of the strands would entirely prevent water from creeping up the cable, as he understood that with continuous current there was always a tendency for the water to travel up the negative cable.

The PRESIDENT (Mr. Charles Pilkington) questioned the feasibility of suspending a three-core cable 1,200 feet long, without risk of serious injury.   He had had some experience with electric cables, and knew how difficult it was to prevent water from getting into them.   The filling described by Mr. Preece seemed to be calculated to keep out water altogether. He quite agreed that many hard bitumen-cables became brittle occasionally.

Mr. ALFRED J. TONGE said that a three-core cable 1,320 feet long had been suspended at the Hulton colliery.   He asked whether Mr. Preece had made any tests for oxidation, or for the effect that atmosphere or temperature had upon the bitumen. In his own experience, nothing surpassed a coat of rubber for the repeated taking-up and laying-down conditions that often prevailed in mines. This cable had been designed to meet the difficulty of moisture; but unless the bitumen was free from cracks there would still be trouble. Mr. Preece had spent a large amount of time in considering this matter, and it seemed reasonable that the members should give encouragement to engineers who were trying to overcome the difficulties specially attached to the use of electricity in mines.

Mr. PREECE, replying to the discussion said that, where a large amount of power was taken down the pit, it certainly might in some instances be better to take three single cables for convenience in handling; it must be remembered, however, that for electrical reasons this method was not so good for three-phase working, especially when large conductors were being used. Further, the single cables would of necessity be unarmoured, and therefore must be protected, say by wooden casing. Armouring on the wires would create a heavy induction-drop. Personally, he would prefer to build up the main cable in units of convenient size where a large power had to be taken down the shaft; this system also allowed of the possibility, in case of a breakdown on one cable, of another taking its load temporarily. He confessed, however, that he had advocated the use of three single cables; that was before this special solid three-core bitumen-cable had been brought out. The special solid strand-filling would not allow any moisture up the strand. There was no fear of any harm arising from the squeezing pressure exerted on the cable when under single suspension; and in such cases, the special construction of the solid three-core bitumen-cable compared favourably with the ordinary three-core cable. The cores were well gripped and protected by the ribbed outer sheathing, which also ensured a uniform distribution of pressure; while in the ordinary type the pressure was exerted locally. The solid three-core bitumen-cable was not soft actually; it was soft in comparison with some very hard types of bitumen-cable, but it was tougher and more resilient, and the latter quality was most important. Further, the hard type of bitumen-cable was not reliable; it would not stand the oxidizing influences present in collieries anything like as well as the type of bitumen used in the solid three-core cable. In cases where great flexibility was required and the cable had to be moved very often, there was nothing better than vulcanized indiarubber; and the resilient type of bitumen in the special cable was the nearest approach to it. Most cables in collieries, especially shaft-cables, were permanently fixed.

DISCUSSION OF MESSRS. W. N. ATKINSON AND A. M. HENSHAW'S PAPER ON "THE COURRIÈRES EXPLOSION."*

Mr. JOSEPH DICKINSON said that it might appear ungracious not to acquiesce in the conclusion arrived at by such high authorities, and after evidently unwearied painstaking enquiry. It was astounding and not readily admissible that a blown-out shot, with only dust to help, should travel through many miles of workings in several different seams and kill 1,100 persons; and, at all events, a little honest criticism might help the consideration of the point.

There were thirteen shafts, varying in depth, with workings connected underground; and the workings of four of these shafts were seriously affected: No. 2 pit, upcast; No. 3 pit, bratticed, part downcast and part upcast; No. 4 pit, upcast; and No. 11 pit, downcast. Loud noise with clouds of smoke and dust came to the surface at Nos. 3, 4, and 11 pits; a cage at No. 11 pit was thrown upwards towards the pulleys; there was no surface-damage at No. 2 pit; the landing-floor was damaged at No. 3 pit; and the covering was blown open at No. 4 pit. Whilst below-ground at No. 11 pit, the 1,257 feet (383 metres) hooking-on place and surroundings were damaged; at No. 4 pit, something similar happened; at No. 3 pit the 919 and 1,070 feet (280 and 326 metres) entrances were affected, some stoppings shutting off a fire being near the 919 feet (280 metres) mouthing; and at No. 2 pit the flame ceased. It required careful study of the several plans outlining the workings in the respective seams to follow the course of the blast. The blown-out shot was in the Joséphine seam, at the face of the Lecœuvre heading. aired by pipes, and one of two headings, about 450 feet in length, driven from an upbrow or rise-working out of the north 1,070 feet (326 metres) winning from No. 3 pit. From this point, the blast

* *Trans. Inst. M. E.*, 1906, vol. xxxii., pages 439, 340 and 507; 1907, vol. xxxiii., pages 124, 303 and 326; and 1907, vol. xxxiv., page 151.

radiated in various directions into various seams, including the north-west district, the No. 2 pit, in the south-east. Another part turned back at an acute angle and then divided, some going to the north and some by the 1,070 feet (326 metres) tunnel to No. 3 pit, up which some went. The blast passed into the south 1,070 feet (326 metres) tunnel and entered the east and west workings, that going west, apparently re-invigorated with fresh gas, extended to Nos. 4 and 11 pits, north and south of which extensive damage was produced. Worked-out districts were not shown on the plans, but only main roadways, and some main air-currents by arrows with feathered tails, and some of the blasts by arrows without tails. The position of the fire in the Cécile goaf near No. 3 pit was shown, also of the seven stoppings shutting it off, the shut-off part including a large area of goaf. Fires in mines were usually dreaded, and suitable precautions taken until the enclosed area became too much fouled to be explosive. This fire, however, discovered in an airway in the goaf four days before the explosion did not cause alarm; work elsewhere went on as usual; and the shutting-off was only just finished before the explosion occurred.

The writers of the paper did not meet with any fire-damp; but their investigations did not begin until long after the explosion, and it might reasonably be supposed that they could not enter many, if any, of the goaves.

Fire-damp had, on a few occasions, been previously found in some parts of the workings. Therefore, in endeavouring to solve the difficult problem, it should be borne in mind that it seemed probable from burst piping that some fire-damp was present in the Lecœuvre heading where the shot blew out. The heading above the Lecœuvre heading was not at work; it was found much shattered and inaccessible after the explosion, and might have been full of fire-damp. Taking the lower heading as the starting-point, aided by gas from the upper heading, flame would extend far away, combining with fresh air and lighting fire-damp in the numerous goaves, and gas distilled from the shut-off fire seems more likely than the hypothesis of dust alone. From below No. 7 stopping also, an unfeathered arrow showed the blast coming out of this goaf; and on the rise near No. 3 pit, two stoppings were blown inwards towards the fire; the blast apparently raked through this goaf.

In his (Mr. Dickinson's) time, excluding explosions from
shots alone, he had investigated between 350 and 400 fatal explosions, large and small, in some of which dust contributed, but
fire-damp played the leading part; and, so far as memory served,
not one was caused from dust alone.   The greatest explosion
was at the Oaks colliery, Barnsley, in 1866, when 361 lives were
lost including 27 rescuers. This explosion followed the firing of
a heavily-charged shot, in about the time that it would take to
count two; but it was in a fiery mine with accumulations of gas,
and found yielding about 550 cubic feet of gas per minute after
the explosion. Dust had long been recognized as a factor in fire-
damp explosions. A few explosions were recorded as having
occurred in strong air-currents, especially in intake air newly
compressed by descent into the mine, the starting-point being a
shot. Caution was, therefore, requisite. Fire-damp seemed the
leading cause, but dust was an accessory. The writers' suggestions in this respect were, therefore, commendable   Dust
should be removed, or watered so far as not to occasion more
harm than it would remedy. Spraying with water might perhaps help in dispersing fire-damp, instead of the now discarded
batting out with cap and jacket.

The PRESIDENT (Mr. Charles Pilkington) said that the details
were exceedingly complicated in the case of the Courrières explosion, and it seemed to him that dust must have carried it on.
He could hardly imagine the presence of a body of gas large
enough to blow up and down the staple pits and cause the havoc
that had been wrought in so great a length of roadway. He
thought that most explosions began with gas, but were extended
by dust. At the same time, Mr. Dickinson had had so large and
varied an experience of mine-explosions, that he deferred to his
opinion, although he could not help thinking that the Courrières
explosion was carried on by dust. He hoped that the dust question would be set at rest by the trials about to be made by the
Government.

Mr. G. B. HARRISON (H.M. Inspector of Mines) could not
fully accept Mr. Dickinson's view, that gas was the predominant
factor and dust a possible accessory in all explosions. In many
explosions, unless all the engineers who had investigated them
were wrong as to the point of origin, the circumstances did not

favour Mr. Dickinson's theory.   He (Mr. Harrison) instanced in
particular the circumstances under which the explosions took
place at Brancepeth and Wingate Grange collieries.   It seemed
to him that at the Courrières colliery dust must have been the
cause of the explosion travelling over such long distances, and
irrespective of the cause and the point of origin, dust and air
must have been the factors that carried on the explosion.

Mr. W. OLLERENSHAW wrote that, some fifteen or more years
ago, he had an opportunity of making investigations after an
explosion which occurred at the Ashton Moss colliery by which
two men lost their lives and several others were injured.   There
were many points in connection with his investigations which
were supported by Mr. Atkinson's and Mr. Henshaw's observations at the Courrières collieries.   (1) No gas had been found
at the point at which the explosion took place previous to the
explosion, nor was any found at that point at any time after the
explosion; (2) the point at which the explosion occurred was
very dry and dusty; and (3) the flames from the firing of the
coal-dust, in both directions in which they travelled, on coming
to points at which the nature of the dust changed, or where the
dust was moistened by water, were arrested as they were at the
Courrières collieries.   The explosion at the Ashton Moss colliery
was proved to have been caused by a blown-out shot, fired at the
junction of the main return-airway.   The roadway in which .
the shot was fired was small, and, as a result of the increased
velocity of the air passing the full tubs of coal, considerable
quantities of dust were blown off the tubs and accumulated along
the roadway; and, so far as this dust extended in both directions,
the flames from the explosion had travelled until arrested at
wet points caused by water coming from the roof.   The airway,
below the point at which the shot was fired and from which
direction the air was travelling, was made through gob or waste;
but no coal-dust was present in the airway, nor could any traces
of the explosion be found in the airway.   The fact that some
coal-dusts were explosive pointed out the necessity of the experimental gallery suggested by Mr. Henry Hall, Mr. John Gerrard,
and others; but in other mines the dust was not of so highly an
inflammable nature as to be explosive. If samples of dust, taken
from the roadways of any mine, were found (on being subjected

to a reasonable test in the experimental gallery) to be of such a nature that danger of a coal-dust explosion might be feared; then, H.M. Inspector of Mines might reasonably demand that remedial measures should be adopted in that particular mine, but it would be unfair to place all mines on the same footing. The dust in some mines, the result of floor-crushings, etc., and not of broken coal, was perfectly safe. He (Mr. Ollerenshaw) suggested two methods which would lessen, if they did not altogether prevent, the occurrence of coal-dust explosions: (1) the tubs used for carrying the coal from the coal-face to the surface should be kept in first-class repair, in order to prevent the leakage of coal-dust on to the roadways; and (2) the haulage-roads should be kept in first-class repair in order to prevent the leakage of coal-a high velocity as to blow dust out of the tubs whilst in transit to the shaft.

He thought that the members would agree that, if Great Britain was to maintain her supremacy as a manufacturing nation, the manufacturer should be supplied with cheap fuel. This fact ought not to be lost sight of in discussing what precautions should be taken to minimize the dangers from coal-dust; and, whilst taking steps to protect the lives of the workmen, care should be taken not to cripple the coal-industry.

Mr. JOSEPH DICKINSON said that he had investigated the explosion that took place at Ashton Moss colliery. Return air, from the deep workings of the mine, was coming through a very small hole and blowing with great pressure into an upper level, and some men fired a shot into the compressed air, which, no doubt, was contaminated with fire-damp.

The PRESIDENT (Mr. Charles Pilkington) said that the composition of dust, its fineness as well as its quality must be considered. Coarse granular dust would not explode, but a very fine one would. Samples taken from the Clifton and Kersley pits had been tested, and the result confirmed this opinion. He had found also that, while floor-dust was comparatively harmless, the dust taken from the sides and roof was more or less explosive. Consequently, in watering any mine it was absolutely useless to water the bottom-dust unless the top and sides were also watered or cleaned. At the same time, he considered that the question of watering was fraught with great danger. Watering was beneficial if it could be done without destroying the roof or lifting the floor; but, in every case where the floor was soft or the roof tender, more men might be killed by watering than would be saved by keeping them from the risk of an explosion.

Mr. JOSEPH DICKINSON said that there were different kinds of blown-out shots.   Holes drilled with an obstruction in the drilling had the effect of giving a twirl to the shot, in the same manner as a rifled barrel did to a bullet or bolt.   He did not think that any drill-hole which was not bored straight should be allowed to be charged; it was dangerous to fire the shot in a drill-hole that was askew, as the flames spread out so widely.

MANCHESTER GEOLOGICAL AND MINING SOCIETY.

ORDINARY MEETING,
HELD AT VICTORIA UNIVERSITY, MANCHESTER, JUNE 28TH, 1907.

MR. JONATHAN BARNES IN THE CHAIR.

Prof. W. BOYD DAWKINS conducted the members through the Geological Museum, and indicated some of the recent additions, including the skeleton of a Plesiosaurus, presented by Mr. W. H. Sutcliffe, and a large slab showing footprints of an enormous reptile found in the Wirral district and presented by Mr. W. Balfour Stewart.

Prof. W. BOYD DAWKINS said that the discovery at Whitby was an interesting illustration of the manner in which those members of the Society who occasionally disported themselves at the seaside, if they kept their eyes open, might make discoveries. Last summer, Mr. W. H. Sutcliffe—a good friend to the Museum— was walking along the cliffs near Whitby, when he saw, some 15 or 20 feet above the beach, what he thought were bones. On making his way up the cliff, he found that they were vertebræ ribs, and portions of paddles, and with the aid of some fishermen he unearthed them. He also found, close by, the skull of a beast which turned out to be that of a Teleosaurus, an order allied to the crocodiles now found in the river Ganges. Mr. Sutcliffe brought the skull and the vertebræ to the Manchester Museum, and Mr. Watson and Mr. Hall, in the geological department (whom, he was sorry to say, they had just lost by his removal to another sphere of activity) put the bones together. Then it was realized that the head had nothing to do with the bones, and the head which did belong to the latter was probably still in the cliff. This Mr. Sutcliffe found on a subsequent visit, and in the specimen in the case it was seen in its natural position at-

tached to the vertebræ of the neck : it belonged to the Pleosauria.
That was the history of one of the most interesting specimens in the
Museum. Continuing, Prof. Boyd Dawkins said that the
Museum had grown to be what it was by the assistance afforded
by members of the Society, who collected the larger and the more
important portion of the geological specimens, notably those
groups of Carboniferous fossils which illustrated so well the
Carboniferous flora. A better collection of the Carboniferous
flora was not to be seen in any other museum in the world. Prof.
Boyd Dawkins also called attention to the slab of stone, recently
placed in the Museum, showing the wonderful " footprints in the
sands of time " of which Longfellow spoke in his *Psalm of Life*.
The footprints, clearly indicated on the slab, were those of a
great Dinosaur, with small fore and large hind limbs, called
Chcirotherium. It was one of the finest slabs found of late years.

Prof. W. BOYD DAWKINS read the following " Introductory
Remarks on the Coal-measures " : —

INTRODUCTORY REMARKS ON THE COAL-MEASURES.

By Prof. W. BOYD DAWKINS, D.Sc., F.R.S., F.G.S.

When the members considered Carboniferous forests, the vegetation covering the land, he thought that they ought to deal with the question of Carboniferous geography. Of course, as was well known, the vegetation followed a zone, more or less definite according to its relation to the sea-level. When they realized the reason why the seams of coal were so remarkably parallel to one another, they grasped the fact that they were vegetable accumulations grown on great tracts of flat land, close to water-level. They must discard the idea of isolated basins, as now represented by the coal-fields. The general idea, presented by Sir John Prestwich and Mr. A. R. C. Godwin-Austen as to the geography of this country in remote ages, was as follows:—A mountainous region in the Highlands of the north, a great tract near sea-level, extending over the lowlands of Scotland, round the Lake District, round North Wales, as far south as Brittany, and extending over nearly the whole of Ireland with the exception of the Wicklow mountains, and to the east into the valley of the Rhine. This great mass of alluvium covered an area more than a thousand miles long and 500 miles broad. The forests, represented by coal-seams, grew on great tracts of flat land near the water-level. In some of the shales and sandstones, there are also remains of driftwood (*Dadoxylon*) carried down from the uplands by floods, and belonging to a flora which grew at a higher level. The character of the water was shown in some cases by shells: marine shells indicated the presence of sea-water, and fresh-water shells, fresh water. The question of the zones for the purposes of classification was now under discussion.

At the close of the Coal-measure period, this great mass of alluvium, no less than 7,200 feet thick in Lancashire, was thrown into a series of folds, and riddled with faults. In later geological times, it was largely destroyed by processes of denudation, and the existing coal-fields are merely those portions of it which had been preserved, by being sunk deep beneath the then surface of the land, and by the protection offered by newer rocks.

———

Mr. DAVID M. S. WATSON read the following paper on " The Formation of Coal-balls in the Coal-measures " : —

THE FORMATION OF COAL-BALLS IN THE COAL-MEASURES.

By DAVID M. S. WATSON, B.Sc.

The Upper-foot Mine of Lancashire and the Halifax Hard Bed in Yorkshire have been known for many years to contain in the actual seam of coal numerous calcareous nodules, enclosing beautifully-preserved plants. It has been found by Miss M. C. Stopes and the author that coal-balls occur in at least two other coal-seams in Lancashire: (a) one of these, well exposed in the banks of the river Tame at Stalybridge, is about 100 feet below the Woodhead Mill rock, and about 20 feet above the top of the Millstone Grit. (b) The other, worked about 100 years ago in a pit near Laneshaw Bridge, about 2 miles north-east of Colne, is in the Millstone Grit, probably in the Second Grit. Exactly similar coal-balls occur on the Continent at Orlau, in Silesia, and in Westphalia, in each case at more than one horizon. In all these occurrences the coal-balls are associated with marine conditions in the roof.

The roof is always a black or dark-grey shale, containing abundant crushed goniatites and *Aviculopecten*. In these shales also occur spheroidal concretions, containing the same fossils uncrushed. These are known as *baum-pots* and yield fine specimens of the contained fossils. Baum-pots have roughly the same composition as coal-balls: differing only in having a higher percentage of clay. The similarity in composition between baum-pots and coal-balls, both being impure dolomites, when taken in connection with the constant occurrence of baum-pots over coal-seams containing coal-balls, suggests some relationship between the two.

The conditions at the junction of the Upper-foot and Gannister coals to form the Mountain Four-foot seam, as seen at the Old Meadows pit, Bacup, renders this relationship almost certain. It is found, on approaching the junction in the Gannister Mine, that isolated coal-balls occur when there is a parting 10 inches

thick between the top of the Gannister seam, which elsewhere
is free from coal-balls, and the Upper-foot Mine with its marine
roof.   When the junction is reached and the parting is only 1
inch thick, numerous large coal-balls are found in the lower
part of the Union seam, which represents the Gannister coal.   It
is found, in certain cases, that stems run through two or more
coal-balls, thus showing that the latter cannot have moved since
their formation.   The often very irregular form of the coal-balls
also testifies to the same fact.

It is consequently certain that the coal-balls must have some
connection with the conditions prevailing during the deposition
of the roof, and the explanation adopted by Miss Stopes and the
present author is as follows:—It has been shown by Mr. H. B.
Stocks* that the bacteria which produce decay are capable of reducing calcium-sulphate to the carbonate, which, being less solu-
able, is deposited. It seems almost certain that the same reaction
would occur with magnesium-sulphate. Now, sea-water contains
appreciable quantities of the sulphates of calcium and magnesium.   It has been shown above that the marine conditions of
the roof have something to do with the formation of coal-balls,
and this occurrence of calcium and magnesium-sulphate in sea-
water suggests that the marine roof had supplied the raw material of the coal-balls.   Another use of the sea-water was as a
preservative.   It has been found by Miss Stopes and the author
that plants are perfectly preserved for a long time by a mixture
of sea-water and peat.

Full information about the occurrence and probable methods
of formation of coal-balls and some similar structures, notably
the dolomite replacing coal at the Wirral colliery, described by
Mr. A. Strahan,† will be found in a paper by Miss M. C.
Stopes and the writer.‡   A short preliminary account has been
published by Miss Stopes.§

* "On the Origin of Certain Concretions in the Lower Coal-measures," by
Mr. Herbert Birtwhistle Stocks, *The Quarterly Journal of the Geological Society of
London*, 1902, vol. lviii., page 46.

† "On the Passage of a Seam of Coal into a Seam of Dolomite," by Mr.
Aubrey Strahan, *Ibid*, 1901, vol. lvii., page 297.

‡ "On the Present Distribution and Origin of the Calcareous Structure
known as Coal-balls," by Miss M. C. Stopes and Mr. D. M. S. Watson.

§ "On the Coal-balls found in Coal-seams," by Miss M. C. Stopes, *Report
of the Seventy-sixth Meeting of the British Association for the Advancement of
Science*, York, August, 1906, page 747.

The CHAIRMAN (Mr. J. Barnes) said that Mr. Watson's paper had been interesting to him, as he had had some share in working out the composition of coal-balls; and he agreed with Mr. Watson's idea as to their being of marine origin, because dolomite was not altogether a marine body, although largely so. The question of dolomitization was not yet definitely settled, but he hoped soon to place something on record that would make the matter as simple as it was now difficult.

Mr. BERNARD HOBSON, in moving a vote of thanks to Mr. Watson, said that he had been specially interested in the relationship that Mr. Watson had pointed out between the coal-seam in the Wirral peninsula and the coal-balls in the coal. In the case of the Dolomite-mountains of the Tyrol, the deposit apparently had no connection with coal-seams. The question of the formation of dolomite was one on which there was great difference of opinion, and every observation which tended to show how dolomite might be formed under certain conditions would, he thought, probably lead to throwing light on its possible mode of formation in other cases. He thought that there was an instance of a dolomite-deposit cementing the pebbles in the river Neckar, at Cannstadt, Germany.* There were various ways in which dolomite could be formed, but the mode, in the case of coal-balls, was of special interest.

Mr. SYDNEY A. SMITH seconded the resolution, which was passed.

———

Mr. GEORGE HICKLING read the following paper on " Carboniferous Flora as an Aid in Stratigraphical Classification ":—

---

* *Die Paragenesis der Mineralien,* by Prof. August Breithaupt, 1849, page 46.

CARBONIFEROUS FLORA AS AN AID IN STRATIGRAPHICAL CLASSIFICATION.

By GEORGE HICKLING.

The main object of the writer is to endeavour to show that fossil plants are more worthy of the attention of geologists, as an aid to the correlation of strata, than has of late been generally admitted.

For a clear discussion of any question of this kind it is essential that one or two fundamental geological terms should be used with greater precision than is commonly the case. " Formation," " system," and " zone " are such familiar terms that their precise meaning and relation are apt to be neglected. They differ not merely in magnitude but in kind. Formation was originally used, and should still be used, simply to denote a mass of rocks of a definite character, which might be traced as a sheet over a greater or smaller tract of country: it represents a particular type of sedimentation. System is defined in a totally distinct manner, although this is not always clearly recognized. While formations for the most part followed one another in unbroken succession, widespread unconformities here and there interrupt the sequence and split the whole series of stratified rocks into larger subdivisions or systems. A prevalent misconception must be corrected here: namely, that a system is a series of deposits defined by a particular assemblage of organic remains. It was perfectly natural that this idea should have been grafted on to the definition of a system in the days of cataclysmic theories; but, in view of the modern belief in the continuity and gradual modification of the organic world throughout geological time, it must be at once recognized that systems can only be defined by natural breaks in the succession of the rocks. A zone, on the other hand, is only palæontological in its conception. Interruptions in the series of deposits which make systems impossible, are the greatest hindrance to the zoning of the geological scale. Single species or groups of organisms may be used to define

zones, as at all periods there are species and groups which attain
the height of their development.   They are the dominant species
of the time, and their abundant remains in a stratum mark it as
having been formed at that time.   Zones, then, but certainly not
systems, are defined by certain dominant groups of organic
remains.

The value of plant-remains for the definition of zones within
a given system depends upon the amount of evolution which the
plants or certain groups of them, have undergone while the rocks
of that system were being deposited; that is, the extent of the
changes in the dominant species.   The first enquiry, then, must
be: what are the chief groups of plant-bearing strata that must
be included in the Carboniferous system, and what, approximately, is their relative position?   A discussion of the intricate
problems of the correlation of the Carboniferous rocks of Great
Britain would be beyond the limits of this brief communication,
and the outline-scheme set forth in plate iv. summarizes the
most probable views, he believed, on this subject. The only points
in this scheme likely to meet with serious opposition are the in-
clusion of Upper Old Red deposits in the Carboniferous system,
and the position assigned to the Devonshire rocks.   These, the
writer hoped, shortly, to make the subject of a detailed paper.
Meanwhile, he might remark that, following the definition of the
system given above, the Upper Old Red must be included, while
he believed that this is further necessitated by the contempor-
aneity of some, at least, of those deposits with part of the Carboniferous Limestone Series.   Below that group of deposits there is
a great unconformity, which will probably be found eventually to
extend over the whole of the British Isles, and may also be traced
on the Continent.

Taking this great break as a natural base, the Carboniferous
system presents three well-known floras, which may very well
indicate the amount of change undergone by plants during the
whole of that period.   Near the true base are found the plant-
beds of Kiltorcan, Kilkenny; about the middle are the cement-
stones and oil-shales of the Calciferous Series; and at the top
are the various Coal-measure horizons.

Of the distinctness of these three floras it is unnecessary to
speak at length.   Kiltorcan has unfortunately yielded as yet
only about ten species, or perhaps less; nevertheless, poor as the

flora is, it is perfectly distinct from that of the Calciferous Sand-stone or Cement-stone, although presenting some affinity with it. Luckily, the knowledge of the plants of this period may be extended from the rich beds of Bear Island, the plant remains from which have been described by Heer, and more recently by Prof. A. G. Nathorst.* *Archæopteris* is there represented by three species, *Bothrodendron* by five, and the peculiar *Sphenopteridium* has likewise five species. These three genera may be regarded as the characteristic forms of the Upper Old Red flora. They are represented in the Calciferous Series, but by almost completely distinct species. *Bothrodendron* and *Archæopteris* only are scantily represented, even as genera, in the Coal-measures. The total absence of many of the commonest upper Carboniferous forms, *Sigillaria, Alethopteris, Neuropteris,* and *Pecopteris,* is no less striking a feature of this earliest flora. So far as one can at present judge, the Calciferous-Sandstone flora occupies a fairly intermediate position between that of the Upper Old Red of Ireland and Bear Island and that of the Upper Carboniferous. It bears a very similar relation to both. Most of the important Upper Carboniferous genera appear in the Calciferous Series, but they are represented mainly by distinct species, while there are several genera peculiar to each group. Of genera peculiar to the Calciferous Series, *Sphenopteridium, Rhacopteris,* and *Adiantoides* may be mentioned, while the complete or almost complete absence of the genera *Pecopteris, Alethopteris,* and *Neuropteris* is still noteworthy, although these begin to appear in slightly higher beds. Thus, the wide distinction between the floras and the great divisions of the Carboniferous is abundantly evident.

The only doubt that could be cast on the value of these distinctions would be on the possibility of their being due to mere local conditions, and not to the evolution of the plant-world. If corresponding plants can be shown to exist at similar horizons in a number of localities, this objection will be overruled. Evidence on this point is steadily accumulating, and it is pointing in the right direction. Telia Quarry, near Prestatyn, contains a plant-bed, referred by Mr. R. Kidston, from its plants, to the Calciferous Series, and by Dr. Wheelton Hind, from its mollusca, to the Pendleside series. Formerly, these two horizons were considered as most widely separated; but, as indicated in the diagram (plate

* *Zur Oberdevonischen Flora der Bären-insel.*

iv.), the writer thought that the evidence pointed to their being comparatively near. Similar plants are associated with similar shells in the Limestone Series at Poolvash, near Castleton, Isle of Man. In the diagram (plate iv.), again, the writer has placed the Culm Series of Devonshire as entirely Upper Carboniferous, and this for purely stratigraphical reasons. From an examination of the plant remains, Mr. E. A. Newell Arber* had concluded that most of the Middle and Upper Culm of Mr. W. A. E. Ussher, is of Middle Coal-measure age. Instances of this kind to illustrate the reliability of plants for determining horizons when they occur under somewhat abnormal conditions, might be considerably multiplied; but that scarcely seems necessary.

The possibility of using plant-remains for the determination of horizons in a broad way would probably be admitted by most geologists with little demur. It was more especially with regard to their applicability to finer zoning in Coal-measure deposits that scepticism seemed to prevail. It was the more unfortunate that this should be the case, as work in this field is of the highest possible value both scientifically and economically; while at present the number of workers is minute. This hesitation to accept results based on plant-remains, may, the writer thought, be traced to one of the very circumstances which appeared to him to make these fossils particularly suitable for the purpose. Certain genera (*Neuropteris* is a good example) have been divided into large numbers of species, which can only be separated with great care. Geologists as a rule have little botanical knowledge, and identifications of species of these genera have been more often wrong than right. Whether results can be trusted when identifications are so difficult is the natural query. The difficulty in these cases was due to the fact that species *do* run into one another. The evolution of a whole genus was in full swing, and each main form was producing numerous varieties. How many species may be recognized is a matter of small importance; the essential fact was that the great number of varieties pointed to their rapid production, and, it may be assumed, to their equally rapid extinction. The conditions are exactly paralleled among the genera of corals and brachiopods which have been used in zoning the Carbon-

* "The Fossil Flora of the Culm-measures of North-west Devon, and the Palæobotanical Evidence with Regard to the Age of the Beds," by Mr. E. A. Newell Arber, *Philosophical Transactions of the Royal Society of London*, 1902, vol. cxcvii., series B, page 291.

iferous Limestone or the ammonites of the Mesozoic rocks, in which it had been found convenient to retain only the main species, and to group the others round them as variants.

Considerable success had already attended the patient and careful work of Mr. R. Kidston, who had been able to show that marked distinctions exist between Lower, Middle, and Upper Coal-measure floras; so marked, indeed, that each had numerous species peculiar to itself, while the common species of one division were rarely common in either of the others, making the actual floras in the field look much more distinct than might appear from lists. With careful collecting, there was little doubt that much finer subdivision and correlation of this important series of rocks will ultimately be possible, and the way will be paved to that accurate and detailed knowledge of their history which must ultimately enable us to predict their position and development in those regions where they now lie buried below later deposits. One great barrier stands in the way of the successful pursuit of this enquiry: the scarcity of specimens, the exact horizon and locality of which are known. In view of the enormous economic value of the results to be attained, though even a single fruitless sinking were saved, the writer felt that he could not better conclude than with an appeal to the members to assist in supplying the necessary materials.

———

The CHAIRMAN (Mr. J. Barnes), in proposing a vote of thanks to Mr. Hickling for his paper, trusted that the members would give Mr. Hickling ample material, so that he could put this matter on a proper footing, just as Dr. Wheelton Hind and others had done in regard to the mollusca.

Mr. BERNARD HOBSON, in seconding the motion, expressed the opinion that in the end the results obtained from mollusca would be found to agree with the results obtained from plants; and, if they did not, it would seem to point to the fact that mistakes had been made in working out the results.

The motion was passed.

on of Mining Engineers.
sactions.1907:1908

To illustrate Mr. G.Hickling's Paper on "Carboniferous Flora," etc   Vol. XXXIV. Plate IV.

FIG. 1.—TABLE OF THE EQUIVALENCE IN TIME OF THE CARBONIFEROUS ROCKS IN VARIOUS PARTS OF GREAT BRITAIN AND IRELAND.

IRELAND: SOUTH-WEST	IRELAND: SOUTH-EAST	BRISTOL DISTRICT.	LANCASHIRE AND YORKSHIRE.	DURHAM AND NORTHUMBERLAND	SCOTLAND: CENTRAL.
UPPER CARBONIFEROUS		UPPER COAL-MEASURES	UPPER COAL-MEASURES	COAL-MEASURES	RED COAL-MEASURES
		PENNANT GRIT	MIDDLE COAL-MEASURES		UPPER COAL-MEASURES
		LOWER COAL-MEASURES	LOWER COAL-MEASURES		ROSLIN SANDSTONE
LIMESTONE		MILLSTONE GRIT	MILLSTONE GRIT	MILLSTONE GRIT	UPPER LIMESTONE
		GAP HERE?	PENDLESIDE		EDGE-COAL SERIES
			YOREDALE	BERNICIAN	LOWER LIMESTONE
CARBONIFEROUS SLATE	CARBONIFEROUS LIMESTONE	CARBONIFEROUS LIMESTONE	CARBONIFEROUS LIMESTONE		OIL-SHALE GROUP
				TUEDIAN	CEMENT-STONE GROUP
COOMHOLA GRITS					
UPPER OLD RED SANDSTONE				UPPER OLD RED SANDSTONE	UPPER OLD RED SANDSTONE

NOTE.—NO ATTEMPT IS MADE TO REPRESENT THE RELATIVE THICKNESS OF THE DEPOSITS IN DIFFERENT

MANCHESTER GEOLOGICAL AND MINING SOCIETY.

ANNUAL GENERAL MEETING,
HELD IN THE ROOMS OF THE SOCIETY, QUEEN'S CHAMBERS,
5, JOHN DALTON STREET, MANCHESTER,
OCTOBER 8TH, 1907.

MR. HENRY BRAMALL, VICE-PRESIDENT, IN THE CHAIR.

FOGGS COLLIERY DISASTER.

Mr. JOHN ASHWORTH moved a vote of deep sympathy with the relatives of the ten men who had been killed, and with Mr. Henry Bramall and Messrs. Andrew Knowles & Sons, Limited, in the disastrous shaft accident which had occurred at Foggs colliery on October 4th.

The motion was carried.

Mr. HENRY BRAMALL said that he greatly appreciated the expression of sympathy.

The following gentlemen were elected, having been previously nominated :—

MEMBERS—

Mr. MALCOLM BURR, Mining Engineer, Sibertswold, near Dover.
Mr. WILLIAM SHERMAN TOPLIS, Electrical Engineer, Novara, Rowan Avenue, Brooklands, near Manchester.
Mr. JOHN T. JONES, Colliery Manager, Foggs House, Little Lever, near Manchester.

The HONORARY SECRETARY (Mr. Sydney A. Smith) read the Annual Report of the Council as follows :—

ANNUAL REPORT OF THE COUNCIL, 1906-1907.

The Council have pleasure in submitting the sixty-ninth Annual Report on the work and progress of the Society during the past year.

Twenty-four new members have been elected during the year, of which 16 are federated members, 4 federated associate members, 2 federated associates, 1 federated student, and 1 non-federated member. Two students have been transferred to the federated members' list. The resignations during the year have been 9 federated members and 5 non-federated members; and, in addition, the names of several members whose subscriptions to the Society were considerably in arrear, have been removed from the list.

The classification of the membership for the year 1906-1907 is shown in the following table : --

Classification.	Non-federated Members.	Federated Members.	Totals.
Honorary members	11	—	11
Members, inclusive of life members	53	197	250
Associate members	—	9	9
Associates		5	5
Students	—	10	10
Totals ...	64	221	285

It is with feelings of deep sorrow that your Council record the death of Mr. Mark Stirrup, who became a member in the year 1880, and in the year 1904 was elected an honorary member in recognition of his devoted service to the Society. Mr. Stirrup was President for the year 1896-1897, and for a number of years previously had held the office of Honorary Secretary. Other members whose deaths have to be recorded are Mr. Henry Jobling, elected in 1883; and Mr. Ronald Gordon Grant, elected in 1904.

During the session, nine ordinary meetings of the Society have been held, a special evening meeting, and an excursion-meeting, all of which have been well attended by the members. There have been twelve Council meetings. The special evening meeting was held at Victoria~University on June 28th by the kindness of the University authorities, and Prof. W. Boyd Dawkins exhibited a number of recent additions to the Museum-collection. At this meeting, short addresses were delivered on

"The Coal-measures," "The Formation of Coal-balls in the Coal-measures," and "Carboniferous Flora as an Aid in Stratigraphical Classification," by Prof. W. Boyd Dawkins, Mr. D. M. S. Watson, and Mr. George Hickling respectively.

At the invitation of the President, Mr. Charles Pilkington, the members visited the Pilkington Tile and Pottery Works at Clifton Junction on June 19th. The excursion was attended by upwards of fifty members. The party was courteously conducted over the company's extensive works by Messrs. Burton (the managers) and their assistants. An enjoyable afternoon was spent in examining the methods of production of tiles, pottery, etc., in various stages, after which the members were hospitably entertained by the directors of the company.

Mr. Charles Pilkington in his Presidential Address discussed a number of present-day problems, including the practicability of coal-cutting by machinery, applied to the general run of collieries; the difficulties arising from the inevitable increase in the depth of pits; the methods to be adopted for the prevention of coal-dust explosions; and the various obstacles to be overcome in working coal under water-bearing strata.

The geological interest has been maintained by communications from Prof. W. Boyd Dawkins, Messrs. John Gerrard, George Hickling, W. A. Ritson, Mark Stirrup, F. J. Thompson, William Watts, and David Watson. The contributors of mining subjects include Messrs. W. H. Coleman, John Galliford, A. M. Henshaw, William McKay, Charles Pilkington, G. G. L. Preece, F. J. Thompson, and T. H. Wordsworth.

The list of papers and short communications, read before the Society during the session 1906-1907, is as follows:—

"The Cook Calorimetric Bomb," by Mr. W. H. Coleman.
"The Coal-measures," by Prof. W. Boyd Dawkins.
"A New Patent Reflector for Safety-lamps," by Mr. John Galliford.
"Demonstration of the Weg Apparatus," by Mr. W. E. Garforth.
"Horizontal and Vertical Sections of Coal-measures from Rishton, Lancashire, to Pontefract, Yorkshire," by Mr. John Gerrard.
"The Courrières Explosion," by Mr. A. M. Henshaw.
"Carboniferous Flora as an Aid in Stratigraphical Classification," by Mr. George Hickling.
"The Boultham Well at Lincoln," by Mr. William McKay.
"Presidential Address," by Mr. Charles Pilkington.
"Recent Improvements in the Design of Electric Cables for Collieries," by Mr. G. G. L. Preece.

"A Comparative Section correlating the Seams in the South and West Yorkshire Coal-fields," by Mr. W. A. Ritson.

"Demonstration of the Aerolith Rescue-apparatus," by Mr. Otto Simonis.

"The New and the Old Geology; and New Ideas of Matter," by Mr. Mark Stirrup.

"The Rock-salt Deposits at Preesall, Fleetwood, and the Mining Operations Therein," by Mr. F. J. Thompson.

"The Formation of Coal-balls in the Coal-measures," by Mr. David M. S. Watson.

"Report of the Delegate to the Meeting of the British Association for the Advancement of Science, York, 1906," by Mr. William Watts.

"Cage-lowering Tables at New Moss Colliery," by Mr. T. H. Wordsworth.

The attention recently directed by the Royal Commission on Mines to the question of rescue-apparatus has resulted in a number of these appliances having been brought forward for discussion during the session, and members have had an opportunity of testing the merits of several types, including the Weg and the Aerolith.

The following papers, printed in the *Transactions* of The Institution of Mining Engineers, have been discussed at the meetings during the session 1906-1907, in addition to the discussion on papers contributed specially to this Society's *Transactions*.

"The Value of Fossil Mollusca in Coal-measure Stratigraphy," by Mr. John T. Stobbs.[*]

"The Courrières Explosion," by Messrs. W. N. Atkinson and A. M. Henshaw.[†]

"Rescue-apparatus and the Experience gained therewith at the Courrières Collieries by the German Rescue-party," by Mr. G. A. Meyer.[‡]

"A New Apparatus for Rescue-work in Mines," by Mr. W. E. Garforth.[§]

Resulting from the discussion on Mr. John T. Stobbs's paper on "The Value of Fossil Mollusca in Coal-measure Stratigraphy," a Mollusca Search Committee has been formed, and a number of valuable specimens (both marine and freshwater) have been placed by Mr. John Gerrard, the chairman of the committee, at the disposal of members interested. The Council commend this work to the consideration of members, believing that, if followed up, it will be of great assistance in the more definite correlation of the Coal-measures in this and other districts.

[*] *Trans. Inst. M. E.*, vol. xxx., page 443; and *Trans. M. G. M. S.*, vol. xxix., page 323.

[†] *Trans. Inst. M. E.*, vol. xxxii., page 439.

[‡] *Ibid.*, vol. xxxi., page 575.   [§] *Ibid.*, vol. xxxi., page 625.

Two general meetings of The Institution of Mining Engineers have been held during the past year. The London meeting was held at the Geological Society's rooms, on June 13th and 14th, 1907, when the President, Mr. Maurice Deacon, delivered an important Address to the members, and the following papers were read :—

"Improvements Required in Inland Navigation," by Mr. Henry Rudolph de Salis.

"A Bye-product Coking-plant at Clay Cross," by Mr. W. B. M. Jackson.

"Notes on Bye-product Coke-ovens, with Special Reference to the Koppers Oven," by Mr. A. Victor Kochs.

"Water-supplies by Means of Artesian Bored Tube-wells," by Mr. Herbert F. Broadhurst.

"Gypsum in Sussex," by Messrs. W. J. Kemp and G. A. Lewis.

"The Application of Duplicate Fans to Mines," by the Rev. G. M. Capell.

"The Reform of British Weights and Measures," by Mr. Austin Hopkinson.

"The Thick Coal of Warwickshire," by Mr. J. T. Browne.

"The Ozokerite or Mineral-wax Mine of the Galizische Kreditbank at Boryslaw, Galicia, Austria," by Mr. D. M. Chambers.

"Notes on the Structural Geology of South Africa," by Dr. C. Sandberg.

"New Rand Gold-field, Orange River Colony," by Mr. A. R. Sawyer.

"Cast-iron Tubbing: What is the Rational Formula?" by Mr. H. W. G. Halbaum.

A number of papers printed in the *Transactions* were discussed, and visits were made to the Park Royal generating-station of the Great Western Railway Company, and the Knight, Bevan & Sturge cement-works, at Northfleet, of the Associated Portland Cement Manufacturers (1900), Limited.

The annual general meeting of The Institution of Mining Engineers was held at Sheffield, at the University, on September 4th, 1907. The annual report and accounts were read, together with the following practical mining papers :—

"The Sinking of Bentley Colliery," by Messrs. J. W. Fryar and Robert Clive.

"Roof-weights in Mines," by Mr. H. T. Foster.

"Deep Boring at Barlow, near Selby," by Mr. H. St. John Durnford.

Excursions were arranged to the works of Messrs. Hadfield's Steel Foundry Company, Limited, and of Messrs. William Cooke and Company, Limited; to the deep sinking at Bentley colliery; the extensive plant at the Silverwood colliery, belonging to the Dalton Main Colliery Company, Limited; Messrs. Samuel Osborne and Company's steel-works, and the Derwent Valley waterworks.

The representatives of the Society on the Council of The Institution of Mining Engineers for the year 1907-1908 are:— Mr. John Ashworth, Mr. Charles Pilkington, Mr. Henry Bramall, Mr. John Gerrard, Col. George H. Hollingworth, and Mr. Sydney A. Smith (Honorary Secretary).

Mr. Charles Pilkington (President) and Mr. Henry Bramall were appointed by the Council as representatives of this Society, and gave valuable evidence before the Home Office Departmental Committee in reference to the proposed Miners' Eight Hours Bill.

Mr. Charles Pilkington and Mr. Sydney A. Smith have been invited to join the committee of the Mining Engineering Section by Mr. Maurice Deacon, the chairman, to represent this Society on the arrangements now being made in connection with the Franco-British Exhibition, to be held in London from May to September, 1908. Both gentlemen have accepted the appointment on behalf of the Society.

Prof. W. Boyd Dawkins was appointed to represent this Society at the centenary meeting of the Geological Society of London from September 26th to September 30th, 1907, and presented a congratulatory Address from the Society.

A new and enlarged edition of the Library Catalogue has been printed, and is available for the use of members. The expense of compilation and printing of this catalogue is considerable, and your Council hope that the benefit to members will be appreciated, and that a still greater use will be made of the valuable collection of books and general geological and mining literature, augmented during the year by several recent works on mining and mineralogy, together with a number of sections. Several new exchanges have been arranged, as a result of which the collection of serial publications will be made still more useful. The number of periodicals devoted to mining has also been increased.

Further additional exchanges have been arranged through The Institution of Mining Engineers, and the publications are added to the Library as issued by the following Institutions, Societies, etc.:—

Annales des Mines de Belgique.
Cuerpo de Ingenieros de Minas del Peru.
Geological Institution of the University of Upsala.
Institution of Mining and Metallurgy.
Massachusetts Institute of Technology.

Mining Society of Nova Scotia.
North-East Coast Institution of Engineers and Shipbuilders.
Rugby Engineering Society.
Sociedad Nacional de Minería de Santiago de Chile.
Maryland Geological Survey. .
Missouri State Bureau of Geology.
Mining and Geological Institute of India.
Comité Central des Houillères de France.
Revue Universelle des Mines et de la Métallurgie.
Lake Superior Mining Institute.

The following direct exchanges have also been arranged during the year:—

Geological Survey of Ireland.   The Engineer.
The Naturalist.   The Electrical Engineer.
Mining Engineering.   Page's Weekly.
The Practical Engineer.   The Machinery Market.
The Mechanical Engineer.

The Honorary Treasurer's accounts show that the financial position of the Society continues in a satisfactory state, despite the considerable expenditure incurred in the compilation of the Library Catalogue, the binding of a large number of *Transactions* of the various Institutes for past years, and general improvements in the library. Eighty-six volumes have been bound during the year. In this respect, the disbursements of the current year may be regarded as covering several sessions, and the bank balance (now £51 10s. 11d.) may consequently be expected to assume a much larger total at the close of the year 1907-1908.

While thanking those members who have during the past session communicated papers and interesting articles at the meetings of the Society, your Council would again urge upon members the desirability of further supporting the Society's work by contributing papers which are likely to benefit the members, and by introducing eligible gentlemen to membership.

———

The CHAIRMAN (Mr. Henry Bramall) moved the adoption of the Council's annual report and accounts.

Mr. W. PICKSTONE seconded the resolution, which was approved and adopted.

———

The Honorary Treasurer's Annual Statement of Accounts was read as annexed.

THE TREASURER IN ACCOUNT WITH THE MANCHESTER GEOLOGICAL AND MINING SOCIETY, FOR THE YEAR ENDING SEPTEMBER 30TH, 1907.

Dr.

	£	s.	d.			
Sept. 29th, 1906.						
To balance in bank ...	89	13	4			
,, ,, in Secretary's hands ...	3	16	2			
	£ s. d.					
Sept. 30th, 1907.						
To members' subscriptions:—						
Arrears	15 8 0					
Current:						
Members ...	368 12 0					
Associate members	18 18 0					
Associates	6 5 0					
Students	8 15 0					
Non-federated members	38 0 0					
Subscribers	2 0 0					
	442 10 0					
In advance	8 4 1					
		466	2			
To dividends:—						
Birkenhead Railway ...	22 16 0					
Lancashire and Yorkshire Railway	20 17 10					
		43	13	10		
To Bank-interest, less commission ...		2	16	0		
,, hire of rooms		5	2	6		
,, sales of *Transactions*		8	17	11		
		£620	**2**	**8**		

Cr.

	£	s.	d.
Sept. 30th, 1907.			
By rent, wages, and expenses of rooms ...	159	2	4
,, printing and stationery	21	15	6
,, Catalogue of library, and books purchased	50	8	11
,, postages, etc.	25	5	9
,, reporters	10	10	0
,, furniture	6	17	3
,, sundry expenses, insurance, etc. ...	4	6	7
,, The Institution of Mining Engineers	250	12	3
,, printing *Transactions*	15	16	0
,, expenses of meetings	6	11	3
,, Honorary Secretary's expenses ...	9	0	0
,, balance in bank	51	10	11
,, ,, in Secretary's hands...	8	5	11
	£620	**2**	**8**

LIBRARY FUND.

Dr.								£	s.	d.
Sept. 30th, 1907.										
To library expenditure	...	...	...	...	...	...	...	1	13	9

Cr.							£	s.	d.
Sept. 30th, 1906.									
By balance	...	...	...	...	...	...	1	13	9

BALANCE SHEET, SEPTEMBER 30TH, 1907.

LIABILITIES.

			£	s.	d.
Outstanding accounts, say	...	...	58	10	3
Balance in favour of the Society	...	...	1,850	1	1
			£1,908	11	4

ASSETS.

				£	s.	d.
Investments:—						
£600 Birkenhead Railway 4 per cent. Consolidated Preference Guaranteed Stock, at £114	...			684	0	0
£733 Lancashire and Yorkshire Railway 3 per cent. Consolidated Preference Stock, at £82½				604	14	6
Library and furniture	...	...	...	500	0	0
Cash in bank	...	...	...	51	10	11
,, in Secretary's hands	...	...	...	8	5	11
Arrears of subscription (estimated value)	...			60	0	0
				£1,908	11	4

The assets of the Society, as shown by the balance-sheet, are £134 11s. 3d. less than at September 29th, 1906: £53 5s. 6d. of this diminution is accounted for by the shrinkage in the market value of the Society's investments in railway stocks, and the balance is almost entirely due to the expenses incurred upon Library account: by the binding of many volumes, the purchase of new books, and the preparation of the Catalogue, which has just been printed and issued.

GEO. H. HOLLINGWORTH, *Honorary Treasurer.*

The Investments of the Society consist of £600 Birkenhead Railway 4 per cent. Consolidated Preference Guaranteed Stock, and £733 Lancashire and Yorkshire Railway 3 per cent. Consolidated Preference Stock. The certificates for these are deposited at Messrs. Williams Deacon's Bank, Limited, St. Ann's Street, Manchester, and were inspected by us on October 8th, 1907.

Audited and found correct, October 8th, 1907.

JON. BARNES,
G. H. WINSTANLEY, } *Honorary Auditors.*

ELECTION OF OFFICERS, 1907-1908.

The following officers were elected for the ensuing year:—

PRESIDENT:

Mr. JOHN ASHWORTH.

VICE-PRESIDENTS:

Mr. W. E. GARFORTH, M.Inst.C.E. | Mr. WILLIAM PICKSTONE.
Mr. GEORGE B. HARRISON, H.M.I.M. | Mr. T. H. WORDSWORTH.

HONORARY TREASURER: Mr. GEORGE H. HOLLINGWORTH, F.G.S.

HONORARY SECRETARY: Mr. SYDNEY A. SMITH, Assoc.M.Inst.C.E.

COUNCILLORS:

Mr. H. STANLEY ATHERTON.
Mr. C. F. BOUCHIER.
Mr. VINCENT BRAMALL.
Mr. LEONARD R. FLETCHER.
Mr. D. H. F. MATHEWS, H.M.I.M.
Mr. W. OLLERENSHAW.

Mr. GEORGE H. PEACE, M.Inst.C.E.
Mr. LIONEL E. PILKINGTON.
Mr. ALFRED J. TONGE.
Mr. JESSE WALLWORK.
Mr. GEORGE H. WINSTANLEY, F.G.S.
Mr. PERCY LEE WOOD.

HONORARY AUDITORS:

Mr. VINCENT BRAMALL. | Mr. GEORGE H. WINSTANLEY, F.G.S.

New Members (Federated):

BEALES, HENRY BATSON
BROOKING, JOHN HENRY CHILCOTE
BROWN, FRANCIS VERILL
CHRISTOPHER, GEORGE ALFRED
CROSS, T. OLIVER
FILES, JAMES
FLETCHER, CLEMENT
HOBBS, WILLIAM LOW-BRIDGE
HUNTER, SHERWOOD
MILLWARD, ALBERT ED-WARD
OLDFIELD, WILLIAM
RABY, GREGORY
STONE, TOM
TAYLOR, HUGH FRANK
WATSON, PERCY HOUSTON SWANN
WILLIAMS, THOMAS

New Associate-Members (Federated):

CUNLIFFE, JAMES
DUBOIS, MARCEL
SCHEMBER, FRIEDRICH
WAINEWRIGHT, WILFRID BENJAMIN

New Associates (Federated):

GALLIFORD, JOHN | WYNNE, GEORGE REYNOLDS

New Student (Federated):

BOLTON, H. HARGREAVES, JUNR.

New Member (Non-Federated):

ACKROYD, ALFRED

Student-Members (Federated) transferred to Members (Federated):

MACALPINE, GEORGE L. | CRANKSHAW, HUGH M.

Member (Federated) transferred to Ordinary Member:

PLATT, S. S.

Hon. Member Deceased:

STIRRUP, MARK

Member (Federated) Deceased:

GRANT, R. G.

Member (Non-Federated) Deceased:

JOBLING, HENRY

The following have ceased to be Members:

MEMBERS (FEDERATED).

BODEN, PETER	EVANS, ROBERT	ROTHWELL, S.
BROOKING, J. H. C.	GAMLEN, ROBERT L.	SCARBOROUGH, G. E.
DEANE, A.	JOHNSON, WILLIAM	SHAW, ALFRED
FLETCHER, ARTHUR	POWER, THOMAS	YOUNG, WILLIAM.
FLETCHER, THOMAS	ROSCOE, THOMAS	(Thurles)

ORDINARY MEMBERS.

AROKSAMY, M. R.	GREGSON, G. E.	SALISBURY, S.
BURKE, HARRY	HEATHER, FRANK	SETTLE, JOEL
COLLIER, REV. E. C.	HUTCHINSON, J. W.	WASLEY, J. C.
DAVIDSON, ROBERT	JOBLING, ALBERT	

LIST OF EXCHANGES.

SOCIETIES WITH WHICH THE SOCIETY EXCHANGES ITS TRANS-
ACTIONS, AND INSTITUTIONS AND JOURNALS TO WHICH A COPY
IS SENT FREE.

(* Exchange through the Institution of Mining Engineers.)

I.—ENGLAND.

London...............British Association for the Advancement of Science, Burlington House, London.

British Museum, the Superintendent, Copyright Office.

British Museum Library, Natural History Department, Cromwell Road, Kensington.

Geological Society, Burlington House; W.

Geologists' Association, University College, Gower Street,
W.C.

*Institution of Mechanical Engineers, Storey's Gate, St. James's Park, Westminster, S.W.

*Institution of Mining and Metallurgy, Salisbury House, E.C.

Iron and Steel Institute, 28, Victoria Street, S.W.

Royal Society, Burlington House, W.

Manchester ....The Manchester Association of Engineers, Grand Hotel, Aytoun Street.

Free Library, King Street.

Literary and Philosophical Society, 36, George Street.

Victoria University.

Bristol ....*British Society of Mining Students, 22, Cromwell Road.

Cambridge ....Geological Museum, University.

Cardiff ....*South Wales Institute of Mining and Mechanical Engineers, Park Place, Cardiff.

Cornwall............Royal Institution, Truro.

Hull..................Yorkshire Naturalists' Union (T. Sheppard, F.G.S., Municipal Museum, Hull).

Leeds ....Yorkshire Geological Society, the Museum, Leeds.

Liverpool............Free Library and Museum.

Newcastle-upon-) North of England Institute of Mining and Mechanical
   Tyne   )   Engineers.

*North-East Coast Institution of Engineers and Shipbuilders, 4, St. Nicholas Buildings, West.

Rugby...............*Rugby Engineering Society, Benn Buildings.

Salford ....Royal Museum and Library.

Wigan ....Free Library.

Mining School.

II.—SCOTLAND.

Edinburgh ....Edinburgh Geological Society, India Buildings, George IV. Bridge, Edinburgh.

Glasgow ....Geological Society.

III.—IRELAND.

Dublin...............Royal Dublin Society.

Geological Survey of Ireland, 14, Hulme Street, Dublin.

IV.—AUSTRALIA.

Melbourne ....Public Library of Victoria.

Sydney...............Department of Mines.—The Government Geologist, New South Wales.

V.—CANADA.

Halifax............*Mining Society of Nova Scotia.

Montreal ....*Canadian Mining Institute, 413, Dorchester Street, W., Montreal.

Ottawa...............Geological Survey of Canada (Director, Museum, Sussex Street, Ottawa).

Toronto ....Canadian Institute, 58, Richmond Street, East.

VI.—INDIA.

Calcutta ....Geological Survey of India.

*Mining and Geological Institute of India.

VII.—UNITED STATES.

Baltimore ········*The Maryland Geological Survey (The State Geologist, Johns Hopkins University).

Boston ....*Massachusetts Institute of Technology.

Cambridge, Massachusetts ; Museum of Comparative Zoology, Harvard College.

Columbus, Ohio...Geological Survey of Ohio. (The State Geologist.)

Houghton, Michigan ; Michigan Mining School, Houghton.

Indiana ....Department of Geology and Natural Resources (W. S. Blatchley, State Geologist, Indianopolis). .

Ishpeming, Michigan ; *Lake Superior Mining Institute.

Minneapolis ......Geological and Natural History Survey of Minnesota.

New York ....American Institute of Mining Engineers (R. W. Raymond, Secretary, United Engineering Society Buildings, 29, West 39th Street, New York City).

Philadelphia ......Franklin Institute.

Rolla, Mo ....Missouri Geological Survey.

Scranton, Pa......Editor of "Mines and Minerals."

Washington........Library of the U.S. Geological Survey.
                Smithsonian Institute.

VIII.—FOREIGN.

Berlin ....Gesellschaft für Erdkunde. Wilhelmstrasse, 23, S.W.

Brussels ....*Annales des Mines de Belgique.

Chile ....*Sociedad Nacional de Minería de Santiago de Chile.

Halle an der Saale........... Verein für Erdkunde.

Liége ....*Revue Universelle des Mines et de la Métallurgie, Rue Bonne-Femme, 18.

Lima ....*Cuerpo de Ingenieros de Minas del Perú.

Paris ....Société Géologique de France.
                *Comité Central des Houillères de France.

Rome.................Reale Accademia dei Lincei.

Upsala ....*Geological Institution of the University of Upsala.

IX.—SCIENTIFIC JOURNALS, ETC.

Iron and Coal Trades Review, 165, Strand, London, W.C.
Mining Journal, 46, Queen Victoria Street, London, E.C.
Colliery Guardian, 30 and 31, Furnival Street, London, E.C.
Science and Art of Mining, 27, Wallgate, Wigan.
Engineer, 33, Norfolk Street, Strand, London, W.C.
Electrical Engineer, 139 and 140, Salisbury Court, London, E.C.
Practical Engineer, 55 and 56, Chancery Lane, London, W.C.
Mechanical Engineer, 52, New Bailey Street, Manchester.
Page's Weekly, Clun House, Surrey Street, London, W.C.
Mining Engineering, Clarence Works, Wallgate, Wigan.
Machinery Market, 146A, Queen Victoria Street, London, E.C.

[For list of publications received from the above Societies, etc., see separate list of additions to the Library.]

The complete list of the Council for the ensuing year is as follows :—

President:
Mr. JOHN ASHWORTH.

Vice=Presidents:
Mr. W. E. GARFORTH, M.Inst.C.E.

Mr. GEO. B. HARRISON, H.M.I.M.

Mr. WILLIAM PICKSTONE.

Mr. T. H. WORDSWORTH.

Vice=Presidents (ex officio):
Mr. JOSEPH DICKINSON, F.G.S.

PROF. W. BOYD DAWKINS, M.A., D.Sc., F.R.S., F.G.S.

Mr. R. CLIFFORD SMITH, F.G.S.

THE RIGHT HON. THE EARL OF CRAWFORD AND BALCARRES.

LORD SHUTTLEWORTH OF GAWTHORPE.

Mr. EDWARD PILKINGTON, J.P.

Mr. HENRY HALL, I.S.O., H.M.I.M.

Mr. JOHN S. BURROWS, F.G.S.

Mr. W. SAINT, H.M.I.M.

Mr. WILLIAM WATTS, F.G.S.

Mr. ROBERT WINSTANLEY.

Mr. JOHN RIDYARD, F.G.S.

Mr. W. S. BARRETT, J.P.

Mr. G. C. GREENWELL, M.Inst.C.E., F.G.S.

Mr. JONATHAN BARNES, F.G.S.

Mr. G. H. HOLLINGWORTH, F.G.S.

Mr. JOHN GERRARD, H.M.I.M., F.G.S.

Mr. HENRY BRAMALL, M.Inst.C.E.

Mr. CHARLES PILKINGTON, J.P.

Honorary Treasurer:
Mr. G. H. HOLLINGWORTH, F.G.S.

Honorary Secretary:
Mr. SYDNEY A. SMITH, Assoc.M.Inst.C.E.

Other Members of the Council:
Mr. H. STANLEY ATHERTON.

Mr. C. F. BOUCHIER.

Mr. VINCENT BRAMALL.

Mr. LEONARD R. FLETCHER.

Mr. D. H. F. MATHEWS, H.M.I.M.

Mr. W. OLLERENSHAW.

Mr. G. H. PEACE, M.Inst.C.E.

Mr. LIONEL E. PILKINGTON.

Mr. ALFRED J. TONGE.

Mr. JESSE WALLWORK.

Mr. G. H. WINSTANLEY, F.G.S.

Mr. PERCY LEE WOOD.

Honorary Auditors:
Mr. VINCENT BRAMALL.

Mr. GEO. H. WINSTANLEY, F.G.S.

Trustees:
Mr. JOSEPH DICKINSON, F.G.S.

Sir LEES KNOWLES, BART., M.A., LL.M., F.G.S.

Mr. JOHN ASHWORTH, in taking the chair, returned thanks for the honour conferred upon him in electing him President for the ensuing year.   He greatly appreciated the honour, and would, he said, do his best for the interests of the Society.

———

Mr. W. A. RITSON moved that the thanks of the Society be tendered to the officers and Council for the past year.

Mr. GERALD H. J. HOOGHWINKEL seconded the resolution, which was cordially approved.

———

Mr. JAMES ASHWORTH read the following paper on " Air-percussion and Time in Colliery Explosions ": —

AIR-PERCUSSION AND TIME IN COLLIERY EXPLOSIONS.

By JAMES ASHWORTH.

The writer, having carefully considered all the published information, and the reports of discussions on the Courrières explosion, has not found a theory which would satisfy what he has
deduced as some of the leading indications, thus:—(1) In the
Lecœuvre gallery (fig. 11, plate xxv.),* there was the clearest
evidence of a developed force at a distance of 43 to 46 feet (13 to
14 metres) from the face of the gallery. Here the floor of the
gallery was lifted, principally on the lower side, the tramroad-
rails were forced towards the higher side of the road, and the
lower-side rail was raised much above its normal level; two iron
air-pipes (No. 139) were pounded into very small pieces (89 being
counted); the roof at this point was much damaged, causing a
heavy fall; one man (No. 129) was thrown outbye, with one leg
and one arm torn off, and these were found further outbye than
the body, which was also entirely denuded of clothes; a force also
radiated from this point inbye and moved the ventilating air-
pipes (Nos. 118, 119, and 120) out of position similarly to the
tram-rails; the tram (No. 66) which was being loaded at the
moment of the explosion, was driven inbye, clear of the end of
the rails, and partly over the handle of a shovel (No. 64), the iron
part of which was in the heap of coal at the face, whilst Leroy
Regis, who had been using the shovel and loading the tram, was
thrown on his back on to the top of the heap of coal, and, though
his trousers were torn to pieces, the lower parts remained around
his legs; his lamp (No. 46), still attached to the band of his hat,
was found close to and not covered with coal; a one-ended pick
(49) was lying on the top of the heap of coal, a hammer (No. 51)
and wedge (No. 68) were in the middle of the level, close to the
tram, but no miner's pick was found near the face; on the heap
of coal was part of the coal-cutter trestle (No. 149), and on this

* *Trans. Inst. M. E.*, 1906, vol. xxxii., page 492.

lay the naked bodies of Arthur and Joseph Lecœuvre, with a prop and other wood on the top of all; many pieces of torn clothing were found between the tram and the lower side, also a plank off the tram, a very short bit of burnt fuze, and a miner's shoe; a torn hat in front of the tram; the lamp (No. 26) of Joseph Lecœuvre was found on the opposite side of the level to his body and 39 feet (12 metres) farther outbye; a small piece of fuze (No. 150) was found on the coal-cutter platform, another and the longest piece (No. 138) where the floor was disturbed, and another short piece (No. 28) between Henri Lecœuvre (No. 128) and his right arm.

(2) The box (No. 81) containing seven cartridges of No. 1 Favier powder, which had been taken out of store on the morning of the explosion, was found intact near the cut-through into the parallel heading; and near to, on the opposite side of the level, a short piece of burnt fuze (No. 25), and about $16\frac{1}{2}$ feet (5 metres) outbye, on the same side, a ring of unburnt fuze (No. 9).

(3) No detonators were found.

(4) The remaining part of a shot-hole, in the face of the gallery, was similar to that of a shot-hole found in the face of the parallel heading.

(5) A comparison of the positions in which the tools in the gallery were found with those in the parallel heading gives a practical idea as to where the coal-cutter platform, the drilling-machine and drills would be placed when not in use, as the same men used them in both headings.

(6) In the Marie seam, north-east workings from the 1,070 feet (326 metres) north bowette of No. 3 pit, safety-lamps were alone used for lighting (fig. 8, plate xxii.).*

(7) These Marie-seam workings adjoined a fault, and no other seam of coal was worked below it on the north side of the pit (fig. 2, plate xvii.).†

(8) The indications of force from the Lecœuvre gallery were not directly towards No. 3 pit (fig. 7, plate xxi.).‡

(9) Where the recoupage or recovery-drift, at 1,070 feet (326 metres) joined the Marie seam north-east workings, there had been a door, the position of which was not exactly known; and therefore it was not possible to determine its projection, as there

* *Trans. Inst. M. E.*, 1906, vol. xxxii., page 492.
† *Ibid.*   ‡ *Ibid.*

was a heavy fall of roof for a long distance on both sides of its
place; to the south-east of the recoupage or recovery-drift, there
was very good evidence of force by the inclination of timbers,
all falling towards the south-east.   To the north-west of the
recovery-drift, there was an indication of force towards the north-
west: the wheel of a tub was thrown 16½ feet (5 metres) away in
that direction.   There were contrary evidences of the same importance farther west in the Marie north-east district.   In the
recovery-drift itself, there was a very heavy fall of roof and no
evidence of the direction of the force (Mr. G. Léon).   The men in
the Marie seam were all killed and burned.

(10) The indications of force emerging into the bowette at ˙
1,070 feet (326 metres) from the Marie north-east district were
both towards the north and the south in the bowette, and into the
west workings of the Marie seam, but principally southwards
towards No. 3 pit.

(11) The wet condition of the 1,070 feet (326 metres) north
bowette did not restrain the flame of the explosion.   .

(12) No. 3 pit was the main downcast pit for all the seams
affected by the explosion; and, therefore, as soon as it became
choked with débris, the whole of the residual force of the explosion, and the deleterious gases produced, were compelled to
attempt to find an exit principally through Nos. 2, 4, and 11 pits.

(13) Very heavy percussive effects were produced on the seams
affected by the explosion.

(14) Two fires were discovered in the Joséphine workings between Nos. 2 and 3 pits,* some days after the explosion, and both
were on the inbye side of goaves.

With these proved facts in mind, and considering the whole
area of the mines affected by the explosion, it is practically certain that the explosion originated in the workings of No. 3 pit;
that the centre of the demonstrated force was about the point
where the return-air from the south-east Marie workings joined
the recovery-drift at its western end; that there was practically
a simultaneous explosion in the Lecœuvre gallery at a point 46
feet (14 metres) from the face, that is, where the air-pipe was
broken into 89 or more pieces (fig. 11, plate xxv.);† that the upheaval or disturbance of the floor at this point indicated, either

* *Trans. Inst. M. E.*, 1906, vol. xxxii., page 451, and fig. 7, plate xxi.
† *Ibid.*, page 492.

an outburst of gas, or the accidental ignition or detonation of an explosive above the air-pipe. Now, as an outburst of gas could not smash the air-pipe in the way described, neither could a blown-out shot do so, and the writer prefers to conclude that an explosive, probably introduced into the pit surreptitiously, was ignited or detonated. There was no surveillance of the miners calculated to prevent them taking any sort of explosive into their working-places, especially in this gallery, where the men fired their own shots' and were therefore uncontrolled. From the position of the drilling-machine and drills, the writer believes that Henri Lecœuvre was making a cartridge, and that the brothers Arthur and Joseph were carrying the drilling-tackle towards the face at the moment of the explosion.*

This supposition, however, does not apparently fit in with the fact that the largest part of a trestle, which was used in connection with the Sullivan coal-cutter, was found partly on the top of one of the brothers Lecœuvre, and the other brother on the top of the trestle, whilst the remaining part of it (No. 62) was found between the tram and the face of the level. The photographs taken by Mr. A. M. Henshaw show the positions of the bodies and materials so clearly, that everyone can decide for himself whether or not the force which piled up the men and materials in this manner went inbye or outbye, from the shot-hole, and whether or not a prop and other pieces of timber had followed the projection of the bodies. It does not seem possible to imagine any sort of force originating at the face which could force the bodies and materials into the positions shown (figs. 18 and 21).†

The writer's supposition that some explosive was accidentally fired whilst in the hands of Henri Lecœuvre would be very much strengthened if it were known what was the course ordinarily pursued by the Lecœuvre brothers, that is to say, whether they drilled the shot-hole and then holed the coal, or whether they holed the coal first and drilled the shot last. The plans show quite distinctly that it was possible to drill a hole whilst Regis was loading the coal, but impossible to hole the coal; and as these men do not seem to have had any other work to do than prepare

---

* After the Fernie colliery explosion in 1902, one of the miners was found with a piece of paper rolled on a pick-elve, to make a cartridge, and yet no hole was drilled ready for a shot.

† *Trans. Inst. M. E.*, 1906, vol. xxxii., pages 469 and 472.

a shot-hole, this appears to be the most practical supposition. In some confirmation of this supposition, there is the fact that the longest piece of burned fuze was found at the point where the air-pipe was smashed into fragments, and the drilling-tackle in the middle of the road.

This explosion conveys the very strong impression that it was of a complex character, (1) that it originated in or near to the north-east Marie workings in No. 3 pit, where there was a single door which might have been open ; (2) that the indications of force were found to radiate from this point, and also to a less extent from the Lecœuvre gallery ; (3) that the explosion in the Marie seam produced a detonating or percussive effect on the Lecœuvre gallery and fired an explosive compound ; (4) that the enormous air-pressure resulting from these almost simultaneous explosions was completely bottled up by the caving in of No. 3 pit ; (5) that the burning effects, coking of dust, and enormous volumes of carbon monoxide were caused largely by air-pressure, and not by what is termed a " coal-dust explosion." Coal-dust doubtless added to the force of the explosion at its initiatory points, but the main cause of the extent and magnitude of the disaster was air-percussion.

The writer does not expect that many students of colliery-explosion phenomena will at first agree with his deductions as to the effect of percussion ; but he is supported by actual facts, and particularly by a disaster which occurred in the Mount Kembla mine in New South Wales in 1902, when 95 lives were lost in a mine free from fire-damp, and where every man used an open light. Shortly stated, that disaster resulted entirely from a huge fall of roof, which so heavily compressed the air along one particular haulage-road as to create enormous volumes of carbon monoxide and damage, which was entirely confined to the haul-age-road. The dust thus lifted off the floor and sides was carried along like a bullet in a gun, and became so highly heated by pressure and friction that, on being disseminated in the outer air, it burst into flame, wrecking the surface plant, setting materials on fire, and burning and killing several persons.*

Probably every large colliery-explosion in Great Britain has demonstrated percussive or detonating effects, and only one of

* *Mount Kembla Colliery Disaster, 31 July, 1902 : Report of the Royal Commission, together with Minutes of Evidence and Exhibits,* Sydney, 1903.

these, at Udston colliery, has received official recognition, namely, in the report of Mr. Joseph Dickinson,* but the writer thinks that this phenomenon has not been followed up by its diagnoser, or by any other investigator excepting himself.

At the Fernie mine in British Columbia, peculiar effects were observed and were attributed to the inexplicable vagaries of colliery-explosions, whereas they became perfectly clear when examined by the aid of the percussive or detonative theory and as resulting from practically simultaneous explosions of nearly every can containing powder in every part of the mine affected by the explosion. This explosion was described in a paper,† contributed to The Institution of Mining Engineers in 1902, as a coal-dust explosion, but it had none of the characteristics of a coal-dust explosion : that is to say, it did not traverse the main haulage-road, but developed its greatest force from the gas in the main return-airway, from which road the force swept broadside on and crossed the main haulage-road through practically every stopping, from east to west. In the case of Fernie, as at Courrières and at Mount Kembla, the agent most destructive to human life was carbon monoxide and not flame.

One other point of the greatest importance in the elucidation of colliery-explosion phenomena is that of " time "; thus, if we follow the popular idea of a coal-dust explosion, time must be allowed for the distillation of gas, the ignition of this gas, and many repetitions of this process; but, unfortunately for the theory, there is no evidence of such a period of time as would be necessary thus to carry flame throughout the ramifications of a mine like Courrières.   Taking the latter as an example, the evidence showed that instead of the force increasing after it had caused the blockage of No. 3 shaft, it actually decreased, was not perceptible at the top of No. 2 shaft, and did very little damage at the tops of Nos. 4 and 11 shafts. Had the precise time of the arrival of the explosion-effects at the tops of, say, Nos. 3 and 4 shafts been taken, very valuable information would have been added to our knowledge of colliery-explosions, and the writer is confident that it would have proved that an explosion is

---

* *Explosions (Udston Colliery): Report to the Secretary of State for the Home Department*, by Messrs. J. Dickinson, H.M. Inspector of Mines, and C. C. Maconochie, Advocate, 1887 [C.–5192].

† "The Fernie Explosion," by Mr. W. Blakemore, *Trans. Inst. M. E.*, 1902, vol. xxiv., page 450.

more or less of an instantaneous character. It has been observed
in all cases where smoke and dust have been projected from
the pit-tops or other openings, that this effect only occupies a few
seconds of time, and that the column of dust is then cut off as if
by a knife, part immediately rushing back into the workings to
fill the vacuum caused by the condensation of the heated gases
and steam, and thus causing backlash and other contrary effects
underground.

The instantaneousness of a colliery explosion is also clearly
demonstrated by the positions in which men are found, many of
them in the precise positions in which they were at the moment,
such as eating food, and the effect on men who have escaped alive
has been of such a temporary character as not to alarm them or
cause them to cease work, although the sudden pressure of the air
has always been noticeable in the stoppage of watches on the
persons of both the living and the dead. Nothing could more
convincingly show the instantaneousness of colliery-explosions
than such facts as these, coupled with the firing of shots and
powder cans at distances of upwards of a mile from the point of
origin, as at Tylorstown and Fernie collieries.

The importance of taking time into account when looking for the point of origin of an explosion, was demonstrated by the explosion at the Albion colliery in 1906.* There
were two falls of roof on a main level at a distance of 240 feet
apart, and the evidence given by the men working at the one
nearest the pit-shaft was that the explosion occurred about 5
minutes after the largest fall took place, and that the volume of air
was 20,000 cubic feet per minute passing along a level which at the
double parting was 13 feet wide. The roughest calculation will
show that any gas given off by this fall must have passed the
second fall long before the explosion took place, and that it was
therefore the second fall that forced flame through a sound
bonneted Clanny safety-lamp placed on the floor immediately
underneath, where it was subsequently found by the first explorers.

* "Report on the Accident at Albion Colliery," by Mr. F. A. Gray, H.M.
Inspector of Mines, *Reports of W. N. Atkinson, H.M. Superintending Inspector of
Mines, and F. A. Gray and J. Dyer Lewis, H.M. Inspectors of Mines, for the
Cardiff and Swansea Districts (Nos. 10 and 11), to His Majesty's Secretary of State
for the Home Department, under the Coal Mines Regulation Acts, 1887 to 1896, the
Metalliferous Mines Regulation Acts, 1872 and 1875, and the Quarries Act, 1894,
for the Year 1906,* 1907 [Cd. 3449—IX.], page 24.

The caving-in of No. 3 shaft at Courrières collieries caused similar effects to that of bursting a boiler by holding down the safety-valve, and as all the haulage-road connections with Nos. 4 and 11 shafts were close to No. 3 shaft on the south bowette, the pent-up force had naturally to expend itself in that direction. Although the Ste. Barbe and Cécile roads were either too dirty or too dusty to carry flame, yet the mechanical effects of the explosion were demonstrated along them all the same.

The writer is of opinion that the Courrières disaster was only to a certain extent a coal-dust explosion, and that the partial coking of dust, burning effects on some of the bodies, and the huge production of carbon monoxide, were due to percussive effects and not to actual flame, excepting only where flame was produced by the firing of stores of explosives at various points, as at Fernie colliery.

The most important lesson taught by the Courrières explosion, as well as by some colliery-explosions in this country, is not that of the danger of working a large number of mines connected together for convenience of ventilating arrangements; but the danger of having shafts so insecurely lined and so filled with material, bratticing especially, that either an explosion or an accident with the winding-arrangements may cause a block in the shaft and entirely preclude all hope of saving the majority of the persons underground. Consequently, if No. 3 pit at Courrières collieries had remained open, and the rescue-parties could have entered the mine without delay, the loss of life would have been comparatively small.

———

Mr. Joseph Dickinson (Pendleton), in moving a vote of thanks to Mr. James Ashworth, said that observations on the time occupied in explosions were rather rare, persons at such times having usually their attention otherwise occupied. To some extent he agreed with what was stated by Mr. Ashworth in his paper, but not entirely. Indeed, of the many explosions that he had investigated, he never but once actually took the time. On one occasion he felt the suck, and, guessing what was coming, he lay down in the gutter until the blast had passed: it seemed to occupy a long time. The time would most likely vary with the quantity of gas, its mixture with air, and the point of ignition. In former times, when gas was burnt out to admit of shots being fired, he had seen the flame travel slowly.

The explosion which he timed occurred in the Limehurst colliery, Ashton-under-Lyne, where three seams of coal were being worked. On August 5th, 1884, the Two-feet seam was giving off fire-damp freely; and, as height was required in the roadways, blasting was resorted to under the restrictions, including ordinary work-persons being out of the mine. At one part fire-damp began to issue very freely, causing the manager to issue a written notice forbidding blasting. Notwithstanding the notice, a shot-lighter fired a shot, and lighted the issuing gas. He tried to flap out the flame, but failed. The manager was called, and came forthwith with help, and they did what they thought best, but without much effect; and soon afterwards the consulting mining engineer arrived. Efforts were continued, the flame kept burning, and gas kept flashing at intervals overhead, tumbling the party about until they became disheartened. Mr. J. S. Martin, then Assistant-Inspector of Mines, arrived; and another effort was made, about twelve fire-extincteurs being applied, but the gas kept burning as it issued, and the packing took fire. He (Mr. Dickinson) visited the colliery, and found that it was being flooded with water, but with air still passing through the workings, and that it would take at least a fortnight before the water could rise high enough to shut off the air from the fire, during which time the fire would be spreading. It was resolved to shut off the fire; but with so much gas it was thought unsafe to do so by building stoppings or by closing the top of the shafts; it was, therefore, arranged to use two wooden air doors which, with a little preparation and some sand, could be quickly closed tightly. Preparations being thus made, all ascended the shaft and a consultation was held. Four volunteers undertook the closing, and descended the shaft. In 10 minutes, having closed the doors, they re-ascended; and, 13 minutes later, the mine exploded, going off like the crack of artillery. Dense black smoke flowed out of the upcast shaft, and continued for nearly half-an-hour. Two of the volunteers were members of this Society, Mr. Robert Winstanley and Mr. Walter Evans. The disobeying fireman was prosecuted, and fined 5s. and costs: a trifling sum compared with the damage done. Had building-off instead of closing-doors been adopted on this occasion, all the persons engaged in the operation would have been lost.

The suggestion made by himself (Mr. Dickinson), as to percussion or detonation of air assisting the explosion at Udston colliery, was, he thought, concurred in by Mr. Ralph Moore, H.M. Inspector of Mines for the district.

He agreed that dust was only a contributor, and not the main factor of the Courrières explosion. He had previously stated his view that it originated from fire-damp, and was helped on by dust and gas in the air and goaves, notably in the just-shut-off area surrounding the fire. He had hoped for some explanation of one of the shutting-off stoppings shown on the plans as being in a main airway; but, so far, none had reached him.

Mr. HENRY BRAMALL (Pendlebury), in seconding the resolution, said that he would like to know the amount of compression that must be produced in air before the temperature became so high as to fire gas: and he could not see how it was possible for the air in a mine to become so compressed as to acquire that heat. He would also like to know how carbonic oxide could be generated by compressing air. No doubt Mr. Ashworth had some explanation, or the terms in question would not have been used by him; but, so far, he was at a loss to understand those points of the paper.

The resolution was carried.

Mr. JOSEPH DICKINSON said that perhaps Mr. Bramall had forgotten or did not know Dr. Angus Smith's invention for testing fire-damp; air mixed with a small percentage of fire-damp became readily ignited by compression. Mr. Dancer made a similar instrument, and with it he (Mr. Dickinson) experienced no difficulty whatever in producing sparks from fresh air, although it fired much more readily with a mixture of fire-damp. It had one peculiarity, that, the air having once been fired, a second spark could not be obtained from the same volume of air.

Mr. HENRY BRAMALL knew the scientific toy referred to by Mr. Dickinson. Granted that one could fire tinder by the compression of the air in this toy (as only a very moderate temperature was needed), even that compression was very much greater than could take place in the airways of a mine. He had had little experience of explosions, and might be wrong. but he could not conceive how so great a compression could be produced in a mine as was required for Mr. Dickinson's experiment.

22

Mr. JAMES ASHWORTH said that he did not think that anyone thoroughly understood percussion or detonation as demonstrated by colliery-explosions; and he purposely brought this matter forward for discussion, because it was a subject which required thoroughly well ventilating. If a mine were watered, and either percussive or detonative effects were produced, then watering could exercise very little protective value. Members would perhaps have noticed on the plan accompanying the report of the Llanerch explosion* of February 6th, 1890, that some curious percussive effects were demonstrated. The men in the lower-side headings, and at the ends of the levels, were found dead just where they were at work, and showed appearances of burning; whilst the men on the higher side, where there was more air-room, were able to run a considerable distance before they succumbed to the after-damp. This explosion occurred in a mine worked by open lights. With regard to detonation or percussion, whichever was demonstrated in fiery colliery-explosions, he understood Prof. H. B. Dixon to say, in his evidence given before the Royal Coal-dust Commission, that no such thing as detonation could occur, and that the propagation of flame in a fire-damp mixture was not very rapid;† but later he noted that Prof. Dixon had spoken about the detonation of mixtures of air and fire-damp. At the Altofts colliery explosion, gas was set on fire in the goaf and he (Mr. Ashworth) thought that it was caused by detonation or percussion; but Mr. W. E. Garforth assured him that it was caused by actual flame from the explosion. Whatever caused the actual ignition, it was certain that its extinction under the most dangerous conditions was only accomplished and a further disaster averted by the cool-headed bravery of Mr. Garforth and his devoted officials and workmen. After the Universal colliery-explosion, the expert witnesses could not agree whether the disaster originated on the east or on the west side of the shafts, nor in which district it commenced; but the plans showed that the force of the explosion came from the different districts towards the main haulage-road. Nevertheless, the only man who came out alive was in a direct line between the east and the west districts, and,

* Report to H.M. Secretary of State for the Home Department upon the Circumstances of the Llanerch Colliery Explosion and the Inquest Consequent thereupon, by Mr. H. D. Greene, Q.C., 1890 [C.-6098], page 10 and plan.

† First Report of the Royal Commission on Explosions from Coal-dust in Mines, 1894 [C.-7401.-I.], vol. ii., minutes of evidence, with appendices and index, appendix xxi., page 91.

consequently, if actual flame passed from either side to the other, it must have gone over him. Did it pass over him? In his (Mr. Ashworth's) opinion, it did not.

It remained to be proved whether there was not another factor to be reckoned with in colliery-explosions, namely, that of electricity. He had thought sometimes that there were electrical conditions which concerned colliery-explosions, but had been assured by electricians that they were impossible; still later he had found that other people were considering the electrical possibilities—for instance, that damp coal-dust in the air carried a charge of negative electricity.

In the case of the explosion described by Mr. Dickinson, it was marked by smoke coming away from the fire. Smoke might deaden the speed, or add to it according to the condition of the fire, but he thought that it would most probably lower very considerably the speed of the explosion.

He (Mr. Ashworth) would be pleased to have the questions that he had raised thoroughly discussed, particularly at the present time, when watering was generally considered to be a necessary safeguard against the extension of colliery-explosions; whereas, if detonation or percussion occurred, it was most probable that the extension of an explosion would be facilitated rather than retarded by watering and spraying.

MANCHESTER GEOLOGICAL AND MINING SOCIETY.

ORDINARY MEETING,
HELD IN THE ROOMS OF THE SOCIETY, QUEEN'S CHAMBERS,
5, JOHN DALTON STREET, MANCHESTER,
NOVEMBER 12TH, 1907.

MR. JOHN ASHWORTH, PRESIDENT, IN THE CHAIR.

The following gentlemen were elected, having been previously nominated :—

ASSOCIATE MEMBER—
Mr. GEORGE EDWARD LOMAX, Fern Hill, Huyton, Liverpool.

STUDENT—
Mr. THEODORE HODSON NUTTALL, Long Port, Freshfield, near Liverpool.

MEMBER, NON-FEDERATED—
Mr. BENJAMIN PALIN DOBSON, South Bank, Heaton, Bolton, Lancashire.

Mr. JOHN ASHWORTH delivered the following " Presidential Address " :—

PRESIDENTIAL ADDRESS.

By JOHN ASHWORTH, C.E.

My first and pleasant duty is to thank you for the honour which you have conferred upon me in electing me President of your Society. I fully appreciate this honour, and it will be my endeavour, with your co-operation, to advance the interests of the Society in every way during my term of office.

This Society was originally a geological society, and it is only of late years that the mineralogist has been admitted into its fold. This is as it should be, for you cannot wisely divide such closely-allied subjects.

Sir Archibald Geikie says, in his Address at the Centenary Meeting of the London Geological Society : —*

> "It has been said that the geologist ought never to forget that the mineralogist was his father. The study of minerals certainly preceded that of rocks ; and it should always be remembered that the early mineralogists were in reality the first geologists, by whom the foundations of the petrographical divisions of geology were laid. But, if the geologist is to own the mineralogist as his father, he must surely acknowledge the miner to be his grandfather.
>
> "For many centuries, and long before the use of scientific mineralogy, most of the current knowledge of the nature and disposition of the minerals and rocks of the earth's crust sprang out of the labours of those engaged in mining operations. It was the business of those men to make themselves practically acquainted with the subject, so far, at least, as regarded the facts that had to be attended to in the sinking and working of mines.
>
> "As a rule, however, they did not trouble themselves with explanations or speculations as to the origin of the rocks with which they had to deal, and when they did so they usually attained to no greater measure of success than other theorists before them. But when the miners established mining schools for the training of those who were to follow their craft, they took an important forward step in paving the way for the creation of a sound geology."

I think, gentlemen, that we can endorse this statement, and the more readily because in a later portion of his Address Sir Archibald Geikie made reference to the admirable assistance given to geologists by the mining institutions, which were originally designed for the practical miner, but eventually developed into training colleges for the geologists.

* *Times*, September 27th, 1907, page 14.

It would ill become me to address you on the vast subject of geology, for this Society has been honoured by many noted geologists, of whom I may mention Mr. Joseph Dickinson, who has been a member of this Society for over 50 years, and has done yeoman service in this branch of knowledge, not only for our members but for the country. We have also Prof. William Boyd Dawkins, who has laboured in the past, and is still an active member amongst us, thus forming the link which binds the early and speculative work of the Society with the more practical and accurate operations of to-day. To the late Mr. Mark Stirrup, whose death took place so recently as June last, we owe also a great debt for the enrichment of the *Transactions* with extracts and original contributions of a very high standard, apart from his labours as Honorary Secretary for a lengthy period. My Address, therefore, will be rather as a mining man than as a geologist.

The various matters relating to the regulations and working of mines are too familiar to you to require anything in the nature of a review from me. We can still congratulate ourselves that this country continues to hold the first place in respect of regulations for the safe working of mines. This point was conceded at the recent Miners' Congress at Salzburg. I am, however, afraid that the proposal of the Belgian representative at that Congress, namely, that coal-production should be regulated by international laws, will not prove a very workable proposition, although it secured the moral support of most of the delegates.

It is also worthy of note that His Majesty King Edward has created a new medal for gallantry in mines and quarries. The warrant states that the King is " desirous of distinguishing by some mark of our royal favour the many heroic acts performed by miners and quarrymen and others who endanger their own lives in saving or endeavouring to save the lives of others from perils in mines or quarries within our dominions and in territories under our protection or jurisdicton." The new medal will be of two classes, to be designated (1) "The Edward Medal of the First Class," in silver, and (2) " The Edward Medal of the Second Class," in bronze. This indicates his Majesty's great interest in the efforts made to reduce the danger and the death-roll in collieries, and such encouragement has been applauded and approved by everyone connected with mining.

Mining men in this country are in many cases looking anxiously for the report of the Royal Commission now taking evidence on the safety of mines. The subjects of rescue-apparatus and the best treatment of coal-dust continue to excite the keenest interest, and a preliminary report on the first subject has already been issued. It is to be feared that far too much is expected to result from rescue-apparatus, the cheapest of which is very expensive. Up to now, there does not appear to be more that one instance of the saving of the life of a miner by such an apparatus, and this occurred in an Australian mine, at Bonnieville, where a miner, having been imprisoned by an inrush of torrential rain, was kept alive for nine days through the daring and cool bravery of two divers, who made periodical visits with food and eventually rescued him, very little the worse for his imprisonment. Doubtless, rescue-apparatus will be found of great advantage in the case of gob-fires and other similar occurrences underground, but it seems curious that a simple. inexpensive apparatus called the Denayrouze, which was introduced into the country many years ago, was not brought before the present Royal Commission.

On the general subject of rescue-work in mines, it seems to me that a great deal more might be done by the Home Office in apportioning a considerable sum of money per annum for scientific investigations, which would materially tend to increase the safety of the persons engaged in our great coal-industry.

With respect to the attitude of capital and labour, I think you will have noticed that there is a strong desire growing up amongst all sections of the community that courts of arbitration should be formed to settle industrial disputes, rather than to be obliged to resort to the barbarous practice of strikes. This point was very clearly brought out in Mr. Enoch Edwards's speech at Southport,[*] when he strongly suggested the settlement of disputes by means of reason and conciliation rather than striking first and adopting reason and conciliation afterwards.

The Miners' Eight-hours' Day Bill will probably become law during the next session, and although the conclusions of the Parliamentary Committee make light of the assumed loss on a production of 25,783,000 tons, calculated on the output for 1906, it is not denied that considerable loss will inevitably follow the

[*] *Manchester Guardian*, October 9th, 1907.

adoption of such a measure; and many other expedients are suggested for mitigating the effects of the proposed reduction in the time underground. The employer's remedy lies apparently in the extended use of labour-saving machinery, such as coal-cutting machines and conveyors, and in the improved mechanical equipment of the mines generally. The Committee admit that certain temporary and permanent relaxations of the rigid rule might be found necessary in the interest of safe working, and that the conditions are variable to such an extent that special regulations would be required for certain districts.

From many of these conclusions it is clear that the operation of law will not remove all the intricacies of the existing systems, and a hard-and-fast limit of working-time will be found detrimental to the industry, both from the point of view of the employer and that of the employed. And, further, is there not some reason in all these considerations for the feeling that many of us have, that these processes of restriction—this law-pervading atmosphere dominating our industrial and social life—are destroying in our race and nationality that individualism, that spirit of self-reliance and of self-restraint which, beyond doubt, have had much to do with the industrial as well as the national development of England.

We move in days of big combines and trusts, where everything is done by concerted moves, and where, except in a few brilliant instances, personality is submerged and operations are performed mechanically. The danger which I apprehend is loss of character by the reduction of the personal element in our undertakings, and if we look upon our national industries not merely as means whereby men may rapidly and selfishly accumulate wealth, but also as means of intercommunication of sympathies and goodfellowship between capital and labour (which, after all, should be one of the principal motives of our commercial aims and pursuits), the danger to which I refer becomes a real one, and one to which philanthropic capitalists might well direct attention.

We cannot be unmindful of the service rendered by the workers in the days when legal ties and binding clauses were almost, if not quite, unknown, and when labour was performed from a sense of duty sufficiently strong to keep the worker at his post and assist in the wonderful development that has characterized modern English history.

Let us for a moment reflect on the great and rapid change made in England by the coal-miner. Has he not completely revolutionized our land, and turned a large part of an agricultural country into a huge workshop? Only one hundred and fifty years ago, Lancashire was the poorest county in England and the most sparsely populated. To-day, if you take an area of 50 miles round this room, you enclose a population more dense than in any other part of the world. You may liken it to one big workshop built on a coal-hole. The miner has mainly effected this change.

This change has brought with it many problems, and one is tempted to ask whether some of us have not lingered too long on the slopes of the Pennine Chain? Is there any outlet for this congested population? Gentlemen, I venture to think that there is; it is not in Old England, but in a New England beyond the sea. A land that covers one-fifteenth of that of the world, and yet has not a total population of London to-day. I refer to Canada.

It seems to me that the miner would do well to turn his attention more assiduously to this vast Dominion, for in my opinion it will prove to be one of the most solid and enduring jewels in the British crown. It is a happy omen that the 1908 summer meeting of the Canadian Mining Institute is to be held, to some extent, under Canadian Government auspices, and promises to be a most important one and worthy of the best representation from this country. For these reasons I shall for a short time take your thoughts away from the mother-country to her offspring across the sea, feeling sure that it is for us, and especially the younger amongst us, whose field of observation is widening out, to mark, learn, and inwardly digest every phase of mining activity, every fresh means adopted to meet a particular contingency, and to apply the beneficial result of our observations to the work in hand. In order to acquire this experience, it is absolutely necessary that fresh fields should be sought and appropriated; the situation is one which should have a peculiar attraction for the young mining engineer or colliery manager whose steps should be directed towards new fields where his energy can find full development. Could he have better prospects than in Canada—a land that can feed an empire? A land that is the best watered, and has the best fisheries in the world; a land that extends from latitude 42 (in a line with Rome) until it is lost in the land of the Midnight Sun.

After every visit I have paid to this vast dominion, I have returned more impressed with its possibilities, and if I can only arouse your interest, I feel sure that you will agree with me that Canada is the country of the future for our surplus population, and that the Lancashire man, especially the miner, will do well to give it his serious attention.

The miner was ever the pioneer in the industrial world, and I therefore venture to bring before your notice some points bearing upon the potential resources of this country. For the sake of brevity I propose to group Nova Scotia, New Brunswick, and Prince Edward Island together, as they are known as " Acadia," or the " Land of Plenty."

Of all the numerous British Colonies, this region presents the strongest family likeness to the mother country, not only in the singular variety of its resources, but also in its proximity to the markets of the world.

Gold, which excels in purity that of Australia and California, is found here. It is here that we have unequalled fisheries, safe harbours, extensive coal-fields near the water's edge, and, above all, a position almost midway on the very highway between the old and the Pacific side of the new world. And I am one of those who believe that Acadia will inherit a full share of that greatness which Britain in her old age must resign.

Though it cannot be said to have yet attained the prosperity predicted for it, this region is to-day by no means an insignificant contributor to Canadian wealth, for of the total number of miners engaged in Canada, 42 per cent. find occupation in Nova Scotia.

Coal is the staple mineral, and the Cape Breton, Cumberland, and Pictou fields provide a fair proportion of the whole output. In Cape Breton, coal is obtained in great quantity from the Sydney and Inverness coal-fields. The Sydney coal-field was the first exploited, operations dating from the year 1785.

In 1863, the output of coal for Cape Breton was 214,812 tons: to-day it is over four million tons. The other fields in this district of Acadia are responsible for a little more than a million tons, or a quarter of the Cape Breton output.

The Coal-measures of the Cumberland area crop out on the sea-shore, and have been worked extensively at the Joggins Mines, where a seam of about 6 feet of coal has yielded an annual output of about 80,000 tons.

The Pictou coal-field, opened up in 1827 by the General Mining Association, is about 11 miles long, 8 miles wide at its broadest part, and covers an area of about 22 square miles. Its structure presents many interesting features, and a few problems, some of which are: the remarkable thickness of the seams (in some cases 40 feet); the extensive deposits of black and brown shale; the marked changes in these deposits noted at comparatively short distances; the nature of the dip, ever chang- . ing, but always considerable; and the numerous large and small faults intersecting the field at many points.

The three coal-producing districts of this field are the Albion, the Westville, and the Yale. The coals vary in character, but are all of the bituminous coking variety. Some tested at the Gaslight and Coke Company's Works, London, have yielded 10,450 cubic feet of gas, of 15 candlepower, per ton. The slack is valuable for blacksmith's purposes.

The fiery nature of the seams has necessitated the use of fans of large capacity and modern construction, in place of under-ground furnaces. Mueseler, Marsaut, and other lamps have been in use for many years.

Quite recently the local papers announced that the Dominion Coal Company, whose annual output is about 3½ million tons, had a representative in this country busily seeking to enlist an army of 2,000 miners for their collieries in Nova Scotia. Lanca-shire, Yorkshire, and Staffordshire men are preferred, it is said, as being more accustomed to the longwall system, which is the method of working adopted in this part of the Dominion.

The area of gold-measures in Nova Scotia has been estimated by various authorities to cover from 5,000 to 7,000 square miles, or from one-fifth to one-third of the area of the province, yet the actual area from which the gold thus far obtained has been won is less than 40 square miles.

In gold, as in other things, this region has not yet realized the great hopes entertained some years back, when it was thought that the saddleback reefs, which bore a close resemblance to the famous Bendigo saddlebacks, would turn out similarly success-ful in the process of extraction. Still, there are many who feel that gold-mining has here a great future before it: and the field is being thoroughly tested, in order to determine the future course of action.

Gypsum is plentiful in Acadia, 435,805 tons having been produced in 1905.

Entering the gulf of St. Lawrence, we reach Quebec with its rocky fortress, and at once come into touch with that great highway of the world, the Canadian Pacific Railway, which extends from ocean to ocean, a distance of 3,077 miles, traversing in its course the rich St. Lawrence valley, the prairie uplands, and the metalliferous Rocky Mountains to Vancouver. The Grand Trunk Pacific Railway, another great highway, is in course of construction, taking a direction to the north of the Canadian Pacific line, and will extend from Halifax to Prince Rupert on the Pacific Ocean, passing through and opening up many mineralized districts.

The Province of Quebec is a most picturesque land of green hills, forests, rivers, lakes, and waterfalls, with the pretty villages and old-fashioned churches and homesteads of the French Canadians, dotted here and there with luxurious orchards, and sleek kine grazing in the rich pastures; everything in this province of 350,000 square miles is agreeable to the view of the traveller. Here agriculture and the lumber-trade loom large in the people's industry, but there is also a steadily increasing mineral-production, which in the year 1905 employed 5,017 persons, 1,650 of whom were engaged in the mining of asbestos. This mineral occurs in considerable quantities, in the form of small veins in intrusive serpentine, in the eastern townships of this province, and also at various points north of Ottawa, in association with serpentine rocks of the Laurentian formation.

In addition to asbestos—the most important of Quebec's minerals—cement-stone, copper-ore, graphite, and mica are obtained in considerable quantities.

Passing on to the Province of Ontario, which is known as the " Garden of Canada," we behold what is in many respects a wonderful province. It extends for 1,000 miles from east to west, is considerably larger than the whole of Germany, has innumerable lakes connected by ship-canals, a multitude of rivers, tremendous water-power plants, a rich soil, an equable climate, and great mineral resources, opening a boundless field of operation for the prospector. The nickeliferous deposits near Sudbury are the richest in the world, and the output of this mineral for 1905 was about £4,500,000 market-value.

Extensive iron-ore deposits are worked at the Helen mine, Michipicoten, along the line of the Kingston and Pembroke and Central Ontario Railways, and smelted at Sault Ste. Marie furnace and steel-works, and at Port Arthur ironworks. Twenty miles north of the famous nickel-district, iron-ore is also found in the Moose Mountains (rocks of Keewatin age). Drilling and blasting are here in full operation, and a railway only is required to make the Canadian iron-ore certain of admittance into all markets. The proportion of iron in the ore is 60 per cent. It is reported that a mineral smelting-plant is to be erected near Toronto, which will provide employment for 15,000 men, treating 1,400 tons of ore daily. No coal had been discovered in this province until quite recently, when lignitic coal was found near Lake Abitibi.

North of Sudbury lies the recently developed rich silver-mining area of Cobalt, which is confined within the small compass of 12 square miles, 5½ miles from south-west to south-east, and only 3 miles across at its broadest part. At the Larose mine, considerably more than 300 feet of good ore has been proved, and a much greater depth of equally rich ore is anticipated.

Natural gas and petroleum are found in Southern Ontario, the annual value of the former exceeding £5,000,000 and of the latter £190,000. At Goderich, a seam of rock-salt of very fine quality, 30 feet thick, is worked, and at the Jarnion mine, Madoc, Hastings county, iron pyrites has been discovered.

A comparatively new industry in Canada, and one allied to mining, is that of clay-working. In the year 1887, no one appeared to realize the importance of this province as a producer of pressed brick, and bricks were imported from Ohio, in the building of a large Toronto bank, at a cost of £8 per thousand. To-day they are manufactured in the province at a price considerably lower than that mentioned, and the importance of this industry is further evidenced by the announcement of a new Canadian organization under the name of the Canadian Clay-products Manufacturers. A valuable handbook on the Clay Industry has recently been published by the Ontario Bureau of Mines.

Further encouragement to mining in Ontario is given in the arrangement of summer classes in connection with the mining colleges of McGill, Toronto, and Kingston : the students who are sent out into the interior to study geology are each provided with samples of the ores that are likely to be found in the various districts.

Travelling westward, we reach the great prairie Province of Manitoba, the indications of the development of which are best given in the figures of population, which, during the last 36 years, has increased twenty-fold, and now stands at something like 360,000. Winnipeg, the midway emporium of the Dominion and the capital of Manitoba, in 1871 had a population of 100; to-day its population exceeds 100,000.

It may be of interest to note that Mr. J. Obed-Smith, the Commissioner of Immigration, is a Lancashire man, and is responsible for all immigrants passing through to the west.

Though primarily the " granary " of the Empire, Manitoba and the North-West Provinces are not bereft of mineral wealth, as they are underlain by rich stores of lignite, which is a useful fuel for the cities and scattered farming population.

Saskatchewan and Alberta, the two new provinces, each have an area of more than twice the size of the British Isles, but only a very small portion of their rich arable land is as yet under cultivation; nevertheless, the population had increased from a few hundreds in 1890 to nearly 500,000 in 1906.

In Alberta there are enormous deposits of lignitic coal. The principal developments of anthracitic and bituminous coals have been made in the neighbourhood of Banff, Frank, and Coleman, and to a similar extent between Calgary and Edmonton.

Considerable deposits of iron-ore recently discovered in Alberta, suggest an industrial as well as an agricultural future for this district. In Southern Saskatchewan, at the northern fork of Willow Creek, there has been found a deposit of ore containing 5 per cent. of manganese, which will doubtless be profitably worked by some enterprising capitalist.

Natural gas is found in many places, the principal supply at present being found at Medicine Hat. After 14 years' continuous use, the volume of gas has shown no shrinkage, and the supply appears to be very far from exhaustion. The town is supplied with light and fuel from this source at a cost of 8½d. per 1,000 feet.

At Lethbridge colliery, in the Belly-river coal-fields, Alberta, a 5½ feet seam, with a fire-clay parting, is worked in a mine equipped for the production of 1,000 tons per day, coal-cutting machines and endless-rope haulage being employed.   Some 60 miles of railway are owned by the company, the line extending from Lethbridge to Coutts.   Coal is delivered in Edmonton at 5s. and 6s. per ton, and along the steep banks of the North Saskatchewan it can be obtained open-cast without being actually mined.

Leaving the prairie provinces, we approach British Columbia, the region of the Rockies, the backbone of Canada, as the Pennine Range is of England.   This region is by far the most important part of Canada as regards mining.   The surface, which extends for 400 miles across a confused mass of high ranges and uplands, has been little more than scratched, in a mining sense.   Its economic development seems only just to have made a systematic commencement, but great advances are now being made.   Last year showed a production valued at about £5,000,000.   The annual output during the last ten years has been more than doubled, and one company—the Granby Mining, Smelting, and Power Company, Limited, the largest and most important, with a capital of £10,000,000, over £3,000,000 of which is issued stock, and held by Americans who control the undertaking—produced 645,000 tons of ore out of a total output of 930,000 tons.

Though not as yet a large producer of iron-ore, British Columbia has proved deposits at Cherry Bluff, Kamloops, and Texanda Island.

Copper, gold, coal, and lead, in the order named, are the most important minerals, the output for 1905 being valued at £1,657,713, £1,055,808, £910,182, and £533,515 respectively.

Of the total lead-output of the Dominion, 98 per cent. is produced by British Columbia, the output being shared largely by the Fort Steele and Slocan districts.   In 1905, more than 56½ million pounds of lead-ore were obtained, the value being something like £480,000.   The Boundary district is responsible for much of this mineral yield ; and, having had an opportunity of visiting the district, I must say that for the variety and value of its ores it will be difficult to find an equal.

The coal-output of British Columbia during 1906 was restricted to the Crow's Nest Pass and the collieries on Vancouver Island.   Coke for smelting is in increasing demand, and, owing to the scarcity of labour and the urgent call for coal, the supply

of coke has been inadequate, as a consequence of which a cargo
of 3,000 tons was imported from Australia during 1906 by the
Crofton smelter.

All this proves that development-work can be very materially
and profitably increased by the influx of more capital and additional labour. Thus the Canadian Pacific Railway Company is
spending £300,000 in opening up collieries in the Crow's Nest
district, and is also a great producer of anthracite at Bankhead,
Banff, Alberta.

At Princeton, a seam of coal, 18 feet thick, has been
found, 49 feet from the surface; and, on the western edge, a bore-
hole, 863 feet deep, has disclosed seventeen seams, aggregating
50½ feet of good coal.

In the Telqua district, a recent telegram states that very
rich deposits of coal have been located, and as the projected line
of the Grand Trunk Pacific Railway runs through the heart of
this new field, some speedy development-work may be expected.

Some novel features have been introduced by the British Columbia Board of Examiners in the rules for the examinations for
the coal-mine officials' certificates. For example, every man is to
be submitted to a sight test, to ascertain his capacity to see a
cap of fire-damp in a safety-lamp, and each candidate will have
to pass this test successfully. The Board of Examiners have also
concluded that a man ought not to be expected to memorize a
bookful of formulæ, and if he can work out the problem when
the formula is set before him, he will thereby satisfactorily
demonstrate his capability. Holders of first-class managers'
certificates must be British subjects. In addition to the examinations for officials for the first, second, and third-class
certificates, monthly examinations are held for miners, all of
whom are required to hold a certificate of competency.

In and about the coastal collieries the proportion of alien
labour is very great, and 774 Chinese, 86 Japanese, and 55
Indians found employment in 1906. Trouble has arisen over
the influx of Japanese into Vancouver, and the feeling of the
whites is very bitter against its continuance. In the Crow's
Nest collieries no yellow labour is employed.

I have now taken you across the continent, from coast to
coast; and, although we have rapidly examined the localities
of the most important minerals, a vast amount of exceedingly

interesting information on mining progress has necessarily been omitted. The principal object of my Address is to endeavour to bring before your notice something of the immensity of the value of our Canadian possessions, with a view to impressing upon you the importance of the direct investment of British capital in Canada, instead of through the New York and other stock-exchanges, as at the present time; for the danger is that Americans will secure the command, through capital, of Canadian mining and other undertakings. If this country is to have and to keep the controlling power, it is essential that she should hold the purse-strings.

You will see, therefore, why the people of the United States of America have secured controlling interest in the Crow's Nest Pass Company. Last year they took 230,000 tons of coal against 150,000 tons sold in Canada, and 53,000 tons of coke against 134,000 sold in Canada. I think that only one colliery in Vancouver Island makes coke—about 23,000 tons in all—of which the United States of America took 8,000. Taking the Vancouver Island collieries into account, the United States took 664,000 tons against 682,000 tons taken by Canada. The export to the United States would have been still larger, if the Californians had not taken to using an increased tonnage of crude oil.

The latest news from the famous Crow's Nest district reports huge beds of coal above the railway, and the Pennsylvanian miner sees in this district—if the British speculator cannot—great beds of coal more than equalling the enormous deposits of his own State.

There was, and is, a great truth underlying the remark made by the Prince of Wales in reference to his Canadian visit, that "England must wake up," otherwise she will fail in her duty to the Dependencies.

Tables I. and II. give interesting particulars relating to the mining industry of the Canadian provinces, and the various mineralized areas are indicated on the map of the Dominion (plate vii.).

Now, gentlemen, it remains for me to thank you for having listened so patiently to the rather cursory treatment of a most important subject; but I trust that what I have said may be found of some little use, and that, as opportunities permit, mem-

TABLE I.—SHOWING AREAS, POPULATIONS, AND MINING STATISTICS OF THE CANADIAN PROVINCES FOR 1905.

Province	Capital	Area (square miles).	Population (1901).	Population per square mile.	Mining Population.	Mining population per 100 square miles.	Proportion of Population engaged in Mining. Percentage.	Death-Rate per 1,000 engaged in Mining.	Total Mineral Production. Dollars.	Total Mineral Production. £ Sterling.	Mineral Production per head of Mining Population. £
Nova Scotia	Halifax	21,88	459,574	21·80	10,780	50·00	2¼	2·00	11,507,047	2,301,409	213·48
Quebec	Montreal	351,83	1,648,898	4·82	5,017	1·42	⅓	0·40	4,405,975	881,195	175·84
Ontario	Toronto	260,82	2,182,947	9·90	11,151	4·27	½	0·35	18,833,292	3,766,658	337·78
British Columbia	Victoria	372,630	178,657	0·48	8,117	2·59	5	3·50	22,378,187	4,475,638	551·38
Prince Edward Island	Charlottetown	2,184	103,259	47·30	—	—	—	—	—	—	—
New Brunswick	Fredericton	27,985	331,120	11·90	—	—	—	—	559,035	111,807	—
Manitoba	Winnipeg	73,732	255,211	3·96	—	—	—	—	—	—	—
Alberta	Edmonton	253,540	} 211,649	0·11	—	—	—	—	11,841,634	2,368,327	—
Saskatchewan	Regina	250,650									
North West Territories		2,130,690									
Totals and Averages		3,745,574	5,371,315	1·48	—	—	—	—	69,525,170	13,905,034	—

Total area of Canadian lakes—125,756 square miles approximately, or 80,483,222 acres, being one-thirtieth of the entire area of the Dominion.

bers will come forward to supplement my remarks by original papers on the various sections in which they may be specially interested.

TABLE II.—GOLD, SILVER, COPPER, LEAD AND COAL-PRODUCTION FOR 1905.

	Gold. Per cent.	Silver. Per cent.	Copper. Per cent.	Lead. Per cent.	Coal. Per cent.
North-West Territories ...	57	—	—	—	30
British Columbia	39	57	78	98	12
Nova Scotia	2	—	—	—	58
Ontario	—	40	18	2	—
Quebec	2	—	3	—	—
Other provinces	—	3	1	—	—
	100	100	100	100	100

Mr. JOSEPH DICKINSON (Pendleton), in moving a vote of thanks to the President, said that Mr. Ashworth had given them a weighty and very comprehensive Address. One must hope that the wish expressed at the beginning would be realized. and that all the members would do their best to aid him in his office and maintain the high standard that the Society had held for so many years.

Mr. JOHN UNSWORTH (Chorley), in seconding the vote of thanks, said that he had listened with the greatest attention to the President's very eloquent Address, which contained much food for reflection. There was no doubt that the Dominion of Canada was a very good field, open to all of them; and to the younger members of the Society in particular, who were looking forward to a successful career, there was something in Canada which was peculiarly attractive.

Mr. JOSEPH CRANKSHAW (London) said that he would like to say a word or two about Mexico, where there was a more equable climate, and mining operations were being carried on with great success and financial profit. The district near the Pacific Coast had been neglected, but it also presented many points of attraction. Near the coast, the climate was hot enough to enable one to dispense with clothes, while further inland they could get to the mountains and to the eternal snow-line. In addition to coal, there was a wonderful salt-deposit. The possibilities were enormous, now that mines were being opened out and lines of steamers were being started. Vessels from British Columbia brought timber, and took back salt.   Here, then, was a field for the employment of British capital.   Already a good deal of Canadian capital was invested, and many of .the stores were run by Canadians.

The resolution was passed unanimously.

The PRESIDENT (Mr. John Ashworth) thanked the members for their patience in listening to his Address.   He was glad to see so many present at the meeting, and hoped that it was an augury of increased usefulness for the Society.   It had occurred to him that interest in the Society would be increased, and its usefulness extended, if they held their meetings occasionally in other places than Manchester.   They might, for example, have meetings at Leigh, at Bolton, and perhaps at Wigan and St. Helens.   This was a matter which perhaps the Council would take into consideration during the session.

TATE'S IMPROVEMENT FOR SAFETY-LAMPS.

Mr. JOHN GERRARD (H.M. Inspector of Mines, Worsley) called attention to Mr. J. W. Tate's arrangement for ascertaining that gauzes are present in bonnetted safety-lamps. If the gauzes were not in the lamp, with this arrangement the light would not burn.   Mr. Gerrard subjected the lamp to several tests, so as to show its effective working (figs. 1, 2, and 3).

Mr. VINCENT BRAMALL (Pendlebury) thought that a portion of the apparatus which should slide up and down freely might stick, and so prevent its perfect working.

Mr. JOHN GERRARD submitted that if the lamp were kept clean, as it ought to be, it would not stick.   It seemed to him to be a very simple check with regard to the use of bonnetted lamps and gauzes.   Of course, there were other arrangements, or inventions, to provide for the gauzes being in the lamps, but this seemed to him to be such a very simple one: he had tried it during the past month, and it had never failed to act.   The idea was simply an extinguisher which prevented the feed of air to the lamp, and caused the flame to die out when the gauzes were not in the lamp.   If the gauzes were in the lamp, they held up the disc or cap and the light would burn freely.

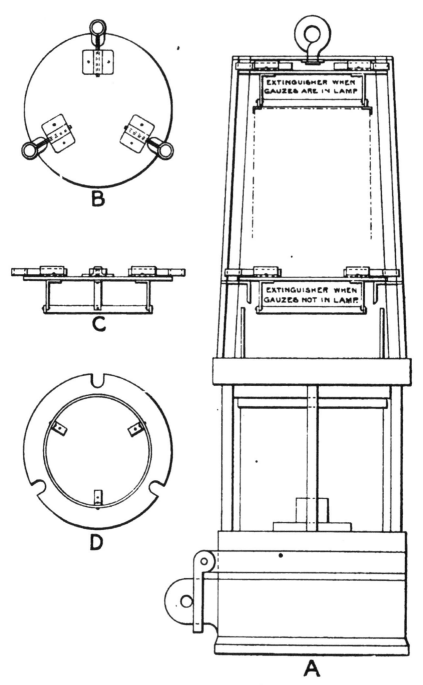

FIG. 1.—DETAILS OF TATE'S APPLIANCE FOR BONNETTED SAFETY-LAMPS:
A, VIEW OF LAMP; B, PLAN OF EXTINGUISHER; C, SIDE VIEW OF
EXTINGUISHER; D, UNDERSIDE VIEW OF EXTINGUISHER.

Mr. J. W. TATE (Tyldesley) said that he had used the lamp daily for a month in the mine, and it had never failed. The appliance could also be adapted to most types of existing safety-lamps; therefore a colliery manager could add this arrangement and have the lamps bonnetted, and could rest assured that before a lamp was given into the hands of anyone the gauze was present, for the light in the lamp would not burn if the gauze were omitted.

FIG. 2.—LAMP WITH TATE'S APPLIANCE HELD UP BY GAUZE.

Mr. PERCY LEE WOOD (Clifton) remarked that before taking up with a new invention, one should be satisfied that it was an improvement on other systems at present in use. While Mr. Tate's arrangement was a step in the right direction, he thought the Patterson arrangement better for several reasons, the principal one being that, with a Patterson bonnet, if either gauze were omitted it would be detected immediately; but he thought that with the Tate arrangement, if one gauze only was in the bonnet, it could not be detected. There were 2,000 lamps with Patterson gauzes at work in the pits under his charge, and he had never had any trouble with them.

FIG. 3.—LAMP WITHOUT GAUZE.

Mr. GERRARD remarked that there might be something to be

said in favour of an invention which secured at least one gauze being in the lamp rather than none.

Mr. WILLIAM HORROBIN (Leigh) said that it was an impossibility for a miner to get into the pit without gauzes in his lamp under the system invented by Mr. Tate.

Mr. JOSEPH DICKINSON (Pendleton) said that Mr. Gerrard's profession brought him into contact with the results of so many mistakes that he naturally desired to know which was the best form of lamp for use in mines.

Mr. JOHN GERRARD (H.M. Inspector of Mines, Worsley) said that he simply brought the lamp before the members for their inspection. He did not ask them to approve of it, or adopt it, but only to look at it and see how far it excelled other forms of safety-lamps.

THE HARDMAN IMPROVED SPUNNEY-WHEEL.

Mr. JOHN GERRARD (H.M. Inspector of Mines, Worsley) next drew attention to Mr. Walter Hardman's improved spunney-wheel, an ingenious contrivance for preventing accidents through the trapping of fingers in the pulley of the ordinary spunney-wheel. It was impossible for a man to get his fingers trapped under this arrangement as the wheel was completely closed in. It was for small jig-brows, where a simple wheel with rope or chain passing through it was in use. Fig. 1 showed the brake free, and fig. 2 the brake in action.

FIG. 1.—HARDMAN SPUNNEY-WHEEL: BRAKE FREE.

Mr. WALTER HARDMAN (Walkden) explained the mechanism of the appliance, which he said had been at work for some time and had proved quite satisfactory. The arrangement for putting on the brake was very simple and positive in action. It applied itself automatically in either direction, excepting when the operator held the lever in the running position. Provision was made, however, whereby the insertion of a pin in the quadrant held the brake in the free position. The brake could be operated from either side, the man standing by preference on the side from which the waggons were running away. In case it became necessary for him to pull the chain in order to assist them, the pull was in the direction away from the wheel, so that there was little chance of his hand being caught. There was nothing complicated about the construction, nor did the parts become unreliable through wear. The method of attachment to the post was by a joint-bolt, or a lashing-chain round the post, which could be fixed in a few moments.

FIG. 2.—HARDMAN SPUNNEY-WHEEL: BRAKE IN ACTION.

FENCE-GATES AND PROTECTORS FOR CAGES IN SHAFTS.

Mr. JOHN GERRARD (H.M. Inspector of Mines, Worsley) further invited attention to a model and photographs of several fence-gates for cages in shafts. The model exhibited was the production of Mr. Henry Houghton, of Skelmersdale. Such fence-gates were in use at Lord Lathom's collieries at Skelmersdale, at the Hulton collieries, and at the Denaby Main collieries. The cage-gates at the latter collieries were automatic, and were the invention of Mr. W. H. Chambers, the arrangement being applicable for the protection of both tubs and men. On arriving at the top or bottom, by certain levers being worked, the gate was raised. Mr. Gerrard explained the points of difference in the several arrangements at these collieries. He favoured the use of gates for cages. There had been a number of accidents in shafts in connection with men riding, and, in his opinion, the use of fence-gates would prevent such accidents. He was glad to note an increasing desire on the part of mine-owners and managers to adopt some simple precaution for the prevention of accidents in the shaft, and thought that it would be interesting to members to have these things brought under their notice.

FIG. 1.—HOUGHTON PROTECTIVE CAGE-GATE AS USED AT LORD LATHOM'S COLLIERIES, SKELMERSDALE: SHOWING METHOD OF OPENING GATE.

The Houghton fence-gate (figs. 1 to 4) would not open by pressure from the inside. To open the gate, it was pushed inwards a little, the lower part raised about its hinges, so as to clear the recess, and then swung outwards. The gate generally fell into its locked position, by gravity, when released. In lieu of hinging the lower part, the same could be made telescopic, and in order to remove the gate from its holding-recess it was simply raised, the lower part sliding on the upper. The gate could, of course, be readily opened from the inside, if desired. When out of use, the gate was slung under the roof of the cage.

The arrangement of the Chambersa'utomatic cage-fence was such that when the cage left either the pit-bottom or top, the fence-bars automatically descended, closing the cage-end, and effectually preventing egress, through accident or otherwise, of

Fig. 2.—HOUGHTON PROTECTIVE CAGE-GATE: SHOWING GATE CLOSED.

FIG. 3.—HOUGHTON PROTECTIVE CAGE-GATE: SHOWING GATE OPEN AND MEN ALIGHTING FROM CAGE.

either man or tubs, in the event of the ordinary stops failing. The fence-bars were raised automatically when at the landings.

The open-end of the cage constituted a great danger; a moment's sickness, causing a sensation of giddiness, might result in the miner so afflicted falling out and meeting a violent death. The cage-gates referred to were ntended to remove this element of danger. The use of such gates would also prevent any sudden attempt being made to enter the cage after the signal for lowering or raising had been given

Mr. VINCENT BRAMALL (Pendlebury) moved a vote of thanks to Mr. Gerrard.

Mr. JOSEPH DICKINSON (Pendleton), in seconding the resolution, said that fence-gates should be automatic in their action,

light in weight, and such as would prevent a man or a tub from falling into the shaft. Much had been done within his recollection to avoid shaft accidents, and it was pleasing to find improvement still being aimed at. Formerly, he had been one of upwards of a dozen persons to descend the shaft with three loops in the coupling-chain, two men each with one leg in each loop, a boy on each of their knees, and some sticking on at couplings. Nowadays

FIG. 4.—HOUGHTON PROTECTIVE CAGE-GATE: GATE OUT OF ACTION AND SLUNG UNDER ROOF OF CAGE.

the usual thing was to step into a cage, where there was a bar or something to lay hold of.

There were few, if any, hoists or man-engines in our colliery shafts, but there were some at metalliferous mines. When the latter mines came within the Mines Regulations Acts, some of these hoists were without sufficient provision for safety. Accordingly, provision was agreed to for the Special Rules in the Manchester and Ireland district, providing for the installation of fend-off boards, or a hinged board to lift, and at the stepping-off places a handle or something to lay hold of where there was any hole to fall into.

MANCHESTER GEOLOGICAL AND MINING SOCIETY.

ORDINARY MEETING,

HELD IN THE ROOMS OF THE SOCIETY, QUEEN'S CHAMBERS,
5, JOHN DALTON STREET, MANCHESTER,
DECEMBER 10TH, 1907.

MR. JOHN ASHWORTH, PRESIDENT, IN THE CHAIR.

The following members were elected, having been previously
nominated : —

MEMBER—
Mr. CHARLES DAVIS TAITE, Electrical Engineer, 196, Deansgate, Manchester.

STUDENT—
Mr. RALPH HAMPSON, Mining Student, Shotton Cottage, Shotton, Queen's
Ferry, S.O., Flintshire.

THE LATE MR. M. WALTON BROWN.

The PRESIDENT (Mr. John Ashworth) reminded the members
that since their last meeting the mining world had suffered a
great loss by the death of Mr. Martin Walton Brown, Secretary
of The Institution of Mining Engineers.

Mr. JOSEPH DICKINSON (Pendleton) thereupon moved the
following resolution : —

"The President, Council, and Members of the Manchester Geological and
Mining Society, in general meeting assembled, receive with profound regret the
intimation of the death of Mr. Martin Walton Brown, the Secretary of The
Institution of Mining Engineers, and desire to express their deep sympathy with
Mrs. Brown and her family in their sudden bereavement. The members also desire
to express their high appreciation of the long and valuable services rendered to
The Institution of Mining Engineers and to the mining profession generally by
the late Mr. M. Walton Brown, whose departure from amongst them is a great
loss to the mining world."

Mr. JOHN GERRARD (H. M. Inspector of Mines, Worsley), in
seconding the resolution, said that he had known Mr. M. Walton
Brown for over thirty years. They all knew how high Mr. Brown

had carried the banner of mining, not only throughout the mining districts of this country, but throughout the mining districts
of the world. The least that they could do, therefore, was to pass
this resolution of sympathy with those who had been overwhelmed
with sorrow in so sudden and so sad a manner. The valuable work
which Mr. Brown had done would be remembered and appreciated
for many years.

The resolution was passed in silence, all the members rising
from their seats.

SPECIMENS OF COAL FROM CALDER COLLIERY.

Mr. JOHN GERRARD (H.M. Inspector of Mines, Worsley)
exhibited some interesting specimens of coal taken from the
Calder colliery, representing the Arley Mine and the Lower Mountain Mine. The distance between the two seams was 6 inches less
than 900 feet. This being the first time that the distance between
these seams had been actually measured, it seemed to him a notable
occurrence, and one worthy of being brought before the Society.

The specimens of marine shells produced came from the two
marine beds associated with these seams.

GAUZES IN BONNETTED SAFETY-LAMPS.

Mr. JOHN GERRARD (H.M. Inspector of Mines, Worsley)
exhibited Mr. John Walshaw's appliance for securing the presence
of gauzes in bonnetted safety-lamps. At the last meeting of the
Society, he had shown a similar invention,[*] and on that occasion
there was some controversy as to the complete utility of the
device. The idea of this particular invention was to have a tab
attached to each gauze, which showed in the glass so that if the
two tabs were not seen there was evidence that the gauzes were
not in the lamp.

Mr. W. H. COLEMAN read the following paper on "Bye-
products from Coke-ovens":—

---

[*] *Trans. Inst. M. E.*, 1907, vol. xxxiv., page 321 ; and *Trans. M. G. M. S.*,
1907, vol. xxx., page 187.

BYE-PRODUCTS FROM COKE-OVENS.

By W. H. COLEMAN.

The exhaustion of our coal supply is proceeding at such a rapid rate that it is necessary to take care that the energy contained in every ton of coal should be utilized to the fullest extent possible, not perhaps that the pinch may be felt in our own time, but because it is our duty to leave as much of our heritage of power as possible to our descendants. The increase in the population of the world is a warning that the question of food supply will be an urgent one in the near future. We have for years been existing upon the stored up wealth of the virgin soils of America and other sparsely-peopled districts, but this cannot go on for ever, and the continued abstraction of plant food from the soil must be compensated for in some way or other. The two most important ingredients permanently removed from the soil by the crops are phosphorus and nitrogen. With the former the writer does not propose to deal in these notes, but he intends to confine his remarks to the latter as being closely concerned with one of the chief bye-products of coke manufacture, namely, sulphate of ammonia, which is at present the only practicable means of returning to the soil the nitrogen which has been abstracted by the crops. No doubt the deposits of nitrate of soda in South America have to a large extent supplied up to the present time the demand, but they are being rapidly exhausted at an ever increasing rate, and until the production of nitrogen-compounds from the air can be accomplished at a reasonable cost, the gap left can only be filled by sulphate of ammonia.

The introduction of motor vehicles has opened an avenue of usefulness for the bye-product recovery-oven, and provides an outlet for the other two of the three chief bye-products, namely, for benzol as a motor fuel, and for tar as a means of combating the dust nuisance, which is made more visible, if it is not entirely caused, by motor traffic.

It may now be desirable to give a few figures showing to what extent the production of coke in recovery-ovens has progressed in this and in other countries. Taking Great Britain first, it was stated in the Report of the Royal Commission on Coal-supplies of 1905* that in the year 1902 only about 10 per cent. of our total output of coke was made in recovery-ovens, and it was then suggested that returns should be obtained showing the annual increase in the use of this kind of coke-oven; and, although owing to the want of powers to compel returns no exact figures can be obtained yet for the last two years, the following figures, published by the Home Office, are extremely interesting :—†

TABLE I.—TONS OF COAL COKED DURING 1905 AND 1906, AND VALUE OF COKE PRODUCED.

Year.	Coal Used.	Coke Made.	. Value.
	Tons.	Tons.	£
1905	33,452,943	18,037,985	10,625,799
1906	35,402,677	19,296,526	12,549,116

TABLE II.—NUMBER AND DESCRIPTION OF COKE-OVENS.‡

Year	Beehive	Simon Carvés	Semet-Solvay.	Coppée	Bauer.	Koppers.	Otto-Hilgen-stock.	Simplex.	Other Kinds	Total.
1905	25,514	726	470	2,233*	52	72	503	—	1,490	31,060
1906	23,454	808	670	2,308	52	108	768	78	1,482	29,728

* Some Otto-Hilgenstock ovens are included in these figures.

TABLE III.—NUMBER OF BEEHIVE AND RETORT COKE-OVENS IN 1905 AND 1906.

Year.	Beehive.	Retort-ovens.	Total.	Percentage of Retort-ovens.
1905	25,514	5,546	31,060	17·85
1906	23,454	6,274	29,728	21·11

From the above tables it will be noted that the number of beehive-ovens has decreased from 25,514 in 1905 to 23,454 in 1906, and the recovery and retort-ovens have increased from

* *Final Report of the Royal Commission on Coal-supplies*, 1905 [Cd. 2353], part i., general report, page 9.

† *Mines and Quarries: General Report and Statistics*, 1905 [Cd. 2974], page 195, and 1906 [Cd. 3774], page 180.

‡ *Ibid.*, 1905, page 196 ; and 1906, page 181.

5,546 in 1905 to 6,275 in 1906, or a decrease in the first case of about 8 per cent., and an increase in the second case of about 13 per cent.

Although no figures are given showing the proportion of coke made in recovery-ovens, for the quantities given in Table I. refer to all coke manufactured, including that made at gas-works, it is possible by ·a somewhat roundabout calculation to arrive at an approximate estimate. Table IV. gives the production of sulphate of ammonia in each year, and the writer considers the figures interesting for two reasons, namely, (1) they show the increase in the recovery of ammonia as sulphate from coke-works for the past nine years, and (2) from these figures the approximate calculation to which the writer has referred has been made.

TABLE IV.—PRODUCTION OF SULPHATE OF AMMONIA FROM 1898 TO 1906.*

Year.	Gas-works.	Iron-works.	Shale-works.	Coke-ovens.	Producer and Carbonizing Works.	Total.
1898	129,590	17,935	37,264	5,403	6,165	196,357
1899	133,768	17,963	38,780	7,849	7,360	205,720
1900	142,419	16,959	37,267	10,393	6,688	213 726
1901	142,703	16,353	40,011	12,255	5,891	217,213
1902	150,055	18,801	36,931	15,352	8,177	229,316
1903	149,489	19,119	37,353	17,438	10,265	233,664
1904	150,208	19,568	42,486	20,848	12,880	245,990
1905	155,957	20,376	46,344	30,732	15,705	269,114
1906	157,160	21,284	48,534	43,677	18,736	289,391

From the foregoing table it will be noticed that the production of sulphate of ammonia from coke-ovens has risen from 5,403 tons in 1898 (the first year for which separate figures are given for coke-ovens) to 43,677 in 1906, or an increase of over 800 per cent., the increase in the last three years being especially noticeable.

The yield of sulphate of ammonia may be taken in round numbers as 25 pounds per ton of coal carbonized, either from gas-works or from coke-ovens (possibly 28 or 30 pounds in many cases may be obtained, but the writer prefers to take the lower figure as being probably the actual yield).

If a calculation be made, on the above assumption, from the quantity of coal carbonized in gas-works, and the figure subtracted from that given in Table I., it will show the quantity of

* *Annual Reports of the Chief Inspector under the Alkali Act.*

coal carbonized in coke-ovens; and, if a further calculation on the same assumption be made, taking the quantity of coal carbonized in coke-ovens where sulphate of ammonia was made and deducting, the remainder will show the quantity of coal carbonized in beehive and other non-recovery ovens. The following table shows the results for the years 1905 and 1906: —

TABLE V.—COAL COKED DURING 1905 AND 1906.

Year.	Gas-works.	Coke-works, all kinds.	Coke-works, Recovery-ovens.	Coke-works, Non-recovery Ovens.	Total Coke Coked.
	Tons.	Tons.	Tons.	Tons.	Tons.
1905	13,973,747	19,479,196	2,753,587	16,725,609	33,452,943
1906	13,991,936	21,410,741	3,914,899	17,495,842	35,402,677

TABLE VI.—PERCENTAGE OF COAL COKED IN RECOVERY AND IN NON-RECOVERY OVENS.

Year.	Non-recovery Ovens.	Recovery Ovens.
	Per cent.	Per cent.
1905	83·9	16·1
1906	81·8	18·2

The above estimated figures agree fairly well with those given for the year 1905 by Mr. Ernest Bury in his most instructive paper read before the Institution of Gas Engineers at Dublin in June, 1907,* wherein, calculating from quite different assumptions, he gives the following figures: —

TABLE VII.—COAL COKED DURING 1905.

Year.	Gas-works.	Non-recovery Coke-ovens	Recovery-ovens.	Total.
	Tons.	Tons.	Tons.	Tons.
1905	14,180,000	16,000,000	3,317,000	33,497,000

The following figures are also taken from the Mines and Quarries Report† : —

TABLE VIII.—NUMBER AND PERCENTAGE OF WORKS RECOVERING BYE-PRODUCTS FROM COKE DURING 1905 AND 1906.

Year.	Number of Works from which Returns were Received.	Number of Works at which Bye-products were Recovered.	Percentage of Works at which Bye-products were Recovered.
1905	271	46	17·1
1906	257	51	19·8

* The Journal of Gas Lighting, 1907, vol. xcviii., page 982.

† Mines and Quarries: General Report and Statistics, 1905 [Cd. 2974], page 196 ; and 1906 [Cd. 3774], page 181.

Turning to other countries, the world's production of coke is given in metric tons in the Bulletin No. 2,706 of the Comité des Forges, published September 30th, 1906, as follows:—

TABLE IX.—WORLD'S PRODUCTION OF COKE.

Country.	1904. Metric Tons.	1905. Metric Tons.
United States of America	21,465,355	29,240,080
United Kingdom (no returns for 1904) ...	—	18,326,593
Germany	12,331,163	16,491,427
France (in part only)	1,673,519	1,907,913
Belgium	2,211,820	2,238,920
Russia	2,402,878	2,374,335
Austria	1,282,473	1,400,283
Italy	607,297	627,984
Spain	605,318	641,689
Canada	493,107	622,154
Australia	173,750	165,576
Hungary	5,103	69,303
Mexico (estimated)	60,000	60,000
Other Countries (estimated)	2,000,000	2,200,000
Totals	45,311,783	76,366,257

N.B.—1 metric ton = 0·9839 English ton.

In reference to the above figures, it should be mentioned that no statement was made to show whether or not they included the coke produced at gas-works. The return for the United Kingdom, as shown by comparison with the figures of the Board of Trade return, included the gas-works production, but that for Belgium did not, as the figure given was the same as that stated in the Belgium Government report as the production of 39 coke-works. The writer has unfortunately been unable to check the other figures.

The United States.—In the year 1901, 5 per cent., and in 1905, 10·74 per cent., of the total production of coke was made in recovery-ovens. In the latter year, there were in the United States of America 3,159 recovery-ovens out of a total of 87,564 ovens of all kinds, or 3·6 per cent.

Germany.—In the year 1901, 53 per cent. of the total production of coke was made in recovery-ovens, and the writer was informed by a competent authority when in Germany last year that 90 per cent. of the coke was at that time being made in recovery-ovens.

France.—The *Colliery Guardian** gave the production in France during the year 1905 as follows:—Gas-works. 2,300,000 tons; coke-works, 2,603,522; total, 4,903,522. There were 2,512 beehive and 1,604 recovery-ovens, the latter being 38·97 per cent. of the total number. The coke produced in recovery-ovens was 1,543,000 tons, being 31·46 per cent. of the total coke produced.

Belgium.—In the year 1904, 1,700,000 tons of coke were produced in 911 bye-product ovens, and in 1905, 2,238,920 tons of coke were produced in 39 coke-works.

Spain.—In the year 1901, about 200,000 tons of coke were produced in coke-ovens of the Semet-Solvay and Simon Carvés types at Altos Hornos and Sestao.

The writer regrets that he has been unable to obtain more recent or more extensive figures, but he believes that enough has been given to show that Great Britain is a long way behind the other nations of the world, except perhaps the United States of America, in the adoption of bye-product ovens. That steps are being taken to remedy this condition of affairs may be seen from Table X., which cannot, however, be considered complete.

TABLE X.—RECOVERY-OVENS EITHER JUST COMPLETED OR AT PRESENT (1907) UNDER CONSTRUCTION IN THE UNITED KINGDOM.

Description.	Number.
Simon Carvés	500
Semet-Solvay	449
Koppers	160
Otto-Hilgenstock	200
Simplex	88
Coppée	353
Total	1,750

The three main products obtained in the process of coking or carbonizing coal, when coal is destructively distilled, are as follows:—Solid, coke; gaseous, heating or lighting gas and ammonia; liquid, tar and benzol or naphtha.

The first of the above, namely, coke, the writer must leave

* 1906, vol. xcii., page 663.

to those who have much more practical experience of the matter than himself. However, he would like to point out that the old objection to bye-product coke is slowly but surely giving way. Recovery-coke is, if properly slacked, quite equal to bee-hive-coke in respect to its appearance, or sulphur content, and its ability to stand the burden of the charge in the blast-furnace. Further, the introduction of mechanically-charged blast-furnaces requires a coke of smaller size than formerly, which will lead in the future to a more extensive use of the bye-product variety.

TABLE XI.—ANALYSES OF VARIOUS KINDS OF COKE (DRY).

Constituent.	Bye-product Coke.	Beehive-Coke.	Gas-works Coke.
	Per Cent.	Per Cent.	Per Cent.
Ash	6·00	8·0	9·90
Volatile matter	1·86	not estimated	1·65
Fixed carbon (by difference) ...	92·14	91·90	88·45
Totals	100·00	100·00	100·00

The second class of products, namely, the gaseous, consists of heating gas and ammonia. At the present day, nearly all coke-works utilize at least a portion of the gas for other purposes than heating the ovens. It has a high calorific value, as will be seen from the following table, taken from that given by Mr. F. L. Slocum[*] :—

TABLE XII.—CALORIFIC VALUES OF VARIOUS GASES.

Name of Gas.	British Thermal Units per Cubic Foot.
Oil-gas	1,350
Natural gas	980
Coal-gas	600 to 628
Coke-oven gas	367 to 686
Water-gas	500
Mond gas	156
Siemens gas	137

In this connection it is interesting to note a suggestion recently made that a producer-gas plant should be combined with a coke-oven plant, especially where both poor and rich coals occur in the same locality. The producer being fed with the poor coal (coal with a small percentage of volatile matter),

[*] "A Comparison of Fuel-gas Processes," by Mr. F. L. Slocum, *The Journal of the Society of Chemical Industry*, 1897, vol. xvi., page 420.

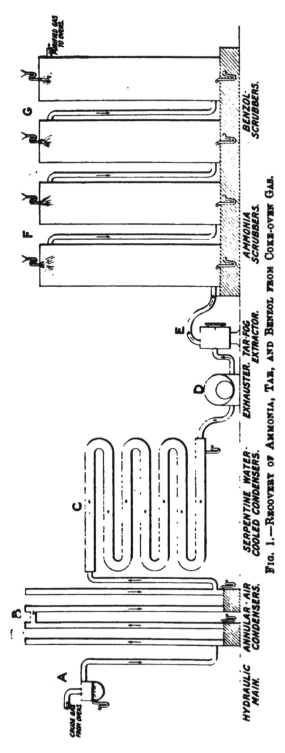

FIG. 1.—RECOVERY OF AMMONIA, TAR, AND BENZOL FROM COKE-OVEN GAS.

even the waste coke, breeze, etc., might be used; and the gas of low calorific power used for heating the coke-ovens; the high-value gas from the ovens being used to produce power by means of gas engines. Coke-oven gas is eminently suited for such a purpose, being of high calorific value and comparatively clean, whereas producer-gas is very difficult to free from dust, and especially from tar, much more so than coke-oven gas. Of course, the great difficulty in the way is the question of storage of power, but consideration of that point is beyond the scope of this paper.

Coke-oven gas is largely used in the United States of America for lighting purposes, and the extension of its use in this direction might with advantage be undertaken in this country. The gas,

especially that evolved during the earlier stages of carboniza-
tion, is of high illuminating value, and it has been proposed
to collect the first portions separately for use as an illuminant.
The remaining products—ammonia, tar, and benzol—now require
consideration. It would perhaps be best to first consider the
question of their recovery from a general point of view, and to
this end the writer invites attention to the rough diagram
(fig. 1) which he has made to illustrate the principles underlying the process. The gas as it comes from the ovens has a
very high temperature, and the first step towards the recovery
of the bye-products is to reduce the temperature to the normal.
From the ovens the gas enters a common main extending the
whole length of the ovens, which is known as the hydraulic main,
A, where the temperature falls to about 212° Fahr. (100° Cent.),
and a considerable portion of the tar and water-vapour is condensed. The latter absorbs some of the ammonia. The gas
passing from the hydraulic main still contains tar in the
state of fog or mist, in addition to ammonia gas and benzol
vapour. It is, therefore, still further cooled by passing through
either annular or serpentine air-cooled pipes, B, and then through
a series of water-cooled pipes, C. By this means, its temperature
is brought down nearly to that of the atmosphere, and a further
quantity of tar and ammoniacal water deposited, which is
allowed to drain away by S-shaped siphon pipes, and conveyed
to store tanks or wells. The gas next passes to the exhausters,
D, which serve to draw the gas from the ovens through the
coolers and to force it through the rest of the apparatus described later. Two different types of exhausting apparatus are
used, the rotary mechanical exhauster and the steam injector.
The latter has the advantage of producing a more constant suc-
tion and of being free from mechanical complications, but has
the drawback of heating the gas so that it has to be passed
through another cooler. The gas from the exhausters, D, still
contains tar-fog, ammonia, and benzol vapour, and is generally
next passed through some form of tar-fog extractor, E. The two
most important types of this apparatus are the Pelouze and
Audouin, and the Mallet extractors, being based on the same
principle, namely, condensation by shock. The tar particles
exist in the gas in the state of little vesicles or bubbles of tar
filled with gas, which are extremely minute and very difficult

to condense. Within the apparatus a number of screens of perforated metal sheets, placed only a short distance apart, are so arranged that the holes in one plate are opposite the blank surface of the next. The vesicles pass through the perforations with considerable velocity, striking against the surface of the plate in front. The shock of the impact breaks up the vesicles, and causes the tarry matter to adhere to the surface of the plate, down which it trickles to the store tank below. Both types have a tendency to get blocked with tar; and to contend with that difficulty the Pelouze apparatus is so constructed and connected to the system of gas-mains that it can be cut out of the path of the gas from time to time and the top removed, so that the dirty plates can be taken out and replaced by clean ones and the apparatus then put in action again. In the Mallet apparatus, the perforated plates, or rather drums, are slowly and continuously revolved, so that the lower portions pass into warm tar contained in the bottom of the casing. The perforations are thus continuously washed by the warm tar. The necessity for the use of the tar-fog extractor will be explained later in the process for the recovery of the benzol vapour. The gas is next passed through the ammonia scrubbers, or washers, F, where the ammonia is absorbed. These consist either of vertical towers filled with chequer-work of bricks, rings, boards, or even with coke; and the gas passes up the towers and is met by a stream of water flowing down. The second type of ammonia-absorber is the rotary mechanical washer of the Holmes or the Standard pattern. This apparatus consists of a horizontal cylinder divided by vertical partitions into a number of compartments, a shaft passing horizontally through the cylinder, a number of brushes or bundles of laths being fixed upon the shaft, one in each compartment. A current of water is caused to flow through the cylinder from one end to the other, from chamber to chamber, each being about half filled with the liquid, and the gas to be washed being caused to flow through the upper part of the chambers in the opposite direction. As the shaft revolves, each portion of each brush dips into the liquid and is afterwards exposed to the current of gas.

The most important points to be borne in mind in carrying out this part of the process are to arrange that the temperature of both gas and water are as low as possible, and that the currents of gas and washing-liquid flow in opposite directions. If tower scrubbers are employed, it is necessary to ensure a good distribution of the washing liquid; and for this purpose the Zschockesche patent distributer is one of the best types, being easy to regulate, as the glass-sight allows the attendant to see if it is working properly.

The next process is the recovery of the benzol, or rather the mixture of hydrocarbons in the gas in which benzol predominates, which is effected by passing the gas through an exactly similar apparatus, G, to that used for ammonia recovery. The benzol is absorbed by a hydrocarbon oil, and for this purpose creosote or anthracene oil is used, which absorbs the benzol vapours, and is afterwards sent to a still where the benzol is separated by fractional distillation. The oil, after cooling, is used over and over again, until it becomes too strongly charged with tar and naphthaline (a crystalline hydrocarbon occurring among the products of distillation of coal, and particularly difficult to separate from the gas) for the purpose, when it is either mixed with the tar or sold separately. In order that this oil may be used as many times as possible, the tar-fog previously mentioned must be extracted completely from the gas. The scrubbed gas is then returned to the ovens and burnt in the flues to effect the coking of the coal, but in most cases a considerable surplus is available for other purposes: this varying, of course, with the quality of the coal, the type of ovens employed, and the care and attention given to the process. It could be used directly in gas-engines for the production of power, or employed for lighting in the works or neighbourhood. For the latter purpose, the gas ought to be further purified by passing it through hydrated oxide of iron, in order to remove the sulphuretted hydrogen which it contains.

Having described how the products are obtained, it remains to say something about their disposal, and this will be considered in the same order as before, namely, tar, ammonia (now in the form of a weak solution), and benzol (in the form of an impure mixture of hydrocarbons known as crude benzol or crude naphtha).

The tar is always sold as such to the tar distiller, who, by distillation and subsequent washing and rectification, pro-

duces the finished first products.   These comprise benzol,
toluol, solvent naphtha, creosote oil, naphthaline, carbolic acid,
anthracene, and pitch, which are used in the colour, rubber, and
explosive industries, as well as for preserving timber, for paving,
asphalting, briquetting coal, preparing disinfectants, making
lamp-black, roofing-felt and many other purposes.   In addition,
the tar itself is now largely used to overcome the dust nuisance,
made so obvious by the rapid increase in the use of motor-cars.
For this purpose the crude tar is not so suitable as the refined
product obtained by distilling off the lighter hydrocarbons,
which render the tar too fluid, and consequently too easily
softened by the heat of the sun, and which also tend to separate
after its application to the roads, especially in wet weather, as a
most objectionable oily film or scum.   The tar could be applied in
two ways, either as a coating spread over a road not in need of repairs, when it acts both as a preventive of dust, and as a
preservative of the road material.   Experiments carried out in
France over several years* proved that not only were the tarred
roads free from dust, but also that they lasted longer, and that
the cost of cleaning was much reduced, so that although the cost
of application might be considerable it was counterbalanced by the
saving effected.   To gain some idea of the quantity of tar that
might find application for this purpose, the writer will quote
figures given by Mr. Maybury in a paper read by him at
Dublin in 1907.†   About 7 tons of tar were required for each
mile of road of the average width, the mileage of roads in
England and Wales being given as follows:—Main roads, 23,826
miles ; other public roads, 95,211 miles ; total, 119,037 miles.   If
these were all tarred, there would be required:—For main
roads, 166,782 tons of tar; for other public roads, 666,477 tons
of tar; being a total of 833,259 tons of tar.

Another method in which tar is used in road construction is
that of paving the road with a mixture of broken stone or slag
with tar or pitch, known as " tar macadam." This is, no doubt,
the more serviceable way, as the painting of the surface is at best
only a palliative, while tar macadam, in addition to the prevention of dust, gives a good hard road which is reported to last'
extremely well.

* Annalen des Ponts et Chaussées, eighth series, 1905, vol. xx., page 260.
† The Journal of Gas Lighting, 1907, vol. xcviii., page 972.

The tar produced in coke-works varies in composition, but approximates to that obtained at gas-works, and contains the same compounds, although in somewhat different proportions. It differs altogether from blast-furnace and producer-tars, which are much less valuable.

Turning to the ammonia, which, as stated above, is recovered as a weak solution in water, the chief compounds present are shown in Table XIII. and compared with the liquor obtained at gas-works.

TABLE XIII.—COMPOSITION OF AMMONIACAL LIQUOR.

Name of Constituent.	Ammoniacal Liquor:	
	From Gas-works.	From Coke-works.
Specific gravity at 60° Fahr.	1·028	1·011
	Grammes per 100 Cubic-centimetres.	Grammes per 100 Cubic-centimetres.
Free ammonia (NH)	1·928	0·841
Fixed do.	0·545	0·102
Total do.	2·473	0·943
Sulphuretted hydrogen	0·510	0·233
Carbonic acid	2·274	0·899

The ammonia compounds are divided into two classes, free and fixed, the former comprising those compounds of ammonia which are dissociated by simple heating, and the latter those which require the addition of lime or of some other alkali to decompose them. The ammonia-water obtained from coke-ovens is generally weaker than that from gas-works, probably because the washed coal contains more water, although it should be possible to improve the strength somewhat by careful attention to the scrubbers. The ammonia water is distilled by steam and the gas passed into sulphuric acid, with which the ammonia combines, forming sulphate of ammonia. It should contain from 24 to 24·5 per cent. of ammonia, and be reasonably dry and free from excess of acid. Its colour should be as white as possible, in order to obtain the best price, and care is required to attain this. If acid containing arsenic is used, the resulting sulphate will be yellow, and may be refused. It may be worth while to give one or two hints as to the cause of discoloration and the means of preventing it. The dirty grey colour often observed is due, not to dark-coloured acid to which it is generally

attributed, but to one of two causes which could with care be easily
avoided.   The first is the priming of the liquor in the still, and the
second is the occurrence of local alkalinity in the saturator or vessel in which the ammonia is absorbed by the sulphuric acid, which
may be due either to a deficiency of steam, which often results in
solid sulphate building up over the perforations in the gas
pipe, or to carelessness of the workman in not feeding the acid
with sufficient regularity.   Either of these causes may also
result in the formation of blue salt, which is very troublesome,
and renders the product unsaleable.   The ammonia-water contains hydrocyanic or prussic acid (which distils over with the
ammonia), carbonic acid, and sulphuretted hydrogen, and which,
in the presence of iron (always found in commercial acid), and
free ammonia, forms ferrocyanide or yellow prussiate of
ammonia.   This compound, as soon as the saturator becomes
acid again, is decomposed, and forms white ferrocyanide of iron,
which rapidly oxidizes into prussian blue and discolours the sulphate.   The remedy is then to use a sufficient excess of steam to
keep the contents of the saturator continually boiling, and to see
that the contents are never allowed to become alkaline for want
of acid.   Care as to these details would save much annoyance.
The carbonic acid and sulphuretted hydrogen, along with some
hydrocyanic acid, escape from the top of the saturator, and as
these gases are not only objectionable but also very poisonous, they
should be treated, and are not allowed to be turned out into the air.
They are, therefore, after cooling, either passed through a purifier or box containing hydrated oxide of iron, when the sulphuretted hydrogen is decomposed into water and sulphur and the
hydrocyanic acid is arrested as prussian blue, or the gases mixed
with a suitable proportion of air and passed through a Claus
kiln or chamber containing oxide of iron, which acts as a
catalytic agent, and the sulphuretted hydrogen is partly burnt
to sulphurous acid, which reacts on the rest of the sulphuretted
hydrogen, forming water and sulphur.   This, by the way, is an
interesting reaction, as the sulphur found near the vents of
volcanoes is probably formed in a similar way.   The waste liquor
from the stills, containing lime-salts and phenols, cannot be
turned into rivers or water-courses; but the various methods
that have been proposed for its treatment cannot be described in this paper.   The writer would, however, refer anyone

interested in the question of the composition of this effluent to the paper by Messrs. Frankland and Silvester,* read before the Society of Chemical Industry, and to the paper by Dr. F. W. Skirrow read before the Manchester section of the same Society.†

The last product to be dealt with is the benzol, which, as was mentioned before, is recovered as a solution of benzol in creosote-oil. The oil containing the benzol is run through a still somewhat similar to that used for the ammonia, and the low-boiling benzol continuously distilled off, while the creosote oil may, after cooling, be used over again. The product obtained, generally that known as 65 per cent. benzol, is a material giving a distillate of 65 per cent. at a temperature of 248° Fahr. (120° Cent.). It is sold to the refiners, who work it up into benzol, toluol and solvent naphtha.

The processes of recovery having been briefly described, it may be worth while to turn attention to the yields from the coal and the possible outlet for the products. The yield of bye-products, of course, depends chiefly upon the amount of volatile matter in the coal carbonized, and also upon the method of coking; but it may be taken as a conservative estimate that the yields are somewhat as given in Table XIV.

TABLE XIV.—YIELD OF PRODUCTS FROM CARBONIZING-COAL IN BYE-PRODUCT COKE-OVENS.

Coke	...	68 to 72 per cent.
Tar	...	6 to 8 gallons per ton of coal.
Sulphate of ammonia ...	...	25 to 30 pounds per ton of coal.
Naphtha, 65 per cent.	...	2 to 3 gallons per ton of coal.

Upon reference to Table V. it will be seen that in the year 1906, approximately 4,000,000 tons of coal were coked in recovery-ovens, about 17,000,000 tons in non-recovery ovens, and about 14,000,000 tons at gas-works. The yield of bye-products at gas-works may be taken as follows:—Tar, 10 to 12 gallons; sulphate of ammonia, 25 to 30 pounds per ton of coal. Calculating from the above data, and using the lowest figure in each case, the estimated production for the year 1906 was as follows,

* "The Bacteriological Purification of Sewage containing a large proportion of Spent Gas Liquor," by Messrs. Percy F. Frankland and H. Silvester, *The Journal of the Society of Chemical Industry*, 190?, vol. xxvi., page 231.

† "The Determination of Phenols in Gas Liquors," etc., by Dr. F. W. Skirrow, *ibid.*, 1908, vol. xxvii., page 58.

assuming that all recovery coke-works recover tar, ammonia, and
benzol, which is certainly not the case so far as benzol is
concerned : —

TABLE XV.—YIELD OF BYE-PRODUCTS AT GAS-WORKS DURING 1906.

Source.	Tar.	Sulphate of Ammonia.	90 Per cent. Benzol.
	Tons.	Tons.	Gallons.
Gas-works	700,000	156,000	1,500,000
Coke-works	120,000	44,000	4,000,000
Other sources ...	?	89,000	......
Totals	820,000	289,000	5,500,000

TABLE XVI.—POSSIBLE ADDITIONAL PRODUCTION IF ALL COKE WERE
MADE IN RECOVERY-OVENS.

Tar ... ... 510,000 tons.
Sulphate of ammonia ... 190,000 tons.
90 per cent. benzol ... 17,000,000 gallons.

Dealing with these separately, it is found in the case of tar
that the present production is about 820,000 tons and the possible increase about 510,000 tons, making together 1,330,000
tons as the possible production. The question as to whether this
could be absorbed or not is a difficult one. The tarring of roads
would, if universally adopted, present an outlet for about 800,000
tons, an amount greater than the possible extra production in
1906, but this is a question for the future. There are two other
possible outlets for additional tar—an increase in the amount of
small coal briquetted, and the carrying out of the suggestion that
poor coals might be successfully coked after the addition of a
certain proportion of tar or pitch.

With sulphate of ammonia the question is simpler: the increase of 100,000 tons in the last nine years has not caused
the price of sulphate to fall, and there is no reason that the writer
knows of to fear that a gradual increase would seriously interfere with the value of this material, as the demand for nitrogen
for plant-food is almost unlimited. The Japanese have adopted
wheat as a food in the place of rice, and are now large importers
of sulphate of ammonia, and the Chinese and other nations may
follow their lead.

With regard to benzol, there is a fairly bright outlook, the
present production in 1906, estimated at 5,500.000 gallons, having been all absorbed either for colour-works or for gas enrich-

ment. During the last year it has come into extended use as a fuel for explosion-engines, and (so far as can be seen at present) this is the only use that can be largely extended. It is estimated that 30,000,000 gallons of petrol were used for motor-vehicles and other explosion-engines in this country last year and that the present rate of increase in its use is 5,000,000 gallons per year. Any large increase in the output of petrol must mean an increase in price, for it only forms a limited proportion of the crude petroleum from which it is made, and outlets have to be found for the residual burning and lubricating oils, paraffin, wax, and pitch, and this does not appear to be easy.

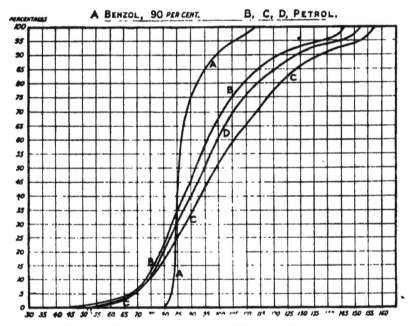

Fig. 2.—Distillation-curves in Degrees Centigrade.

The writer was informed by a very good authority that the supply of petrol is gradually approaching its limit, and that any further increase in the demand for motor fuel must be met from other sources ; moreover, it is stated that the use of the heavier oils or paraffins in spray-carburetters is being seriously considered. In view of these considerations the demand for, and consequently the value of, benzol for this use is more likely to be increased than otherwise. Benzol is a very efficient fuel for motors, as, volume for volume, about 20 per cent. greater efficiency can be obtained from its use. Its range of temperature on distillation is only about

86° Fahr. (30° Cent.), against 212° Fahr. (100° Cent.) for most samples of motor-spirit; and this is well illustrated by the curves (fig. 2), which show the temperature at which each 5 per cent. distils. This means that its efficiency is more constant, and that no trouble owing to the formation of stale spirit need be feared. It is a home-product, and the supply would not be interfered with in the event of political complications. If all the benzol possible were recovered, it does not equal the present demand for motor fuel, so that no fall in its value need be anticipated.

In conclusion, if the writer has interested the members sufficiently to induce them to consider the advisability of investigating for themselves the question of the adoption of recovery-ovens as a less wasteful process than the old one, he will have been amply repaid. His thanks are due to the proprietors of the various coke-ovens and to firms who have kindly supplied information given in the paper.

On the motion of the PRESIDENT (Mr. John Ashworth), seconded by Mr. SYDNEY A. SMITH, a cordial vote of thanks was accorded to Mr. Coleman for his interesting and most valuable paper.

The discussion of the paper was adjourned.

CONGRATULATIONS TO MR. JOSEPH DICKINSON.

The PRESIDENT (Mr. John Ashworth) and all the members present offered their congratulations to Mr. Joseph Dickinson, who, on November 24th, completed his eighty-ninth year.

"I am sure," the PRESIDENT remarked, "we are all exceedingly pleased to see our worthy friend still amongst us and so active."

Mr. JOSEPH DICKINSON (Pendleton), in reply, stated : " I entered upon my ninetieth year on Sunday, November 24th. I am very much obliged to you for your kind congratulations. Long life is a great gift, but it has its drawbacks. We who attain to great length of days have many losses to bear. I stand before you as a member of a large family and the last survivor. Still, we have the memory of those who have gone to cheer us, to guide us, and to lead us, and there is a gain, taking it altogether, in a long life."

SECTION OF BORE-HOLE AT MESSRS. MARK FLETCHER AND SON'S WORKS AT MOSS LANE, WHITEFIELD, LANCASHIRE.

By JOSEPH DICKINSON, F.G.S.

As this part of the Lancashire coal-field is covered with deep drift, and as there are no colliery-workings in the immediate neighbourhood, the section is likely to prove of use to geologists and mining engineers.

The section was proved by Messrs. Chapman and Sons, Salford, while rope-boring for water, the actual strata being of secondary importance; but a few cores were taken as likely to be useful.

These cores were viewed by the writer and Mr. William Pickstone on September 18th, 1907, which resulted in the original section being thus detailed. The cores show the dip of the strata to be about 1 in 4½, and that the strata first met with appear to correspond to the red measures under the Permian Sandstone proved in the section at Heaton Park, as communicated by the writer to the Society, and printed in the *Transactions.**

The site of the Whitefield bore-hole is at the north side of Moss Lane, leading to Unsworth, about 1,200 feet from Four Lane-ends, and close to the western side of Messrs. Mark Fletcher and Son's works.

SECTION OF STRATA BORED THROUGH AT THE WORKS OF MESSRS. MARK FLETCHER AND SONS, LIMITED, MOSS LANE, WHITEFIELD.

Description of Strata.	Thickness of Strata. Ft. Ins.	Depth from Surface. Ft. Ins.	Description of Strata.	Thickness of Strata. Ft. Ins.	Depth from Surface. Ft. Ins.
Loamy clay	15 0	15 0	Light - red clay, with		
Boulder-clay	109 0	124 0	bands of light-		
Quicksand	18 6	142 6	coloured red shale	12 0	195 0
Dark clay	5 6	148 0	Beds of flag-rock	5 0	200 0
Gravelly clay, with			Red clay, with thin beds		
stones	5 0	153 0	of shale	8 0	208 0
Strong clay	9 0	162 0	Red shale, with bands		
Gravel	16 0	178 0	of sandstone	50 0	258 0
Mixed clay and gravel	5 0	183 0	Red shale	68 0	326 0

* "Heaton Park Bore-hole, near Manchester, with Notes on the Surroundings," by Mr. Joseph Dickinson, *Trans. M. G. S.*, 1902, vol. xxviii., page 69.

Description of Strata.	Thickness of Strata. Ft. Ins.		Depth from Surface. Ft. Ins.		Description of Strata.	Thickness of Strata. Ft. Ins.		Depth from Surface. Ft. Ins.	
Fine sandstone ...	1	0	327	0	Hard grey shale ...	3	6	757	0
Soft red shale	55	0	382	0	Grey shale	17	0	774	0
Black shale	10	0	392	0	Black and grey shales ...	42	0	816	0
Mixed various-coloured shales ...	37	0	429	0	Shale, with thin hard band ...	2	0	818	0
Various gritty shales ...	35	6	464	6	Grey shale	4	0	822	0
Close-grained hard sandstone ...	10	6	475	0	Gritty grey shale ...	7	0	829	0
Gritty shale	2	0	477	0	Grey shale	27	0	856	0
Close sandstone-rock ...	5	0	482	0	Mixed grey, black, and red shales, with bands of fine sandstone ...	3	0	859	0
Red gritty shale ...	18	0	500	0	Red shale, with beds of sandstone	4	6	863	6
Red shale, with thin beds of fine sandstone ...	24	0	524	0	Grey grit-rock	6	0	869	6
Coarse sandstone ...	4	0	528	0	Mixed red and grey shales	8	6	878	0
Shale, with thin bed of sandstone	3	0	531	0	Dark-red clay	1	0	879	0
Gritty shale	5	0	536	0	Mixed red clay and shales	12	0	891	0
Coarse sandstone ...	14	0	550	0	Grey shale	10	0	901	0
Gritty shale	5	0	555	0	Grey rag-rock	2	0	903	0
Sandstone-rock ...	35	0	590	0	Grey shale	9	6	912	6
Various coloured shales	14	0	604	0	Red clay and shale ...	2	0	914	6
Black soft shale ...	1	6	605	6	Black shale and clay ...	1	0	915	6
Grey shale	2	0	607	6	Grey rag-rock ...	10	0	925	6
Grey shale, with bands of fine grey sandstone	3	0	610	6	Grey rag-rock, with shale	2	0	927	6
Hard grey sandstone ...	2	0	612	6	Grey rag-rock	1	0	928	6
Rag-rock	2	6	615	0	Black shale	0	6	929	0
Grey shale	7	0	622	0	Grey shale, with bands of rag-rock ...	5	0	934	0
Hard band of grey rock	0	6	622	6	Grey shale ...	4	0	938	0
Fine flag-rock	4	0	626	6	Grey and black shale ...	2	6	940	6
Gritty shale	29	0	655	6	Soft black shale ...	2	0	942	6
Very hard band of fine stone	0	6	656	0	Mixed shales	2	0	944	6
Light-grey shale ...	15	0	671	0	Grey shale	12	0	956	6
Black shale, with thin seam of coal ...	5	6	676	6	Mixed grey and black shales	17	6	974	0
Grey shale	38	6	715	0	Rag-rock ...	1	0	975	0
Black shale	2	0	717	0	Grey shale, with thin beds of rag-rock ...	2	0	977	0
Mixed shale	1	6	718	6	Red shale	1	0	978	0
Red gritty shale ...	3	6	722	0					
Very hard gritty shale	2	0	724	0					
Very hard fine sandstone	2	0	726	0	Total depth from surface ...			978	0
Hard gritty red shale ...	0	6	726	6					
Red and black shales ...	1	0	727	6					
Black shale	26	0	753	6					

The details of the above proving by rope-boring are as follows :—
   Well to 27 feet.
   Borehole to 978 feet.
   Tubing 20 inches in diameter, from 26½ feet to 93 feet below the surface.
      ,, 16 ,, ,, ,, ,, 26½ ,, ,, 226 ,, ,, ,,
      ,, 13¼ ,, ,, ,, ,, 40 ,, above bottom of the 16-inch tubing to 467 feet below the surface.
      ,, 11 ,, ,, ,, ,, the surface to 479½ feet.
   The remainder of the hole was untubed.

A cordial vote of thanks was accorded to Mr. Joseph Dickinson and to Messrs. Chapman and Sons, for this addition to the information relating to the geology of the district.

Vol. XXXIV, Plate VII.

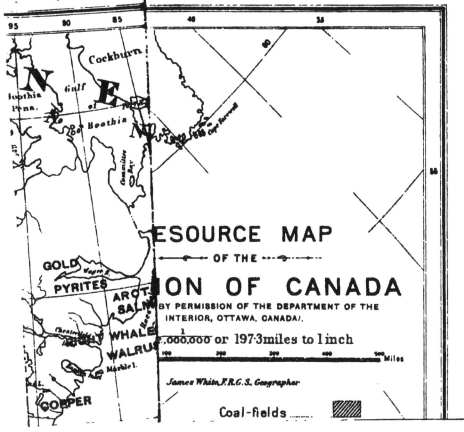

ESOURCE MAP

OF THE

ION OF CANADA

BY PERMISSION OF THE DEPARTMENT OF THE
INTERIOR, OTTAWA, CANADA).

$\frac{1}{,000,000}$ or 197·3 miles to 1 inch

James White, F.R.G.S. Geographer

Coal-fields

MANCHESTER GEOLOGICAL AND MINING SOCIETY.

ORDINARY MEETING,
HELD IN THE ROOMS OF THE SOCIETY, QUEEN'S CHAMBERS,
5, JOHN DALTON STREET, MANCHESTER,
JANUARY 14TH, 1908.

MR. JOHN ASHWORTH, PRESIDENT, IN THE CHAIR.

The following members were elected, having been previously nominated:—

MEMBERS—

Mr. EDWARD T. DAVIES, Colliery Manager, Wynnstay Colliery, Ruabon, North Wales.

Mr. A. GHOSE, Mining Engineer, 42, Shambazar Street, Calcutta, India.

Mr. GEORGE BOOKER HACKETT, Mining Engineer, P.O. Box 3117, Johannesburg, Transvaal.

STUDENTS—

Mr. GEORGE BENNETT ECCLESHALL, 34, Bradford Terrace, Worsley Road, Farnworth, near Manchester.

Mr. THOMAS EMRYS EVANS, 5, Kelnerdeyne Terrace, Rochdale.

Mr. EDWARD FIELDEN PILKINGTON, The Headlands, Prestwich, Manchester.

DISCUSSION OF MR. W. H. COLEMAN'S PAPER ON "BYE-PRODUCTS FROM COKE-OVENS,"* OF MR. W. B. M. JACKSON'S PAPER ON "A BYE-PRODUCT COKING-PLANT AT CLAY CROSS,"† AND OF MR. A. VICTOR KOCHS' "NOTES ON BYE-PRODUCT COKE-OVENS, WITH SPECIAL REFERENCE TO THE KOPPERS OVEN."‡

Mr. CHARLES PILKINGTON (Manchester) said that he felt that those who were colliery proprietors ought to be able to discuss these papers, and he was rather ashamed that he knew so little of the subject. He thought that they would all have to go thoroughly into this question, if they were to make the coal-trade pay in the future, when it became more expensive to get coal;

* Trans. Inst. M. E., 1907, vol. xxxiv., page 331; and Trans. M. G. M. S., vol. xxx., page 197.

† Trans. Inst. M. E., 1907, vol. xxxiii., page 386.   ‡ Ibid., page 398.

26

and he hoped that, in three or four years' time, if this question came up again, they would be able to say something about it, or, at all events, that the younger members would.

The PRESIDENT (Mr. John Ashworth) called attention to the Armstrong vertical bye-product coke-oven, and explained that this oven was constructed on a totally different plan from that adopted generally in the construction of bye-product ovens. Its vertical construction admitted of considerable pressure during the coking period, and its circular form allowed of a larger output than was possible with the horizontal ovens; it was also cheaper in construction. It was efficient in the utilization of the heat generated in the combustion-flues—as might be gauged from the fact that, although the oven was internally very hot, the outside brickwork remained just sensibly warm to the hand.

The strong pressure brought to bear upon the coal in the vertical oven produced denser coke, and the heavy hydrocarbons from the escaping gas passed through the coke above, the result being that a very fine coke, free from gas, was produced. The attention required to be bestowed upon this vertical oven was, he understood, less than was necessary in the case of horizontal ovens; and, in fact, he (Mr. Ashworth) considered that it possessed quite a number of advantages over existing types.

A report upon the new oven had been prepared by Mr. W. Carrick-Anderson, in the course of which that gentleman stated that—

"The coke produced was of a very superior quality. Most of it was of a fine, silver-grey colour; for a 40 hours coke it was hard and strong, and the porosity was very even. Dark-coloured and soft pieces were very few; and little difference, if any, could be detected between the stuff drawn from the upper and lower parts of the oven, showing very uniform heating throughout the flues. The average length of the pieces was considerably greater than can be got from the ordinary horizontal ovens; and, as the width of this oven could, in my opinion, be yet further increased for use with very strong coking coals, it would appear to be possible by its means to obtain bye-product coke at a length hitherto only to be got by the use of the beehive oven."

APPENDIX TO MR. W. CARRICK-ANDERSON'S REPORT ON TRIAL OF ARMSTRONG PATENT COKE-OVEN, NOVEMBER 13TH TO 15TH, 1907.

	Tons.	Cwts.
Weight of washed Durham coal charged into oven on November 13th, 1907, at 6 p.m.	4	16·0
Deduct adherent moisture, 5·74 per cent.	0	5·5
Net charge of air-dry coal	4	10·5

		Tons.	Cwts.
Coke drawn from oven on November 15th, at 10·30 a.m., weighed in waggon		3	13·00
Specimens additional		0	0·75
Total weight		3	13·75
Deduct moisture, 0·62 per cent.		0	0·50
Net dry coke		3	13·25

Dry coke drawn   = 80·9 per cent. of air-dry coal.
Laboratory results = 71·92   „   „   „

ANALYSIS OF COAL.

Volatile matters :	Air-dry. Per Cent.	Moist as charged. Per Cent.
Gas, tar, etc.	27·09	25·52
Water	0·99	6·73
Coke :		
Fixed carbon	66·78	62·91
Ash	5·14	4·84
Totals	100·00	100·00

	Per Cent.	Per Cent.
Total sulphur	0·928	0·874
Caking index	—	18
Coke	—	Swollen and bright.
Ash	—	Reddish-white.

ANALYSIS OF COKE.

	Per Cent.
Moisture	0·62
Gas and tar	1·37
Sulphur	0·776

November 27th, 1907.

W. CARRICK-ANDERSON.

MANCHESTER GEOLOGICAL AND MINING SOCIETY.

ORDINARY MEETING,
HELD IN THE ROOMS OF THE SOCIETY, QUEEN'S CHAMBERS,
5, JOHN DALTON STREET, MANCHESTER,
FEBRUARY 11TH, 1908.

MR. JOHN ASHWORTH, PRESIDENT, IN THE CHAIR.

DISCUSSION OF MR. DAVID M. S. WATSON'S PAPER ON "THE FORMATION OF COAL-BALLS IN THE COAL-MEASURES."*

Mr. JOSEPH DICKINSON (Pendleton) said that the paper described properly the Upper-foot coal-seam and its junction with the Gannister coal-seam near Bacup, forming northwards one seam known as the "Mountain Four-foot." It further described certain peculiar accompaniments, notably that "coal-balls occur in at least two other coal-seams in Lancashire," one of them at Stalybridge, the other near Laneshaw Bridge, about two miles north-east of Colne, probably in the Second Grit. Both of these statements were more than questionable. A reference to Sheet 105 of the 6-inch Geological-Survey map showed that at Stalybridge it was the Upper-foot and Gannister seams, with fossils, which cropped out there. This was confirmed by extensive workings in the Gannister coal, up to about the year 1858. With regard to the coal-workings at Laneshaw Bridge about 100 years ago, he would mention that the mine had since been reopened, and was known as the Mountain Four-foot (consisting of the Upper-foot and the Gannister). The supposition of three similar fossil-horizons therefore disappeared, the supposed three being one and the same.

This horizon was a sure guide, and was well defined in the eastern parts of the Lancashire coal-field. Some of its fossils

* _Trans. Inst. M. E._, 1907, vol. xxxiv., page 177; and _Trans. M. G. M. S._, vol. xxx., page 135.

were found in other parts of the coal-field, both above and below, but not with the peculiarly complete series of calcareous and ferruginous concretions, nor with the same mine-water, which possessed the property of turning a person's skin brown as if with nitric acid.

It was unfortunate that the error had been made in the paper, and he should like the correction to be made in the *Transactions*.

Mr. WILLIAM WATTS (Wilmslow, near Manchester) said that the thanks of the members were due to Mr. Dickinson for his remarks, which were simply an explanation of where coal-balls had been found, apart from the places mentioned in Mr. Watson's paper. Mr. Dickinson wished to give a wider area for the location of these particular nodules or balls.

Many years ago he (Mr. Watts) had found a coal-ball in one of the lower coal-seams, under Gorse Hall, Stalybridge, which had the appearance of a round-headed hammer at both ends, and was composed internally of gritty material with an external film of carbonaceous matter. Both ends were practically equal in size, and evenly narrowed down in the middle.

———.

Mr. JOHN McARTHUR JOHNSTON read the following paper on "Stirling and other Water-tube Boilers as applied to Collieries and Coking-plant " : —

STIRLING AND OTHER WATER-TUBE BOILERS AS APPLIED TO COLLIERIES AND COKING-PLANT.

BY JOHN McARTHUR JOHNSTON.

The writer does not propose to deal with the historical aspect of the water-tube boiler, but to confine himself to the modern water-tube boiler, and the results obtained from its use within recent years.

The success which attended well-designed water-tube boilers led to the invention and production of a large variety of boilers of this type, some of which were possessed of useful and desirable features.   Three of the best known water-tube boilers are the Babcock and Wilcox (fig. 3), the Stirling (figs. 4, 5 and 6), and the Climax; the first-mentioned having straight tubes, with which the members no doubt are familiar, the Stirling slightly bent tubes, and the Climax curved tubes.

The Climax (or Morrin) boiler consists of a vertical drum, into which are expanded a large number of loop-like tubes. These tubes extend practically from top to bottom of the drum, and form the chief portion of the heating-surface.   A firebrick-lined casing surrounds the boiler and forms the combustion-chamber, which is circular in form like the firegrate, the water-level being at a point below the top rows of tubes.   This boiler occupies small floor space, and gives dry steam, but the circulation has no definite path to follow, and the drum is subjected to considerable variations in temperature.

In dealing with water-tube boilers in this paper, the author's object is to place before the members results obtained in actual practice, not merely test-figures showing what has been done during an hour's test here and there; and to show the economy of water-tube boilers when compared with cylindrical boilers under the same conditions, especially where fairly large installations are concerned.

A few comparisons of the work done by water-tube boilers

against those of the cylindrical or internally-fired type will serve to illustrate the advantages to be obtained by using boilers of the water-tube type.

Lancashire, Cornish, egg-ended, and other externally-fired boilers, have for many years been in extensive use for steam generation in iron- and steel-works, coal- and iron-mines; but,

FIG. 1.—WATER-TURBINE   FIG. 2.—CHAIN-SCRAPER
    TUBE-CLEANER.   TUBE-CLEANER.

with the march of progress, water-tube boilers have in recent years been installed in a large number of such works. The increasing number of boilers, of the water-tube type, which have been put down at iron-works and mines have testified to the advantage which they have over the older type of boiler, and call for more serious consideration than they have usually received from those who have hitherto shunned them as being doubtful or troublesome.   The experimental stage has long since been left behind, and hundreds of steam users are now convinced that they have made a big step forward by installing water-tube boilers when extensions of plant had to be considered.

The writer would refer to Mr. Bryan Donkin's work on " The Heat Efficiency of Steam Boilers,"[*] which is valuable as containing a record of 400 carefully made boiler-tests under various conditions.   This book contains a large number of tests for each type of boiler, and shows the average results as representing fairly well the efficiency of the different types of boilers.

The latest edition of this book is dated 1898, and although the cylindrical type of boiler has been improved very little since the book was compiled, the water-tube boiler has made very considerable advancement, in numbers, in detail, and in design.

Of the 107 tests of Lancashire boilers recorded in this book, the efficiency averaged 62·4 per cent.   Unfortunately, Mr. Donkin does not give an equal number of tests of the water-tube boilers; but at the head of the list were shown water-tube boilers with an efficiency of 77·4 per cent.

In comparing the Lancashire boiler with a properly designed water-tube boiler, it is safe to say that the efficiency of the latter represents a saving in coal of about 15 per cent.   Without a careful study of all the causes and effects that lead to this result, the statement appears to be rather a startling one; nevertheless, many large works, where both types of boilers are in everyday use, have shown conclusively that such saving was effected by means of water-tube boilers.   As an example, the writer may mention that a boiler which burnt 10 tons of coal per day, with coal at 10s. a ton, would represent a saving of £170 per annum, and in a very few years' time the water-tube boiler would have paid for itself.

Circulation.—The circulation in a cylindrical boiler of the Lancashire or Cornish type is by convection currents.   In the Stirling boiler the circulation is rapid and follows a definite and pre-determined path, and is so arranged that the dirt, sediment, and matters held in solution, are deposited in a drum placed as far as possible from the hottest gases or furnace.   One

* *The Heat Efficiency of Steam Boilers: Land, Marine, and Locomotive,* by Bryan Donkin, 1898.

very important feature of the externally-fired water-tube boiler being that the combustion-chamber is enclosed on three sides by firebrick walls, the initial temperature in the combustion-chamber is consequently maintained at a very high rate, which ensures practically complete combustion before the gases come into contact with the heating surfaces of the boiler. The temperature in the combustion-chamber is, therefore, much higher than in the internally-fired boiler, in which the fire is surrounded by the cooling surfaces of the boiler. On the grate of the firebrick-lined furnace, very low-class fuel can be burned with high efficiency, a point of very great importance, as will be seen from the figures given below, which are taken from actual practice.

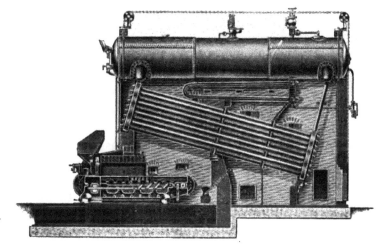

FIG. 3.—BABCOCK AND WILCOX WATER-TUBE BOILER.

In a works with economical mill-engines of 500 indicated horsepower, and Lancashire boilers, coal costing 12s. per ton was used, with a consumption of 2 pounds per indicated horsepower, which amounted to £2 8s. 3d. per day, or £723 per year of 300 working days.

Water-tube boilers were substituted, with special grates for burning " gum " or " smudge " costing 4s. per ton. The consumption of this fuel was 2½ pounds per indicated horsepower, the cost per day of 10 hours being £1 2s., or £330 per annum, showing a saving of over 50 per cent. for the same amount of steam produced. Naturally, to burn fuel of this class ample boiler-power is required, but the increase in capital expenditure would be in a very short time repaid by the saving in running-costs.

Choice of Boiler.—The choice of a boiler is a business proposition, and sentiment should not be allowed to enter into it. The purchaser naturally takes into account the business aspect of the transaction, and the following points require due consideration:—(1) Safety, (2) cost of maintenance, (3) cost of cleaning, (4) fuel economy, and (5) first cost. The author has placed these points in the order named, that being the order of their importance in the minds of most users.

In choosing a boiler the steam user would probably ask: "What steam-generator will earn the most money, at the least cost of coal and maintenance?" The Lancashire boiler, which is a popular and well-tried steam-generator, is probably better known than any other form of boiler. Generally speaking, when new plant is under consideration, it becomes a simple matter to fix the number of Lancashire boilers required; but it is equally wellknown that this type of boiler has also its faults, and it may be assumed that its defects have come to be taken as a part of the boiler itself, and therefore, owing to custom, do not receive any serious consideration when making choice of the boiler to be installed.

It is often stated in favour of the Lancashire boiler that all of the internal surfaces are accessible for cleaning and examination, that it has a large volume of water, and large steam space; but it also has a small ratio of heating-surface to grate-area, the ratio being about 25 to 1; consequently the gases pass away to the flue at a high temperature, and therefore the efficiency is low. The ratio of heating-surface to grate-area in a water-tube boiler is frequently about 65 to 1.

Design.—In the cylindrical type of boiler, owing to its design there is a large area of heating-surface present for the collection of deposits, which constitutes a defect. The design of a Lancashire or Cornish boiler permits of various stresses and racking strains being set up in different parts; when the various temperatures applied to the several parts of the furnace-tubes and shell are considered, it will be seen that totally different rates of expansion take place, resulting frequently in grooving, and other defects.

It might be noted that if this boiler be worked at low rates of evaporation, say from 4,000 to 5,000 pounds per hour, the efficiency rises to from 55 to 65 per cent.; but nowadays boilers are rarely bought or sold to work at a low rate of evaporation, the tendency rather being to overload the boiler: consequently, the gases leave the boiler at a high temperature, which has the effect of reducing the efficiency to about 50 per cent.

In the event of overheating of the furnace-tube of the cylindrical type of boiler, due to excessive scale, or shortness of water,

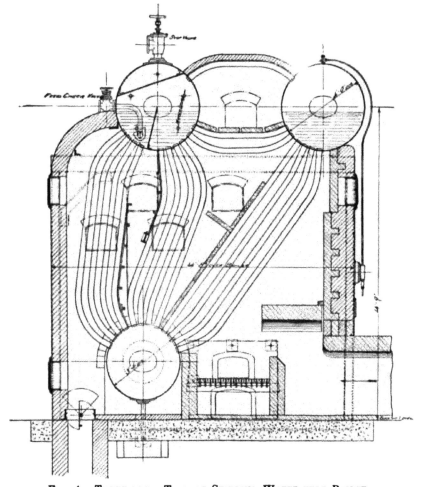

FIG. 4.—THREE-DRUM TYPE OF STIRLING WATER-TUBE BOILER.

from neglect, the tube becomes red-hot and pliable; and, being unable to withstand the pressure, it collapses, and probably ruptures, with serious results. The boiler is then necessarily laid off for a considerable time for repairs, which moreover necessitate the services of a skilled boilermaker.

One special feature of the water-tube boiler is safety, and

should a rupture in a tube occur, it is purely local in its action, and only a limited amount of water and steam contained in the tube can immediately escape, and consequently, no serious damage to property occurs; while in the cylindrical boiler there is a large volume of water, which, if suddenly released, might cause serious loss of life and damage to property.

If ever a Lancashire boiler, 30 by 8 feet, working at, say, 160 pounds per square inch (the usual pressure nowadays), were to rupture in the shell, terrible wreckage and loss of life might occur. How disastrous Lancashire boiler explosions have been, and might be, is wellknown, and may be illustrated by the fact that some years ago a Lancashire boiler, 7 feet in diameter, and working under a pressure of 80 pounds per square inch, exploded at Batley (Yorkshire), causing the loss of 23 lives, and doing immense damage.

In the cylindrical type of boiler the ratio of heating-surface to grate-area is practically a fixed quantity: the grate may be lengthened, though not increased in width; but in a water-tube boiler the ratio can be varied at will, to obtain the best economy from the different classes of fuel.

In the event of repairs being necessary in a boiler of the water-tube type, the works' mechanics, not necessarily skilled men, can do all that is required. At the worst, supposing that a tube becomes split, or bulged, through scale or neglect in other ways, such as oil being allowed to come into contact with the heating-surfaces, all that need be done is to cut out the defective tube and put in a new one, expand both ends, fill up the boiler, and go ahead again. This can be done in a short space of time.

In the cylindrical type of boiler the surfaces in the external flues rapidly became coated with soot, which is an exceedingly bad conductor of heat, and there is only one means by which these surfaces can be cleaned, namely, by entering the flues and cleaning by hand. In the water-tube boiler, all deposits of flue-dust, fine ash, etc., can be removed at any time by means of the steam-lance, applied through sooting-holes and doors provided for that purpose, and without interfering with the work of the boiler.

Combustion.—The space for combustion in the cylindrical type of boiler is very limited; the gases rising from the fuel

(freshly fed) are cooled on the furnace-plates, and consequently perfect combustion is not attainable, and black smoke issues from the chimney.   The above remarks apply especially when the boiler is working under forced conditions.   With hand-firing much excess air passes through the thin places in the fire, and

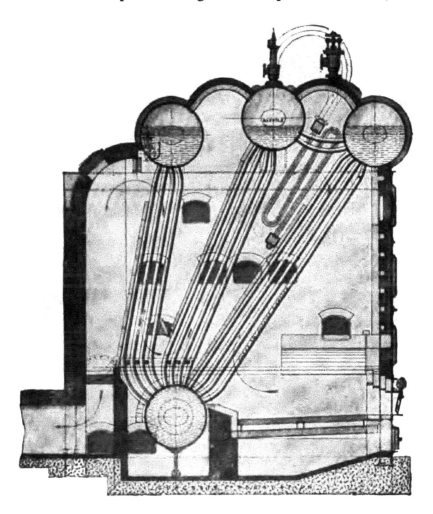

FIG. 5.—FOUR-DRUM TYPE OF STIRLING WATER-TUBE BOILER.

complete mixture of the gases does not take place.   Unless a proper mixture of the gases takes place, the efficiency must be low.

The design of the water-tube boiler exhibited before the members permits of a very spacious combustion-chamber, which, as before mentioned, is lined with firebrick, with a fire-brick arch immediately over the fire; this arch becomes white-hot, and serves various useful purposes, namely, assisting the ignition of the freshly-fed fuel, igniting the gases (requiring a high temperature), also heating the air which rushes in when the firing door is opened, and thus preventing any of the dele-terious effects resulting from the sudden impingement of cold air on the heating-surfaces. It will be readily admitted that the firebrick arch is entitled to special attention, and that it is a most useful factor in the economy of the water-tube boiler; being maintained at a great heat it ensures practically perfect com-bustion, as the gases are completely consumed before they come into contact with the heating-surfaces of the boiler.

It is admitted that perfect combustion can only take place in a properly-designed firebrick-lined furnace. It is also well-known that no type of boiler attains perfection in its work, but reason dictates that we may safely choose the boiler which most closely approaches perfection.

In modern practice, it is desirable to work with high-pressure steam-engines, which necessitates boilers capable of supplying a fixed quantity of high-pressure steam economically. In such a case the cylindrical type of steam-generator is required to be built of plates of great thickness, in order to withstand the high pressure, and thus the initial cost is considerably increased.

The writer would mention that he has selected the Stirling boiler for comparison with other types, because he is more familiar with that type of water-tube boiler.

This boiler (fig. 5) consists of three upper steam and water-drums, and one lower or mud-drum, with various modifications (figs. 4 and 6), connected by three banks of tubes slightly bent so as to enable them to enter the drums radially; the bends also serve to take up the expansion and contraction due to the varia-tions of temperature at the different parts. The front and central drums are connected by bent tubes in both steam and water-spaces, but the rear drum is connected to the central drum in the steam-space only. The three upper drums rest on a steel framework, which constitutes the whole support of the boiler. This is a very important feature, as the arrangement allows free play to the lower part of the boiler, when subject to varying temperatures, and consequently there can be no racking strains at any part.

Course of Gases.—The gases, after leaving the region of the grate and arch, travel up the first bank of tubes, guided by a baffle-wall, cross over to the second bank of tubes, and travel a downward path to the bottom of the second baffle-wall, thence crossing to the third bank of tubes, along which they are made to pass upwards and to the chimney.

Course of Water.—The feed-water enters at the rear drum, and flows down the rear bank of tubes to the mud-drum ; during

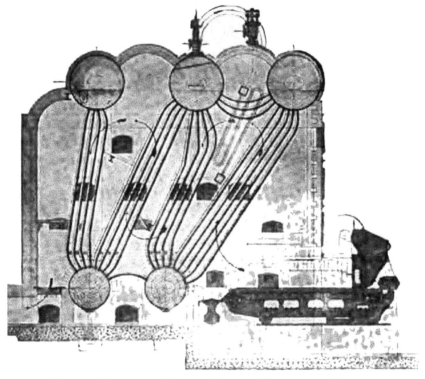

FIG. 6.—FIVE-DRUM TYPE OF STIRLING WATER-TUBE BOILER.

this passage sufficient heat is taken up by the water to render insoluble any lime-salts contained therein, which with the mud are deposited in the bottom drum   From the bottom drum the water and steam rush up the front bank of tubes to the front drum, from which the water then flows to the central drum, and thence downwards to the mud-drum, thus creating a flow in a circular tour at a rapid rate.

Economy and Circulation.—Theoretically, one pound of fuel of a known heat value is capable of evaporating a certain fixed quantity of water, but, in practice, the theoretical evaporation is never reached, for obvious reasons; and it follows that the most efficient boiler is that which will absorb or utilize the most heat from the fuel.   This, expressed as a percentage of the theoretical amount, varies with different types of boilers from 40 to over 80 per cent.

Rapid circulation is a potent factor in the economy of heat-transmission from the furnace gases to the water in the boiler, the evaporation depending on the difference in temperatures between the furnace gases and the water in the boiler; the greater the difference between the two temperatures, the more heat is transmitted through the plate; consequently more heat is absorbed by the water, and more water is evaporated.   When the circulation is sluggish, the bubbles of steam adhere too long to the plate, the difference of temperature is less, and evaporation is slow.   By increasing the circulation, more heat is transmitted, as the water and steam do not remain so long in contact with the metal; economy is therefore obtained by increasing the circulation, an advantage that the water-tube boiler has over the cylindrical type of boiler, in which, as is well known, the circulation is bad.

Safety.—Returning to the design and construction of the Stirling boiler, the writer would point out that as all the surfaces are either cylindrical or spherical, no stays are required.   No castings are employed, and the whole arrangement is practically elastic; trouble from leaky joints or tube-plates, therefore, should never exist.

Another important feature is that the riveted joints are not exposed to the action of the fire.   There is ample room for expansion and contraction of the parts where the greatest heat is applied.

Assuming that an installation of eight Lancashire boilers, 30 by 8 feet, is required at a working pressure of 200 pounds with an economizer, these would occupy, including firing platforms, 5,760 square feet of ground.   Water-tube boilers to take the same duty would occupy a space of 2,436 square feet of ground, including economizer and firing platforms, being half the area.

The approximate price of such Lancashire boilers and economizer would be £4,500, while the water-tube installation would cost approximately £3,500, including economizer.   The net saving in capital outlay would be £1,000, and in addition there would be a saving of at least 50 per cent. on the expenditure in steam and feed-piping.   Estimating the cost of coal at 10s. per ton for a period of 50 weeks of 60 hours each, the saving per annum would represent 24 per cent. on the capital outlay.   The saving in fuel, capital expenditure, piping, and rates and taxes for floor space, would represent a considerable one over the Lancashire type of boiler.

FIG. 7.—STIRLING BOILER FED BY BLAST-FURNACE GAS.

Feed-water.—The writer is of opinion that all feed-water to boilers should be free from matter which might be deposited inside; but in many cases this state of affairs cannot readily be obtained.   Very many Lancashire boilers are seldom, if ever, properly scaled, because of the long time necessarily occupied in the operation.   For removing scale from the tubes of water-tube boilers, the author exhibits two sets of apparatus.   One is a simple mechanical apparatus by which the internal surfaces can be cleaned in a very short time and at a small outlay (fig. 1). It consists of a small water-turbine connected to cutters and chippers, operated by water-pressure of about 100 pounds per square inch. The bends in the tubes present no obstacle to this machine. The second device (fig. 2) is a scraper for removing comparatively soft scale, and drawn through the tubes by means of a chain. The difficulty of cleaning water-tube boilers is often a point urged against this class of boiler, but only by those without experience of them, or by those who have handled some of the earlier types of water-tube boilers; but there is no reason why this simple mechanical contrivance should not clean quickly and thoroughly all the internal surfaces. The turbine should be connected by a $1\frac{1}{2}$-inch hose-pipe to the feed-pumps or water-mains, and entered at the top ends of the tubes. The turbine, rotating at a high speed, causes the arms to fly out and round, so that the cutters remove the scale which is washed down in advance of the turbine into the bottom drum, whence it can be easily removed by hand.

Water-tube boilers can be designed to work at high pressures with very moderate thickness of plates and tubes, also in very large units, up to 40,000 pounds evaporation in one boiler. This involves only one set of mountings, piping, etc., and occupies correspondingly less space than a number of cylindrical boilers to give a like duty. Another feature which may be noted is that cylindrical boilers, with their attendant economizer (an early form of water-tube boiler), require the services of more men than one large water-tube boiler, and running costs are therefore higher.

The first cost of a water-tube installation is lower, the attendance-costs under many conditions are also lower, and, the coal-bill being less, the total saving is therefore considerable.

Removing the scale from the heating-surfaces of a cylindrical boiler is frequently a tedious and laborious task, attended with serious expense, and has to be effected by hand-chipping. The narrow water-spaces between the flue-tubes and the shell are so difficult of access, that it is almost impossible to clean them properly.

Water-tube Boilers in connection with Coke-ovens.—During recent years a considerable number of water-tube boilers have been installed to utilize the waste heat from coke-ovens, and the results obtained have warranted their more general adoption, these results as regards efficiency not having been equalled by any other type of boiler.

The hot gases, on entering a water-tube boiler, become split up into a large number of small streams, and the thin metal of the tubes permits of rapid transmission of the heat to the water ; and, as the rapid circulation prevents the steam-bubbles from remaining any length of time on the heating-surfaces, the efficiency is maintained at a high rate.   It will be seen on reference to figs. 4, 5 and 6, that the gases enter at the bottom of the front bank of tubes.   An ordinary firegrate is fitted for hand-firing, for use when the ovens are shut off.

Assuming the average amount of water evaporated in a Lancashire boiler per pound of coal coked at from 1 to 1·4 pounds, the average in a water-tube boiler is about 1·3 to 1·7 pounds, being an increase of about 30 per cent.

The mean evaporation over a four days' test of a water-tube boiler fired by waste-heat from coke-ovens (as shown below) was 1·7 pounds of water for each pound of coal coked.   The temperature of the gases leaving the ovens was 2,000° Fahr., and the temperature of the gases entering the boiler 1,700° Fahr. ; the boiler-inlet was 29 feet from the coke-ovens, and, in consequence, there were large radiation-losses in the flue.

The following is a comparison of the work of one water-tube boiler with that of two Lancashire boilers under similar conditions, at the Victoria Garesfield colliery, Rowlands Gill, near Newcastle-upon-Tyne :—

	Stirling Boiler.	Lancashire Boiler.
Description of boilers	1 No. 12, Class A ...	2 of 28 by 8 feet.
Number of coke-ovens	22	37
Heating-surface	1,611 square feet ...	1,796 square feet.
Water evaporated from and at 212° Fahr. per pound of coal coked	1·7 pounds	1·33 pounds.
Temperature of gas entering boiler	1,720° Fahr.	1,700° Fahr.

Owing to reasons before mentioned, the water-tube boiler has the advantage of being able to abstract more of the heat from the gases than the cylindrical boiler.

In 1907, a test was made over a period extending from May 21st to 28th, at Chopwell colliery, County Durham, on a water-tube boiler fired by waste-gases from coke-ovens, and the following results were obtained :—

Duration of test	168 hours.
Heating-surface of boiler	4,170 square feet.
Coke-ovens (beehive) ... . per oven	134·5 square feet.

Form of furnace :	47 coke-ovens gas-fired.
Heating-surface of superheater	295 square feet.
Average boiler-pressure by gauge	157·4 pounds.
,, temperature of steam at the above pressure	370° Fahr.
,, of superheated steam	476·2° Fahr.
,, amount of superheat	106·2° Fahr.
,, temperature of feed-water entering the boiler	61° Fahr.
,, of flue-gases leaving the boiler	500° Fahr.
,, ,, of gases entering the furnace ...	2,228·1° Fahr.
,, ,, of the external air	50° Fahr.
Total quantity of coal as fired	1,389,696 pounds.
Quantity of coal as fired per hour	8,272 pounds.
Draught	0·5 inch.
Total weight of feed-water to boiler (actual)	2,136,700 pounds.
Equivalent water evaporated from and at 212° Fahr.	2,734,976 pounds.
Factor of evaporation	1·28.
Water evaporated per hour	12,718·4 pounds.
Equivalent evaporation per hour, from and at 212° Fahr.	16,279·5 pounds.
Evaporation under actual conditions per pound of coal carbonized	1·537 pounds.
Equivalent evaporation from and at 212° Fahr. per pound of coal carbonized	1·968 pounds.

Blast-furnaces.—Water-tube boilers have also been fired by blast-furnace gas (fig. 7). The gas enters at a point immediately under the incandescent arch, the arch being dropped at the back to ensure better mixture of the air and gas. Any accumulation of dust can easily be removed by means of the steam-jet applied through the cleaning-doors, without shutting down the boiler.

Explosion-doors are sometimes fitted to relieve the brick-setting of any strains, should an explosion of gas occur.

At an iron-works in Scotland, comparative tests were made in 1907 of Lancashire and water-tube boilers, fired by waste-heat from puddling-furnaces (fig. 7), showing the following results :—

	Lancashire Boiler.	Water-tube Boiler, No. 1.	Water-tube Boiler, No. 2.
	Pounds.	Pounds.	Pounds.
Pressure per square inch	50	50	110
Water evaporated per pound of coal ...	5·171	5·65	5·66
Water evaporated per pound of coal from and at 212° Fahr.	5·420	5·93	6·02
Water evaporated per hour	2,923	3,032	3,016
Water evaporated per hour from and at 212° Fahr.	3,069	3,183	3,212

In this instance the water-tube boilers were Stirling boilers.

Taking the "from and at" evaporations per pound of coal, from these figures it will be seen that when the water-tube boiler was working at 50 pounds per square inch the evaporation was 9·4 per cent. higher in the water-tube boiler than in the Lancashire boiler.   When working at 110 pounds per square inch, the evaporation of water per pound of coal in the water-tube boiler was 11 per cent. higher than in the Lancashire boiler.

Coal-refuse.—When designing steam-plant for a new colliery, it would be advisable to bear in mind the ability of a water-tube boiler to burn and generate steam from the unsaleable materials taken out of the pits.   In a properly-designed grate of a water-tube boiler, almost any refuse from coal-mines can be used, including old pit-props, other damaged and rejected timber and road-sweepings, and quite a noticeable saving can be effected.

In conclusion, the writer trusts that he has been successful in showing the advantages which may be claimed for the water-tube boiler over the older type, also that a boiler with these advantages can be designed possessing great safety with simplicity of construction, and an economy unrivalled by that of any other type of stationary boiler.

———

Mr. G. B. HARRISON (H.M. Inspector of Mines, Swinton) moved a vote of thanks to Mr. Johnston for his paper.

Mr. CHARLES PILKINGTON (Manchester), in seconding the motion, asked whether the use of the chain-stoker did not cause a great loss of fuel.   He knew that such was the case in many instances.   With regard to the method of cleaning the tubes, it seemed to him that the wearing process might be very severe.

Mr. J. McARTHUR JOHNSTON (Manchester), in acknowledging the vote of thanks, said that with regard to the chain-grate his experience showed that fuel was saved and not lost by its use. The coal was, by its use, laid in more regularly, and used more economically than could be done by hand-firing, and any coal that fell between the bars could be recovered.   He was also of opinion that a much better mixture of the gases was obtained in this way than was possible by hand-firing.   With regard to the method of cleaning, the wear of the tubes by this process was slight and negligible; indeed, the methods of chipping and cleaning a Lancashire boiler were more destructive, in his opinion, than those in use for cleaning the tubes of a Stirling water-tube boiler.

The further discussion of the paper was adjourned.

MANCHESTER GEOLOGICAL AND MINING SOCIETY.

———

ORDINARY MEETING,
HELD IN THE ROOMS OF THE SOCIETY, QUEEN'S CHAMBERS,
5, JOHN DALTON STREET, MANCHESTER,
MARCH 10TH, 1908.

———

MR. JOHN ASHWORTH, PRESIDENT, IN THE CHAIR.

———

The following gentlemen were elected, having been previously nominated : —

MEMBER—
Mr. J. E. H. LOMAS, Mechanical and Mining Engineer, 32, Great St. Helens, London, E.C.

MEMBER, NON-FEDERATED—
Mr. WILLIAM CLIFFORD, Mining Engineer, Jeannette, Pennsylvania, United States of America.

———

Mr. GEORGE H. WINSTANLEY read the following paper on " Accidents in Winding, with special reference to Ropes, Safety-cages, and Controlling Devices for Colliery Winding-engines" : —

ACCIDENTS IN WINDING, WITH SPECIAL REFERENCE TO ROPES, SAFETY-CAGES, AND CONTROLLING DEVICES FOR COLLIERY WINDING-ENGINES.

By GEORGE H. WINSTANLEY, F.G.S., M.Inst.M.E., Lecturer in Mining, Manchester University.

The official statistics for the year 1907, relating to accidents in mines,* show that at the collieries of Great Britain during last year there were seventy-one shaft accidents, which resulted in the loss of ninety-seven lives—a figure considerably in excess of the average of the last twenty-five years, and one which has only been exceeded three times, and equalled twice, during that period. These records do not take into account those occurrences—serious enough in themselves—which fortunately were not attended by personal injury or loss of life.

That the subject of shaft accidents and safety in winding has attracted the attention and engaged the interest of the mining community at home and abroad, is evinced by the evidence submitted to the Royal Commission on Safety in Mines, now engaged upon its deliberations, as well as by the Report, published at Pretoria in 1907, of the Transvaal Commission, which was appointed to enquire into the use of winding-ropes, safety-catches, and appliances in mine-shafts.†

The writer therefore feels that no apology is needed for bringing the subject before the Society, and he does so in the hope that its consideration may be productive of valuable discussion, and possibly beneficial results.

When the Rope Breaks.—At the outset, attention may be directed to a somewhat common but mistaken impression that many winding accidents are due to the breaking of the rope. The prevalence of this idea may to some extent be gauged by reference

---

* *Mines and Quarries: General Report and Statistics*, 1907 (advance proof, subject to correction), page 4.

† *Report of a Commission appointed by His Excellency the Lieutenant-Governor [of the Transvaal] to enquire into and report upon the Use of Winding-ropes, Safety-catches and Appliances in Mine-shafts*, 1907, with Minutes of Evidence.

to the records of new patents, published in the technical press, from which it is abundantly evident that the habitual inventor invariably includes in his repertoire a " Safety-device for Mine-cages, Hoists, and the Like," a fearful and wonderful contrivance to accomplish all manner of marvellous things " when the rope breaks."

The writer has observed close upon a hundred of these appliances, but quite the most weird is one referred to in the following extract from a patent specification, published some little time back in one of the mining newspapers : —

9013 (1906). — " Safety-appliances for Miners Descending into or Leaving the Shaft. — This invention refers to a safety-appliance to be used with ascent and descent cages for mine-shafts, of the kind wherein the cage is attached to a rope which is coiled and uncoiled on a drum by the winding-engine. On a supplemental drum, which in certain circumstances may be connected with the main drum for the cage-rope, a separate safety-rope is provided, which the miner fastens around his body. This rope is uncoiled to exactly the same length as the cage-rope. If the latter should break, the miner who has the safety rope attached to his body will be suspended, and can be let down or drawn up as desired."

The Transvaal Commission investigated between eighty and ninety " safety-"appliances of one sort or another, of which seventy-two were of the " when-the-rope-breaks " type. Of these latter, out of seventy-two, only two were reported upon with anything like definite approval; and even these, on careful examination, are not likely to commend themselves to practical men, nor are the views of the Commission with regard thereto very convincing.

The fact of the matter is that the breaking of a colliery winding-rope, whenever it does occur, is more likely to be the result than the cause of an accident. The writer has been unable to discover a single instance in which the breaking of a winding-rope, in the ordinary course of working, was positively shown to be, primarily, the cause of an accident—except in those cases where it was only too clear that the rope had been neglected or wrongfully treated.

It will no doubt be readily agreed that a rope cannot and will not break without a cause, and that if it is made of the best material, properly proportioned to the load which it has to carry, and fairly treated, it will not break, unless it be subjected to a strain exceeding its tensile strength. It is only too obvious, for example, that a rope having a tensile strength of 100 tons, must have at least that amount of strain put upon it before it will break, unless it has been seriously weakened by improper use. It is equally obvious, too, that if such a rope does actually break, either the normal conditions of working must have been very greatly exceeded, or the rope must have been neglected or sub-jected to improper treatment.

The most effective course to pursue in order to prevent ropes from breaking is that which will prevent the normal conditions of working from being seriously exceeded, at the same time giving to the rope that constant and careful attention which its import-ance demands.   The adoption of a so-called " safety-cage " is not the remedy.   Safety-cages do not prevent serious strains upon the rope, neither do they ensure against neglect; but on this point more will be said hereafter.

It appears probable that many occurrences described as acci-dents are, strictly speaking, not so. An accident is an occurrence of an entirely unforeseen character, usually the result of a com-bination of unexpected circumstances which it was beyond human power to anticipate and provide against.   Any occurrence which might have been prevented, either by the exercise of reasonable foresight or precaution, or the possession of reasonable know-ledge and skill, cannot, the writer ventures to suggest, truly be regarded as an accident.

It is, of course, quite possible that some of those seventy-one occurrences recorded as shaft accidents were of this description, and might perhaps have been obviated by the exercise of greater care on the part of someone, or even a slight knowledge of the laws of motion, the elementary principles of which should be understood by all those having charge or control of machinery in motion, especially colliery machinery.   The fact that so many accidents in which, amongst other things, the rope breaks, are regarded as accidents caused by the rope breaking, is, to the writer's mind, convincing that the real or primary cause of the accident is often overlooked.

Of course, shaft accidents include other occurrences than those associated with the breaking of a rope ; but the scope of the present paper is limited to the matters referred to in the title, the writer's desire being to consider the causes which may lead, or tend to lead, to the breaking of winding-ropes, and the precau-tions to be taken to prevent such occurrences.

The Factor of Safety in Colliery Winding-ropes.—In good British colliery practice it has become almost universal to allow a factor of safety of 10 for colliery winding-ropes. The Transvaal Commission considered this question, and appears to have come to the conclusion that a factor of safety of 6 was a sufficiently liberal allowance.* The Transvaal Commission may be right in its conclusion, at least so far as conditions in South Africa are concerned; but the writer would be sorry to think that a similar conclusion would be accepted in this country, or that the more liberal figure of 10 were to be reduced. Indeed, it would not be amiss, since legislation is likely to result from the Royal Commission's deliberations, if the factor of safety in colliery winding-ropes were to be defined by law, just as the strengths of boiler-plates and other materials are regulated by the Board of Trade. In speaking of the factor of safety of 10, it should be clearly understood that the intention is to provide, at the outset, a rope which shall have a strength of not less than ten times the gross load to be raised, this gross load being made up of the weight of the cage, the tubs with their full weight of coal, and the weight of the rope itself.

The steel wire used in the manufacture of the best qualities of winding-ropes has usually a breaking strength of from 100 to 120 tons per square inch of sectional area. The writer offers for what it is worth a simple rule, by which the weight in pounds per yard, and the size of a winding-rope suited for a given load and depth of shaft, can be calculated with fairly accurate results. It must, however, be understood to refer only to round wire ropes of the usual construction, made of high-grade wire having a tensile strength of about 120 tons per square inch. It is based upon the fact that in round numbers the weight of 20,000 yards of such a rope, of any size, is practically equal to its breaking strength, and that the weight of 2,000 yards therefore represents its safe working-load. The rule is as follows:—Take the maximum suspended load in pounds (that is, the total weight of the loaded cage) and divide by the difference between 2,000 and the depth of the shaft in yards. Thus, take a load of 8 tons, and a shaft 600 yards deep:

$$\frac{8 \times 2,240}{2,000-600} = \frac{17,920}{1,400} = \text{say, } 12\tfrac{3}{4} \text{ pounds per yard.}$$

* *Report of a Commission appointed by His Excellency the Lieutenant-Governor [of the Transvaal] to enquire into and report upon the Use of Winding-ropes, Safety-catches and Appliances in Mine-shafts, 1907, with Minutes of Evidence, page xxx.*

The nearest rope to this, in the list of a wellknown manufacturer, under the head of " improved plough-steel " is one weighing 25 pounds per fathom, of which the breaking strength is stated to be 119 tons; the working-load will therefore be 11·9 tons.

Six hundred yards, at 25 pounds per fathom, amounts to 3·35 tons, giving a gross load of 11·35 tons, showing that the rule in this case gives a correct result. The square root of the weight in pounds per fathom is equal to the circumference of the rope in inches.

Certificates of Test.—When a new rope is purchased, the maker should be required to furnish a certificate of test, which should include both the wire used and a sample-length of the finished rope. The tests should be of as comprehensive a character as possible, embracing tensile, torsional, bending, and fatigue tests; and all the details should be fully set forth on the certificate.

The demand for so-called cheap ropes—that is to say, ropes which can be bought at a low figure per ton, but in practice prove really to be costly on account of their comparatively short life—has led to the manufacture of ropes that can be purchased at a price which, if the purchaser would only reflect for a moment, is very little higher than the cost of best-quality steel bar, and therefore leaves nothing for the various processes intervening between the raw material and the finished rope.

It is surely patent to the most casual observer that, if a wire rope is to be manufactured from the best material, and by the most approved processes, the cost and the value of the finished rope must be considerably in excess of the value of the steel in the earlier processes through which it has to pass. Nevertheless, the writer has been assured by a rope-maker of high repute that he is actually able to sell a certain quality of steel (from which wire is made) in the form of octagonal bars or tool-steel at a price not less than that which is asked for certain makes of steel-wire ropes, and reference to the quotations of the metal market will confirm this. When one remembers that the actual cost of making a rope, after the wire-drawing stage is completed, is the same no matter whether wire of good or of inferior quality is employed, we realize the true significance of this comparison, and wonder what kind of material the cheap rope is made of, and by what processes it has been converted into wire.

All the operations which tend to make good material into good wire are operations which add to the cost, and if the purchaser insists upon a low cost, that is, a so-called " cheap " rope, he cannot in reason expect that he is getting the advantage either of good material, or of the best processes of manufacture. For example, the best steel for the manufacture of high-grade wire is made from Swedish iron, and.one has only to look at the current market prices to see how this variety compares with others in point of cost. The subsequent processes may or may not include forging under the steam-hammer. If the hammer is dispensed with, an element of cost is eliminated, whilst a degree of quality is sacrificed. Cogging, rolling into rods, annealing, and cleansing precede the all-important process of wire-drawing, that extraordinary yet apparently simple process by which a steel rod is forcibly drawn through a hole smaller than itself, resulting in a rod or wire of reduced sectional area, but of greatly increased strength.

It is at this stage that expense may again be avoided by sacrificing quality in the finished article. A rod of given section may be drawn down to a wire of the desired tensile strength either by several gradual steps, or in a smaller number of steps with a greater proportionate reduction of sectional area at each stage. The resulting wire may have the same tensile strength, it may give the same torsional and bending tests, and it will obviously be cheaper to produce; but its life, under ordinary working-conditions, will be shorter, and it will do less work before broken wires begin to appear in the rope into which it has been made, the consequence being that the rope has to be discarded.

One of the first principles, and one of the most important essentials in connection with the efficient, economical, and safe working of winding-ropes, is this question of quality. It is important that purchasers and users should recognize what would seem to be a sufficiently obvious and self-evident fact, namely, that ropes of high quality cannot be purchased at an absurdly low figure. And further that there will always be found, in connection with all trades, makers who will undertake to find something sufficiently inferior to exchange for the price offered, no

matter how low.　Indeed, this unfortunate tendency to insist on
extreme cheapness in first cost, which never yet resulted in real
economy, has had, and is having, a most deplorable effect upon
manufacturing trades generally, and certain branches of the
engineering industry in particular.　It is not suggested that purchasers should pay fancy or exorbitant prices for what they require, but a little consideration will generally serve to show
whether the price at which goods are offered can possibly permit
of the use of the best materials and the most approved processes
of manufacture.

As already pointed out, a rope may be manufactured from
wire which has cost much less to manufacture than another
wire, and yet the two samples of wire will satisfy the same tests
as to tensile strength, torsion, and bending, but not fatigue
Formerly the only test for fatigue was that imposed by the
ordinary conditions of working, and it is too late to discover, when
the rope is done, that its capacity for work was not what was
desired. Messrs. J. A. Vaughan and W. Martin Epton, however,
have invented an appliance—illustrated in the Report of the
Transvaal Commission*—which enables the fatigue-test to be
applied.　Briefly, the object of this machine is to enable a wire
to be put under a load, say one-fourth of its tensile strength, and
worked continuously round pulleys, bending in opposite directions, until fracture takes place.　The pulleys, it must be remarked, are not small relatively to the diameter of the wire, but
of ample diameter, so that the cause of final fracture is actually
fatigue due to working stresses, and quite distinct from the
ordinary bending test in which a wire is bent backwards and for-
wards with a sharp bend of small radius.

The application of this, or a similar test effected in an equally
satisfactory manner by some other form of appliance, would
seem to be the only way to ensure a proper standard of quality.
and if a Government test were to be established, it ought
certainly to embrace tensile, torsional, bending, and fatigue-tests.

The Care of Ropes in Use.—To ensure satisfactory and
economical results, a reasonable working life, and to avoid accidents, the rope in use must be properly cared for, not only as

* *Report of a Commission appointed by His Excellency the Lieutenant-Governor
[of the Transvaal] to enquire into and Report upon the Use of Winding-ropes,
Safety-catches and Appliances in Mine-shafts,* 1907, annexure No. 6, page lxxxix.

regards the strains to which it is subjected in working, but also with a view to avoiding excessive or rapid wear, internal abrasion and corrosion, matters of paramount importance which are only too frequently neglected. Assuming the rope to be made of suitable material, it will stand a considerable amount of bending and unbending on the drum and over the pulley, provided that the diameter is not too small in comparison with that of the rope. Nothing tends to shorten the life of a rope so much as bending and unbending on drums and pulleys which are too small. As a rule, however, this fault is more usually met with in haulage-drums and pulleys than in winding arrangements. The drums of winding-engines and the headgear-pulleys are generally more than 100 times the diameter ·of the rope, which, according to many rope-experts, is the minimum size of pulleys for six-stranded seven-wire ropes.

. At the same time, so far as the headgear-pulleys are concerned, one must avoid the opposite extreme, and, if the pulleys exceed 18 feet in diameter, they become unnecessarily heavy, and lead to trouble and excessive local wear, in consequence of their inertia and momentum.

There can be little doubt that side-friction, caused by the rope rubbing against the cheeks of the headgear-pulley and against the coils of rope already on the drum, is responsible for a good deal of wear tending to shorten the life of the rope. The acuteness of the angle contained between the rope hanging in the shaft and the line of rope from the drum to the pulley (that is, the angle in the vertical plane), does not in itself seriously affect the question—provided that the pulleys are of the proper size; but the lateral angle (that is, the angle contained between the line of the rope from pulley to drum, when the cage is at the surface, and the line of rope from pulley to drum when the cage is at the bottom of the shaft) has undoubtedly an important bearing on the subject. And inasmuch as an acute vertical angle is generally caused by the drum being too near (measured horizontally) the pulleys, the acute vertical angle is often erroneously regarded as the cause of the mischief, whereas it is really the lateral angle caused by the amount of lateral travel on the drum. This will, of course, be greater for deep shafts than for shallow ones, and greater on small drums than on large drums. To avoid excessive side-friction, therefore, the relative position of the drum and pulleys

must be carefully proportioned, and the size of the drum must
bear some relation to the depth of the shaft.

The greatest care must be taken, too, that the rope does not
rub, or even occasionally knock, against anything in working.
One frequently sees, on the sides, top, and bottom of the opening
in the engine-house wall, through which the ropes pass, clear and
abundant proof that they do at times rub against these surfaces,
and such rubbing is not calculated to do them any good or prolong their life.   Indeed, in collecting information for this paper,
the writer has been surprised at the many different ways—simple
and trifling enough in themselves—in which good winding-ropes
are injured, even at collieries where there is an evident desire
to do things properly and well.   Doubtless it is the fact that
these details are simple, and apparently trivial, which leads to
their existence and their effect being overlooked.   Bolt-heads in
the drum protruding above the surface of the lagging, in consequence of wear, have been the cause of broken ropes.

Corrosion and internal abrasion generally go together, and
are due to the neglect of proper cleaning and lubrication of the
rope.   The writer makes use of the word " lubrication " as quite
distinct from " greasing."   In the ordinary course of working,
the wires in a rope move slightly and rub against each other, and
efficient lubrication is therefore just as necessary as it is in the
case of the moving-parts of a machine.   To be effective, the
treatment, however, must be that of lubrication and not mere
greasing.   The rope must be properly cleaned from time to time,
and a suitable lubricant used.

By " greasing " the writer refers to that practice sometimes
observed in which the rope is periodically plastered over with
some vile and filthy compound which attracts both dust and
moisture, and ultimately does more harm than good.   In shafts
where there is great liability to corrosion in consequence of bad
water, good results appear to have been obtained by the use of
galvanized ropes, but even if these are used, the lubrication must
be attended to with the same strict care.   Galvanizing appears
to have the effect of very slightly reducing the strength of the
rope, but tends to prolong its life in places where the water is of
a particularly corrosive character.   The use of copper wire in
cappings is to be avoided as being a possible cause of corrosion,
due to electrolytic action.

The Life of Colliery Winding-ropes.—In good practice a winding-rope is not retained after it has begun to show outward and visible signs that its working-life is approaching completion. It is not easy to find a means of satisfactorily expressing or comparing the life of a colliery winding-rope. To measure its life merely in terms of months or years is obviously unsatisfactory. One rope may do more work in a few months than another one in as many years. Neither is it altogether satisfactory to express the life or work of the rope in tons of coal raised, although this is, of course, the usual way of arriving at its cost for purposes of comparison with other ropes. For example, we may be told that two ropes have respectively raised one million tons, and half-a-million tons. At first sight the former would be considered as having done twice as much work as the latter. If, however, we are further told that the former raised its million tons from a depth of 600 feet, whilst the latter worked in a shaft 2,400 feet deep, it will be apparent that really the rope which has raised 500,000 tons through a distance of 2,400 feet has done twice as much work, in foot-pounds or foot-tons, as the one which has raised 1,000,000 tons through a distance of 600 feet.

But even a record of tons of coal, or other materials raised, and the depth of the shaft, does not furnish a complete statement of what the rope may have accomplished. The average speed of winding has an important bearing on the question, whilst the weight of coal raised does not take into account what may possibly amount to a far greater total weight lifted during the life of the rope, namely, the weight of the cage and tubs and the weight of the rope itself. It is quite certain that in the generality of cases the work done by a winding-rope is three times as much as that represented by the coal raised. Regarded in this way, the work which is safely and satisfactorily accomplished at a very large number of British collieries by high-class winding-ropes, runs into astonishing figures. A rope which raises 500,000 tons from a depth of 1,600 feet runs up a total of 800 millions of foot-tons in coal raised; and, taking into account what is done in raising its own weight and the weight of the cage and tubs, a total of certainly not less than 2,500 million foot-tons.

There is, therefore, no reason for lack of confidence in steel-wire winding-ropes. It is possible to procure ropes which can be absolutely relied upon to do, without fear of breaking, what they are required to do, and breakage will only result either from serious neglect or from strains so greatly in excess of working-conditions that something must go, and if the rope does not, something else will.

Rope-cappings and Attachment to the Drum.—Ample strength in the rope itself is, of course, of no avail if its attachment to the drum on the one hand, or to the cage on the other, is not proportionately strong. The former attachment to the drum is comparatively easily provided for, by taking advantage of what is known as coil-friction. It is usual to purchase a rope of considerably greater length than is represented by the distance from the drum to the bottom of the shaft, so that when the cage is at the bottom there are still three or four complete coils left on the drum. These extra coils not only provide spare rope to make up for the pieces which should be cut off at each re-capping,· but they also serve to secure the rope to the drum.

In the light of the experiments recently reported and commented upon‹by Mr. John Gerrard, H.M. Inspector of Mines, with regard to rope-cappings, demonstrating that the form of capping which was perhaps most popular, and generally considered to be the best, really possessed a strength of from only 50 to 60 per cent. of the strength of the rope,* there is something very like irony in the situation. There can be little doubt that until Mr. F. L. Ward, of Bradford colliery, Manchester, and Mr. Gerrard investigated the matter, most of us were fondly deluding ourselves with the appearance of strength possessed by a form of capping in very general use. To be perfectly frank, most of us were in ignorance as to the true state of affairs; and it is equally certain, from the elaborate provision that one often sees for securing the rope to the drum, that the peculiar property of coil-friction is not generally recognized. Everyone has observed the apparent ease with which a sailor is able to check the movement of a large vessel, by simply giving the rope two or three turns round a post or " bollard." Four turns round a post, assuming a co-efficient of friction of 0·4, would enable a man

* *Trans. Inst. M. E.*, 1905, vol. xxix., page 177; and *Reports of John Gerrard, H.M. Inspector of Mines for the Manchester and Ireland District (No. 6), to His Majesty's Secretary of State for the Home Department, under the Coal-mines Regulation Acts, 1887 to 1896, the Metalliferous-mines Regulation Acts, 1872 and 1875, and the Quarries Act, 1894, for the Year 1906*, pages 17 and 18.

exerting a force of 20 pounds to resist a pull of something like 200 tons, that is, if the rope were capable of sustaining such a strain.

The co-efficient of friction between a greased wire-rope and a wood-lagged drum will probably be about 0·35, which means that one coil on the drum would enable a pull of 1 pound to resist a strain of 9 pounds, two coils 9×9, three coils $9 \times 9 \times 9$ and four coils $9 \times 9 \times 9 \times 9$, or 6,561 pounds. With four coils on the drum, therefore, and the end passing through the lagging secured with a force of, say, 50 pounds, the rope is securely held, even if the load amount to as much as 140 tons.

One frequently finds that the rope-end which is passed through the drum-lagging is given two or three turns round the drum-shaft, in which case the security is increased to an enormous extent, and there is not the slightest necessity for the two or three strong clams which generally complete the arrangement. Yet at the other end of the rope we have been satisfied with an attachment only giving half the strength of the rope.

The experiments already referred to demonstrate, however, that too much care cannot be exercised in the selection, design, and application of rope-cappings. The forged steel conical socket, in which the prepared end of the rope is secured by means of white metal, when properly and carefully made, possesses a greater strength than the rope to which it is fitted. It must, however, be properly made, and the writer cannot do better than refer the members to Mr. John Gerrard's Report for 1906 for instructions on the most important points.*

Another excellent type of capping, which the writer believes to possess the necessary qualification of gripping-strength greater than the strength of the rope, is the Becker capping, which has the important advantage that it can be taken to pieces at any time and the rope inspected right up to its actual extremity.

Re-capping Winding-ropes.—The importance of re-capping the winding-rope, at regular intervals, cannot be too strongly urged; and the reasons for so doing are not difficult to recognize. First, there is the fact that the wires immediately above the

---

* *Reports of John Gerrard, H.M. Inspector of Mines for the Manchester and Ireland District (No. 6), to His Majesty's Secretary of State for the Home Department, under the Coal-mines Regulation Acts, 1887 to 1896, the Metalliferous-mines Regulation Acts, 1872 and 1875, and the Quarries Act, 1894, for the Year 1906, pages 17 and 18.*

capping, at the point where the rope loses its flexibility as it enters the socket of the capping, are subjected to a certain amount of bending of a different character from that which other parts of the rope are called upon to bear. It would also appear that the strains set up in the rope towards the capping, in ordinary working, are of a more trying character than those in other parts of the rope.

In recapping the rope, which should be done three or four times a year, a suitable length of rope should be cut off with the old capping, a length not less than one-half of the circumference of the headgear-pulleys. There are certain points in the cycle of operations in winding which result in greater wear in the corresponding places on the rope than in other parts. For example, that portion of the rope which is in contact with the headgear-pulley when the cage is at the top and the bottom of the shaft respectively. The starting of the wind, and the manipulation of the engine to facilitate the loading and unloading of the various decks, induces greater frictional wear at these points. Another point where wear takes place, especially in quick-winding, corresponds with the point where the speed of the engine is checked, the point where the brake is applied to the drum, and the rope in turn acts as a brake to check the revolution of the headgear-pulley.

For these and other reasons it will be seen that only by cutting off a proper length of rope at each re-capping can the excessive wear at these points be prevented from becoming too local. The piece cut off should be carefully examined and tested, all the tests previously enumerated being employed, and the results compared with the original certificate.

Excessive Strains in Winding.—It will be evident, from what has already been said, that the breaking of a colliery winding-rope may be due, either to excessive local wear and corrosion, or to excessive strains set up during winding operations. It is to cover the latter that the liberal factor of safety already alluded to is provided, and it must not be supposed that this liberal factor of safety is established so as to enable less care to be taken of the rope, or for the sake of avoiding a little extra trouble.

We have now, therefore, to consider the various causes, and probable extent and effect, of the excessive strains which may be set up in the rope in the regular course of working. Some of these strains are unavoidable, as, for instance, the extra strain in starting when the inertia of the loaded cage at the bottom of the shaft has to be overcome, as well as the extra strain during the whole period of acceleration. This may amount, and probably does in some cases, to very nearly double the weight of the suspended load. The strain at starting depends very largely upon the care and skill of the engineman; a sudden "snatching" sort of start imposes very great strains, the extent of which it is scarcely possible to calculate, but which may dangerously approach the limit of strength of the rope.

(*a*) *Acceleration.*—The strain due to acceleration, that is, uniform acceleration, is as nearly as possible 70 pounds per ton for each foot per second in the rate of acceleration. Thus, supposing the gross load to be 10 tons and the rate of acceleration attained at one moment or another to be 25 feet per second, the extra strain would be $10 \times 70 \times 25 = 17,500$ pounds or nearly 8 tons, making for that particular moment the total strain 18 tons instead of 10.

(*b*) *Retardation.*—Another unavoidable extra strain is that which is occasioned when retarding the movement of the engine by the application of the brake. In this case the descending rope is subjected to a strain greater than that due to the actual suspended load; but here, again, if the engine is properly controlled and the brake is properly applied, the extra strain is amply covered by the factor of safety. There are, however, cases on record where the sudden application of a powerful brake has most probably been the cause of a serious accident. When one enquires into this matter closely, the result is at first sight somewhat startling and unexpected, although a little consideration will serve to show that it is the only result that really could be expected. Briefly it is this, that in the case of the sudden application of a powerful brake, whilst the engine is travelling at a fairly high speed, it is the rope attached to the ascending cage, and not that which is descending, which runs the greater risk of being broken.

A simple example will suffice to make this clear. Suppose we have a shaft where the ascending cage weighs 10 tons and the descending cage weighs 6 tons, and that the maximum speed is

80 feet per second. The ascending cage, at this velocity (neg-
lecting friction), would continue to ascend for exactly 100 feet
if the raising force were suddenly discontinued; that is to say,
if such a thing were possible as the application of a brake which
would stop the drum dead, the ascending cage would continue
to ascend for a further distance of 100 feet before it came to
rest, after which it would, of course, fall back.

The writer finds it necessary to say, at this point, that he does
not suggest that it would actually be possible to stop the drum
dead, or even in one revolution of the drum. A letter in one
of the mining periodicals, criticizing a former paper on a somewhat similar subject, made it evident that the present writer's
meaning had not been understood. His desire is to show that
to stop a drum running at a high speed in a very short distance
is not only impossible, but that to attempt to do it means
destruction.

Coming back, then, to the example, let us suppose that the
drum is brought to rest in a distance of, say, 90 feet, that is, 90
feet travel of the descending cage. The extra strain on the descending rope would—if the retarding force were uniform during
the whole period—amount to rather more than double the actual
weight of the suspended load, and this would doubtless be amply
covered by the factor of safety.

The effect on the ascending rope, however, would be quite
different: the strain would take the form of a sudden jerk, of
sufficient force to break the rope. The ascending cage, it will
be remembered, would, by virtue of its kinetic energy, continue
to ascend for a total distance of 100 feet after the first application of the brake, so that it would finally be free to fall back
through a distance of 10 feet, at the end of which it would have
stored up kinetic energy equal to 100 foot-tons, which in all
probability the rope would be incapable of resisting. It will
thus be evident that whilst a powerful brake is a necessary and
desirable detail in the equipment of a winding-engine, its use
must be regulated with an intelligent knowledge of the laws
of motion, momentum, and inertia.

The following figures (taking the acceleration due to gravity
as 32) show the minimum theoretically safe distance in which
the cages may be stopped, if the retarding force is perfectly
uniform :—

Speed when brake applied. Feet per second.	Equivalent distance. $s=\frac{v^2}{2g}$ Feet.	Speed when brake applied. Feet per second.	Equivalent distance. $s=\frac{v^2}{2g}$ Feet.
10	1·56	60	56·20
20	6·25	70	76·60
30	14·07	80	100·00
40	25·00	90	126·60
50	39·00	100	156·09

In actual practice, it is not possible to exert a retarding force so perfectly uniform that we can reduce the speed of the drum and the cages at a perfectly uniform rate of 32 feet per second, which would be necessary in order to bring the cages safely to rest in the distances named above. It is therefore necessary to spread the retarding force of the brake over a much longer distance, and devote a larger portion of the space of time occupied in each run, to the slowing down and stopping, than is represented in the above figures.

The Use of Cage-supports or Keps.—A good deal of difference of opinion exists as to the use of cage-keps. Personally, the writer does not like them, if for no other reason than that they tend to shorten the life of the rope, and occasion great and jerky strains. With the cage resting on the keps, and 6 or 12 inches of slack rope or chain, it will be seen that quite a sudden and serious jerk can be given to the rope in snatching the cage from the supports. The writer has noticed that where keps are regularly used, there is always a great deal more banging and bumping about of the cages than where they are not used.

Tests carried out under working conditions have shown that with less than 1 foot of slack chain the strain on the rope in picking up may be not far short of three times the actual load. A more doubtful arrangement is the mechanical kep, of which there are several types in use, by means of which the banksman can remove the keps from under the cage without troubling the engineman to raise it. By this appliance the banksman can let the cage actually drop on to the slack rope. The use of these appliances should, the writer considers, be prohibited, unless special arrangements are made for accurately adjusting the length of the rope, and extreme caution observed in avoiding slack rope when the keps are released.

Jerky Winding.—Jerky winding, with consequent sudden variation of strain on the rope, is either due to unskilful manip-

ulation, to a wrongly-proportioned engine, or to unsuitable controlling-appliances on the engine. That there are different
degrees of skill amongst enginemen will be readily admitted.
We have all experienced the difference in riding a cage when
one or other of two enginemen has control. One man will produce a smooth and almost imperceptible gliding motion, whilst
another, in charge of the same engines, will jerk the cage about
until one has to grip the hand-rail with both hands to avoid being
pitched out. But even a skilful engineman cannot avoid jerky
action, if the engine is wrongly proportioned. If it is not so
proportioned that either cylinder can start the load, with a
reasonable rate of acceleration, there is a temptation to start
with a run or jerk, so as to get over the dead centre.

The control of the engine should be such as to entail the least
possible amount of physical exertion on the part of the engineman. His duties are such that, if he is to carry them out effectually, he must be free from constant demand upon his physical
energy to operate and control the engine, so that his mental
faculties may be properly concentrated upon his work. To this
end the engine should have an easily-operated steam valve, quick
to open and close; a steam-actuated reversing-gear, of which
there are several excellent types in use; and a powerful brake,
which, whilst capable of holding the engines under the most
unfavourable circumstances, also permits of its being applied
gradually so as to avoid the effects previously discussed.

If these appliances are unwieldy and heavy, and if the engineman has to struggle with them to control the engine, there is a
risk of physical fatigue interfering with mental alertness, and
serious accidents may be the result. Personally, the writer has
nothing but praise for the colliery winding-engineman as a class.
On the whole, they are men of exceptional ability in their particular work, and one has only to reflect upon the nature of that
work, at the average modern colliery, to recognize the fact.

The colliery engineman may have charge of a pair of engines
capable, it may be, of developing more than 2,000 horsepower.
He has to set in motion by means of these engines perhaps a total
weight of 100 tons; not merely to set in motion, but to impart a
velocity of 60 miles an hour in a few seconds, and in less than a
minute to bring the whole moving mass once more to rest, with
the cages adjusted to practically within an inch of the same spot

every time. These operations he may have to repeat three or four
hundred times a day, and the man who can do this, day after
day, week after week, year in year out for many years, without
ever making a mistake or committing an error of judgment, is,
indeed, possessed of skill and ability of more than average character. As a matter of fact, it is too much to expect that an engineman
will never make a mistake; and it should therefore be made incum-
bent upon colliery-owners to provide their engines with such controlling appliances as will relieve the engineman of undue physical
exertion, and also to apply an automatic controlling-device which
will positively prevent overwinding, running away, or starting
in the wrong direction.

Controlling Devices for Winding-engines.—At a number of
Lancashire collieries, for some years now, an appliance has been
in use which goes a long way in this direction. The "Visor" is
too well known to call for a description of its working; suffice
it to say that the Visor makes running-away well nigh impossible, and absolutely prevents a quick overwind. In other words,
it ensures that the speed of the engines, when approaching the
end of the wind, shall be so far reduced that the question of final
stopping is one of no difficulty.

A later appliance, and one which presents several interesting
features, is the Whitmore controller, which is used in conjunction with the Whitmore steam-brake. In the Whitmore controller there are two long vertical screws, one having a right-
hand and the other a left-hand thread. There is also a fly-ball
governor which, as well as the two screws, receives motion from
the engine. On each screw there are two nuts with projecting
wings, and these nuts or cursors, separated by a few inches,
travel up or down the screw as the engine works. The position
of the nuts or cursors may, as a matter of fact, be regarded as
indicating the position of the cages in the wind. The upper
cursor in each case is to prevent overwinding even slowly, or
starting in the wrong direction, and the lower cursor is to prevent
the engine exceeding a certain predetermined speed at any point
in the wind after the maximum speed has been attained. It
not only limits the maximum speed, but it checks the proper and
gradual reduction of speed as the engine approaches the end of
its run.

The brake is of ingenious and yet simple construction, and is actuated by steam; the steam, however, is not the force which applies the brake, this being done by weights. The steam assists in putting on the brake, but it has to raise the weights in taking the brake off. The great advantage of this brake appears to be that the retarding force may be regulated to any degree between full on and full off, that the amount of retarding force is always the same for a corresponding position of the brake-handle, and that it may be very gradually increased until the full braking effect is produced, or, if need be, it can be applied quickly.

In conjunction with the Whitmore controller, its especial advantage is, that in the event of the controller coming into action to check the engine when running away, the brake is not applied too suddenly so as to produce the result that has been already indicated, but gradually, until the full retarding force is developed and the cages are brought to rest within a safe distance. An appliance of this kind is calculated to prevent an accident, not to come into operation afterwards with doubtful results. It does not and cannot interfere with regular and ordinary working, nor come into action when it should not.

Disengaging-hooks.—The use of disengaging-hooks has become very general, but it will doubtless be admitted that their sphere of usefulness is very limited. A detaching hook of the most perfect type affords no protection to the descending cage, whilst the protection that it affords to the ascending cage is limited to comparatively slow overwinds—since in the case of a quick over-wind the cage crashes into the framing which carries the detaching-hook bell or cylinder, and smashes it up, so that it is no longer able to play its part. The writer is not saying a word against detaching-hooks, so long as they are properly applied and regularly taken apart for cleaning and lubrication, and the D-links properly re-annealed every time when the cage-chains are re-annealed, which should be at regular intervals, say when re-capping the rope. Whatever type of detaching-hook is employed, secondary catches should be fixed in the headgear to hold the cage, in the event of the bell or framework being damaged by the ascending cage.

Safety-cages.—In conclusion, the writer would venture the opinion that the safety-cage, as it is called, is the wrong direction in which to approach the subject of safety in winding. It has too much the appearance of condoning carelessness or neglect in other directions, and it is perfectly certain that in nearly all cases where safety-cages could possibly be of service (even if there were any which were reliable), the accident would be due to neglect or carelessness. At the same time, the writer would not express this view so strongly, if it could be shown that such an appliance existed which could really stop the cage gradually and carefully. So far, there are none which can be relied upon to do so.

Briefly, safety-cages may be classed under two heads:—First, those in which the inventors have entirely lost sight of the fact that kinetic energy has to be reckoned with, and that it is just as serious to attempt to stop a cage suddenly as it is to break the rope to which that cage is attached; secondly, there are those, very few in number, in which the inventor appears to have recognized that an attempt must be made to bring the cage to rest gradually. Those coming under the first head are not worth further thought; they are positively dangerous, and many of them would more frequently cause worse accidents than those which they are supposed to prevent. Those coming under the second head are too complicated or too unreliable for practical use. One of these consists of an arrangement for securely gripping the conductors if the rope breaks, but this does not immediately stop the cage. The cage is attached to the gripping device through the medium of a wire-rope coiled on a drum secured to the top of the cage. This drum is provided with a sort of automatic brake, which is applied with gradually increasing force as the safety-rope is paid out. The writer is inclined to the opinion that people who will not take the trouble to look after their winding-ropes will not give much attention to a complicated device of that kind, and if it did not cause an accident itself, in all probability it would be out of order when required to operate.

One of the safety-cages favourably reported upon by the Transvaal Commission* is a device consisting of a steel cylinder containing liquefied carbonic-acid gas, which, in the event of the rope breaking, operates two rams or pistons working in other

* Report of a Commission appointed by His Excellency the Lieutenant-Governor [of the Transvaal] to enquire into and Report upon the Use of Winding-Ropes, Safety-catches and Appliances in Mine-shafts, annexure No. 7, page ciii.

cylinders.   These rams are designed to actuate brake-blocks which bear upon the rigid guides in the shaft, and bring the cage to rest more or less gradually.   Again, the complication is too great, and again the writer finds himself asking the question whether it is likely that people who are going to neglect their ropes or treat them so that they will break, are the sort of people to keep in good working order a complicated apparatus of that kind.   But in this particular case we have an object-lesson at first hand, the full particulars of which will be found in the report of the above-mentioned Commission.

On September 7th, 1906, seven members of the Commission, two officials, and one representative of the inventor conducted a practical test with this apparatus at the " Marcus " shaft, a shaft which had been placed at the disposal of the Commission for the purpose of such experiments.   The first test, allowing the cage to fall from rest, was apparently successful.   Then followed a more severe test, in which the cage was detached whilst descending.   The cage dropped to the bottom of the shaft, there was no evidence of any retarding action, and it was found that there was no gas in the cylinder.*   If this can happen in a specially conducted test, what could happen, one ventures to ask, in ordinary work, when there are no experts standing around to look after it ?

But the last word about safety-cages has surely been uttered by the inventor of this appliance himself.   The writer does not speak disrespectfully nor does he venture to express an opinion ; he contents himself with quoting Mr. K. Schweder's own words in the concluding paragraphs under the heading " Advantages claimed for this Safety-device."†   Mr. Schweder was a member of the Commission, and is the inventor of the appliance in question.   He says :—

"In conclusion, I do not think I have to point out especially that the value of a really reliable safety-catch is not alone that of safety, but of economy as well, as the mine manager is entitled to use a rope a little longer if he can rely on the safety-device.

Another consideration is that the margin in the factor of safety might be reduced for very deep shafts."

* *Report of a Commission appointed by His Excellency the Lieutenant-Governor [of the Transvaal] to enquire into and Report upon the Use of Winding-ropes, Safety-catches and Appliances in Mine-shafts*, page lxviii.

† *Ibid.*, annexure No. 7, page ciii.

The writer commends the above expression of opinion, without comment, to his fellow-members for their most serious contemplation and meditation.

In conclusion, the writer feels that he may appear to have expressed some of his views very strongly, but in a matter of such great importance there is nothing to be gained by beating about the bush. He has no interest in any mining appliances, either connected with winding or other operations, except in so far as they lead to economy, efficiency, and safety; and wherever efficiency and safety are found there also will economy be found. Economy is not measured by the amount of expenditure incurred, but by the degree of efficiency and the standard of safety which can be maintained.

The PRESIDENT (Mr. John Ashworth, Manchester) thought that Mr. Winstanley's paper was of a most instructive character; the members present had been fully interested in it from first to last, and it had been illustrated by excellent slides. The paper showed how important it was, not only to have good laws, but to exercise the utmost possible care and one's best judgment in all circumstances.

Mr. HENRY HALL (H.M. Inspector of Mines, Rainhill, near Liverpool), in moving a vote of thanks to Mr. Winstanley, said that the pleasure of listening to this most excellent paper was enhanced by the fact that it came at a very opportune moment, as several very bad accidents had happened quite recently, due to causes mentioned by Mr. Winstanley. The first part of the paper might, however, lead to some misconception. The author had said a great deal about the qualities of wire used, and what bad ropes there were in the market, and left it to be implied that colliery managers might be willing to purchase some of those ropes of bad quality. In his judgment, colliery managers were not likely to do anything of the kind. If there was one matter as to which they should be careful, it was this question of the quality of the winding-ropes, because a bad rope might lead to serious damage to the mine in addition to the danger to life.

Mr. T. H. WORDSWORTH (Audenshaw, Manchester), in seconding the vote of thanks, said that the paper must have occupied

a good deal of time in its preparation, and that Mr. Winstanley
had laid before them much valuable information.

The resolution was unanimously passed.

Mr. W. OLLERENSHAW (Denton) said that Mr. Winstanley
had previously read many excellent papers before the Society,
but he thought that on this occasion he had excelled himself.
He admired his courage in referring to defective ropes. Mr.
Hall thought that colliery managers did not buy bad ropes; but
the fact remained that manufacturers were selling them, and,
therefore, somebody must be buying them. He believed with
Mr. Winstanley that cheap ropes were not economical but expensive; yet no matter how good a rope was when first put to work,
unless it received proper attention, was regularly cleaned and
properly lubricated, rapid deterioration would soon result.

Mr. LEONARD R. FLETCHER (Atherton) thought that the factor
of safety for winding-ropes need not be 10 in all cases.   It
seemed to him that there was a sufficient margin of safety for
heavy loads with a factor of about 7. Assuming the loaded
cage to weigh 10 or 12 tons, if the rope were made on the principle of being ten times stronger than the weight that it had to
bear, a rope of large diameter would be required. If the cage
were only 3 tons or thereabouts in weight, taking a factor of 10
they would get a rope that would not break under a less strain
than 30 tons, thus allowing a margin of 27 tons. But, assuming that one was content with a factor of 7 for heavy loads, and
had a cage that weighed 10 tons, they would then get a rope
that would stand a strain of 70 tons, and the margin would be
60 tons. Given a rope of the best quality, he thought that he would
be able to sleep at night under such circumstances.   Mr.
Winstanley had referred to the use of keps or catches. As to
the value of these appliances there was, of course, some diversity
of opinion.   He was accustomed to use them, and was heartily
in favour of them.   Mr. Winstanley had referred to some patent
keps or catches which the banksman at the pit-top could draw
without having the engine reversed or the cage lifted, and he
condemned them in what he (Mr. Fletcher) thought was unnecessarily strong language. Personally, he did not like that
condemnation, for at Atherton collieries they had had these
patent keps in regular use for two or three years with very satis-

factory results.   He thought that Mr. Winstanley was labouring
under a misapprehension in stating that a strong reason against
using these patent keps was that the slack chain which might
be resting on the top of the cage would go down with a jerk
when these catches were drawn away.   As a matter of fact, no
slack chain would remain on the cage at the pit-top, as it would
immediately be taken up by the weight of the other cage-rope
hanging in the shaft. · He had been very much interested in
the paper, but thought that it would be a pity if anyone were prevented from using these patent keps by what had been stated
by Mr. Winstanley; and it was only fair both to the makers and to
those who used them, to state what had happened in actual
practice.   Personally, he would be sorry if, as Mr. Winstanley
had suggested, any law were passed to compel a 10 to 1 factor
of safety for winding-ropes, or to prohibit the use of these patent
keps. The laws which they had at present to observe were as many
as they could well remember, without adding to them
unnecessarily.

Mr. S. ECKMANN (Manchester) said that no reference was made
in the paper to the safety-devices for electric winders.   Mr.
Winstanley had rightly emphasized the necessity for providing
winding-engines with such controlling-appliances as would
relieve the driver of undue physical exertion and would
automatically prevent overwinding, running away, or starting
in the wrong direction.   The control-devices of electric winders,
especially those after the Ilgner, Ward Leonard, or similar
systems, were far superior to any of the appliances mentioned
in the paper.   They were well known, and had been dealt with
in a previous paper.*   The superiority of electric control had
been recognized by Continental authorities by allowing the maximum speed of 33 feet per second for winding men with electric
winders, as against a maximum of 18 feet with steam winders; indeed, when testing a new electric winding-engine, gentlemen had
allowed themselves to be wound at a maximum speed of 60 feet
per second.   In another test, the driver of an electric winder
was advised to start the engine and then go away, and to leave
the retarding and stopping entirely to the automatic devices,
with the result that the cage was stopped automatically at bank-
level.   The smooth running and, consequently, the reduced wear

* "The Application of Electricity for Winding and other Colliery Purposes,"
by Mr. Maurice Georgi, *Trans. M. G. M. S.*, 1904, vol. xxviii., page 455.

of the ropes and the greatly increased safety against accidents connected with electric winding-engines, were points well worth taking into account when the question of steam or electrically-driven winding-engines was under consideration.

Mr. A. RUSHTON (Abram, near Wigan) said that at the Abram pits they used cage-keps whilst the men were ascending and descending, after which they were pegged back for the rest of the day, and thereby wear-and-tear of the cages was saved, the cages always being suspended, and the ropes kept tight during the winding of coal   In these pits where the catches or keps were used and heavy weights were raised, a severe strain would be put upon the ropes by using the catches continually when the engine raised the load from the bottom, as the cages rested upon the catches and the rope was slack.   At one of their pits, the weight raised from the bottom was 13 tons.   He was in favour of using the catches for men, but did not deem them needful when drawing coal.

Mr. JOHN LIVESEY (Bolton) urged the desirability of demanding a certificate with every rope purchased. Ropes could be obtained at any price, but Mr. Winstanley had not told them which was the best way to secure a good one. At present, they could only tell whether a rope was a good or a bad one after it had been in use.

Mr. JOHN GERRARD (H.M. Inspector of Mines, Worsley) said that it was not sufficient for a coal-owner to trust a manufacturer to produce a good rope by leaving the price unlimited. He agreed with Mr. Livesey as to the importance of having a certificate of test, and shared Mr. Winstanley's objection to greasing—covering the dirty rope with a villainous paste. Ropes ought to have fair treatment, not only as to the size of the pulleys and drums, and the angles at which these were placed, but also as regarded proper cleaning and lubrication. He was much interested in Mr. Fletcher's observations regarding the weight of ropes; his view was that by improving the quality, giving full consideration to obtaining great strength together with less weight, the factor of safety now prevailing might continue for deep shafts. It should be remembered that human agents were liable to err; and in connection with the taking up of the load and the acceleration to full speed, there were reasons for requiring a considerable margin. He strongly supported the resort to mechanical arrangements to prevent undue speed, or the starting of the engines the wrong way, and to arrest the engines in case the engineman failed to do so. Detaching-hooks had made winding in shafts much safer, but they did not secure safety at other than moderate speeds, for every detaching-hook hitherto brought out had failed at full speed.

Mr. GEORGE H. WINSTANLEY (Manchester), in acknowledging the vote of thanks, said that, with the President's permission, he would defer his reply to the various points raised until the adjourned discussion; but he would like to say, in answer to Mr. Fletcher, that he was willing to admit that the reference to mechanical keps might perhaps with advantage have been differently worded. As a matter of fact, in revising the proof he had made an addition which he feared did 'not appear in the advance-copies printed for use at the meeting.

What he had desired to convey, with regard to mechanical keps was, that perfect as such appliances might be mechanically, they were liable to very serious abuse, and if not applied and handled with the greatest possible care, their undoubted advantages might be converted into disadvantages.

He advocated in all mechanical appliances, so far as was reasonably practicable, arrangements which would refuse to act altogether unless properly operated. Mechanical keps—keps which obviated the necessity for raising the cage to effect their withdrawal—might be, and no doubt were, excellent appliances in themselves. But whilst on the one hand, when carefully used with regular adjustment of the ropes, they might be very convenient and give most satisfactory results, on the other hand, their actual operation was independent of the adjustment of the rope and the manipulation of the engine, and they could, as a matter of fact, be withdrawn when the rope was slack. For this reason, he had desired to introduce a reference that would call forth discussion, to emphasize the importance of extreme care in the application and use of these otherwise valuable appliances.

With regard to the factor of safety in winding-ropes, he was glad that that point had been taken up, and hoped that more would be said upon the subject at the adjourned discussion.

The further discussion of the paper was adjourned.

MANCHESTER GEOLOGICAL AND MINING SOCIETY.

***

ORDINARY MEETING,
HELD IN THE ROOMS OF THE SOCIETY, QUEEN'S CHAMBERS,
5, JOHN DALTON STREET, MANCHESTER,
APRIL 14TH, 1908.

***

MR. JOHN ASHWORTH, PRESIDENT, IN THE CHAIR.

***

The following gentleman was elected, having been previously nominated :—

MEMBER—

Mr. SYDNEY ARTHUR CHAMBERS, Mining Engineer, 96, Gresham House, London, E.C.

Mr. CHARLES F. BOUCHIER read the following paper on " Enlarging an Upcast Furnace Shaft from 10 to 15½ feet in diameter, whilst Available for Winding Men and Coal."

ENLARGING AN UPCAST FURNACE SHAFT FROM 10 TO 15½ FEET IN DIAMETER, WHILST AVAILABLE FOR WINDING MEN AND COAL.

By CHARLES F. BOUCHIER.

Introduction.—The arrangement recently installed at the Strangeways Hall colliery, near Wigan, belonging to Messrs. Crompton & Shawcross, Limited, for the stripping and enlarging of an upcast shaft from a diameter of 10 to 15½ feet inside the brickwork, may be of interest to the members.

The shaft was already sunk to a total depth of 2,160 feet from the surface, being 10 feet in diameter down to 1,620 feet, and 14 feet in diameter for the remaining 540 feet.

The first thing to be done was to get a correct centre-line of the bottom length, and before commencing the work there were certain points that required careful consideration, namely:—

(1) There are four seams dependent on this shaft for ventilation; this is effected by means of a furnace placed in a mouthing 1,200 feet from the surface, which would not allow of the shaft being filled up and enlarged in that way.

(2) The shaft acts as the second outlet from the mines, and consequently must be in constant readiness for any emergency. The first 1,620 feet is being increased from 10 feet in diameter to 15½ feet, instead of 14 feet, for reasons which will be explained later.

There are two pairs of engines (figs. 1 and 2, plate xiii.) fitted up for winding; (1) a pair of cylinders 22 inches in diameter by 3 feet 6 inches stroke, with a drum 7 feet in diameter, which were originally used for winding coal, and afterwards for the sinking of the last 540 feet; and (2) a pair of cylinders 36 inches in diameter by 4 feet 6 inches stroke, with a drum 20 feet in diameter, which have been erected to wind coal when the shaft has been stripped to the entire length. The smaller pair of engines are now used for the stripping operations, and the larger pair for winding coal, examining below the tube, and are available for drawing the men out from the mines below in case of emergency. There are three conducting-rods, $1\frac{1}{2}$ inches in diameter, for the cage to travel up and down, and two $\frac{3}{4}$-inch ropes as conducting-rods for the hoppet.

The stripping operations were first commenced by working three shifts of 8 hours each; but when more coal was wanted, it was decided to wind coal for 8 hours, namely from 6 a.m. to 2 p.m., and to strip from 2 p.m. to 6 a.m. The men working in the mine from which coal is wound are lowered and raised by another shaft, and coal only wound during the 8 hours.

The first 600 feet in depth has been sunk for more than 50 years, and has twice collapsed and completely run in; on the first occasion about 45 feet from the surface, and on the second about 240 feet, which rendered the work of enlarging both difficult and dangerous, as the old curbs had given way and were very much out of the centre-line of the shaft below.

Commencing from the surface, the old shaft was excavated to 24 feet square, and pitchpine timbers 18 inches square put in every 7 feet in depth, with 7-inch by 3-inch planks behind these timbers. As the loose ashes and soft ground were taken out, the old brickwork was left standing to act as a fence and to keep the smoke from the men, being taken down as the sinking proceeded, about 7 feet each time, until solid ground was reached. Where the shaft had previously collapsed and had been filled in with loose ashes, the ground was very bad, and great care had to be taken to prevent it from again giving way for a depth of 45 feet to the metal, when a ring $15\frac{1}{2}$ feet in diameter was laid, and the shaft bricked up solid to the timbers, which were taken out as the bricking up proceeded.

The ashes were wound in barrows, by means of a small steam-winch, which was very handy for getting round the sides, as also for letting bricks and mortar down for the bricking.

After the metal was reached and the sides became stronger, the shaft was excavated in a circular shape, making it 17 feet in outside diameter, in order to leave a finished inside diameter of $15\frac{1}{2}$ feet. Where the metal had to be blown, precautions had to be taken to prevent it from falling down the shaft by slinging a tube of $\frac{1}{2}$-inch plates 18 feet 6 inches long by 8 feet 3 inches in diameter, by means of two crabs on the surface attached by

1-inch ropes, with the large winding-rope as an additional safety. The tube was 8 feet 3 inches in diameter instead of 10 feet (the size of the shaft), in order to allow room for the hoppets to land and give more room for the men to move about, also to allow room for the shots to lift. In order to prevent metal from falling down the opening, a segment was bolted to the bottom of the tube. Double hand ratchet-drilling machines are used; they are first fixed in a horizontal position against the tube to drill holes about 18 inches deep, in which plugs are inserted for the machines, when the machines are fixed in a vertical position for drilling the shot-holes, which average 4 feet to 4 feet 6 inches in depth, according to the nature of the ground. Care has to be taken that these holes are not drilled below the bottom of the tube, and also to prevent the shots from simply blowing out at the bottom and not lifting the metal.

Nine and ten holes, which are all bench-shots, are drilled and charged, and fired by an electric cable and battery from the surface, which blow the metal against the tube, thus preventing it from falling down the pit. The débris is filled out and wound in the ordinary way. Skeleton rings and boards are placed every 4 feet apart to protect the side of the shaft, and are taken out as the brickwork lining is constructed. The tube is always not less than 4 feet below the bottom of the holes to be fired. When the sinking has proceeded to within this length of the bottom, the tube is again lowered, by means of the crabs and the winding-engine, to within 3 feet of the bench on which the sinkers are working.

The depth sunk before being bricked up varies according to the nature of the ground, the greatest length having been 120 feet, a week being occupied in completing the bricking. As an additional support, a bricking-ring was laid in the middle of this length. When suitable ground for plugging is reached, the drilling-machines are set against the tube and the holes drilled ready for the plugs. So as to prevent any damage to the bricking-ring when it is set, by the firing of the next round of shots, the shaft is sunk 6 feet deeper and the last round of shots left in until the bricking is finished. When it has been decided to brick up, the tube is lowered below the ring.

Bricking-scaffold (figs. 3, 4, 5, and 6, plate xiii.).—Owing to the conducting-rods being in the shaft, the bricking-scaffold is made in four parts, and each has to be lowered separately every time that it is used. It is put together on two baulks placed in hangers, 18 inches long, from the bricking-ring and bolted by shot-bars. An opening is left in the centre (fig. 6, plate xiii.), 7 feet in diameter, with sheeting round it 4 feet high, to act as a fence and to allow the ventilation to pass through. It is raised by means of four 1-inch ropes attached to the main winding-rope as the bricking proceeds. When a length of bricking is completed, it is again lowered on to the timbers, there taken to pieces, and then sent up the pit. When coal was being wound, the scaffold was fastened to the skeleton-rings by means of a pulling-jack on each side, in order to liberate the winding-rope. The depth already completed and bricked is 681 feet, and has averaged 24 feet a week.

Ventilation for Sinkers.—During the stripping of the first 300 feet, the ventilation was taken down two ranges of 18-inch galvanized air-pipes, the fresh air going down the pipes and returning up the centre of the shaft. No more pipes have been put in, and the air now goes down the sides of the enlarged shaft and returns up the centre with the ventilation current from the other mines.

———

A vote of thanks was accorded to Mr. Bouchier for his paper.

———

Mr. CHARLES F. BOUCHIER (Wigan), replying to a question by Mr. George B. Harrison with reference to the broken ground where the collapse or " run in " of the shaft had previously taken place about 240 feet down, said that a certain amount of difficulty was experienced in getting through at that spot, which was overcome by putting down piles, and so keeping the soft ground back, and that the shaft had been sunk at various times, and was not absolutely perpendicular; the tube, therefore, could not be taken down the centre throughout.

Mr. ALFRED J. TONGE (Bolton) asked Mr. Bouchier whether they were able to continue bricking all the time that they were winding coal, as he should think it would be necessary to uncouple the rope and leave the main cage at the bottom while the bricking-scaffold was raised, which would shorten their working hours. He considered the arrangement to have been on the whole very well thought out.

Mr. BOUCHIER replied that coal was not wound while the brick-ing was going on. As stated in the paper, coal was wound from 6 a.m. to 2 p.m., and the bricking took place from 2 p.m. to 6 a.m. Replying to Mr. James Ashworth, Mr. Bouchier said that the cage was detached at the surface, and that only about 10 minutes was required to change from coal-winding to sinking. No damage was done by the firing of shots. They were fired simultaneously.

DISCUSSION ON MR. JAMES ASHWORTH'S PAPER ON "AIR-PERCUSSION AND TIME IN COLLIERY EXPLOSIONS."*

Mr. W. N. ATKINSON (H.M. Inspector of Mines, Bridgend) wrote that the object of Mr. Ashworth's paper appeared to be to prove that "percussion-effects" played an important part in colliery explosions. This theory had been advanced or discussed on several previous occasions, but the evidence and arguments put forward to support it failed to carry conviction. In the large number of explosions which he had investigated, he had found no indication to support the theory, nor any reasons for suppos-ing such effects had occurred.

So far as the argument was based on the Courrières explosion, he thought that it was altogether erroneous. In the first place, it might be asked on what evidence or authority it was stated that "the floor of the [Lecœuvre] gallery was lifted,"† and "that the upheaval or disturbance of the floor at this point indicated, either an outburst of gas, or the accidental ignition or detonation of an explosive above the air-pipe."‡ The first of these statements was contrary to the fact; the floor was not lifted, and therefore the deductions in the second statement had no such basis. It was true that the air-pipes were shattered, as stated, and the rails displaced, and this occurred near where the prolongation of the axis of the shot-hole at the face would strike the floor of the gallery.

With reference to the remaining statements in the first para-graph of the paper and those in paragraphs 2 to 7, it was not apparent what bearing they had on the percussive theory advanced in the paper.

* *Trans. Inst. M. E.*, 1907, vol. xxxiv., page 270; and *Trans. M. G. M. S.*, 1907, vol. xxx., page 158.

† *Trans. Inst. M. E.*, 1907, vol. xxxiv., page 270; and *Trans. M. G. M. S.*, 1907, vol. xxx., page 158.

‡ *Trans. Inst. M. E.*, 1907, vol. xxxiv., pages 272 and 273; and *Trans. M. G. M. S.*, 1907, vol. xxx., pages 160-161.

With reference to paragraph 8, wherein it was stated that: "The indications of force from the Lecœuvre gallery were not directly towards No. 3 pit,"* the reason of this was explained on pages 467 and 468 of the paper on the "Courrières Explosion" by Mr. Henshaw and himself,† and on page 15 of the official report by Mr. Cunynghame and himself,‡ namely, that on the two more direct routes to No. 3 pit, the explosion was arrested for lack of coal-dust.

Paragraph 11 of Mr. Ashworth's paper stated that "The wet condition of the 1,070 feet (326 metres) north bowette did not restrain the flame of the explosion."§ The wet condition of this road was discussed on page 16 of the official report above referred to. As steam from the Cécile fire passed through it for many weeks after the explosion, the roof and sides were naturally wet or damp when examined; and the water-course by the side of the road being blocked through falls, and a pump being out of action, there was much more water on the floor after the explosion than before. The roof and sides of the bowette "were blackened with the usual coating of black dust found after explosions on dusty roads, and at one place coked dust was found." He had no doubt that the portion of this bowette traversed by the explosion had contained sufficient dry dust to account for its passage.

No. 3 pit was not "choked with débris" (paragraph 12, page 272). There was always a passage through the débris, first for the smoke of the explosion which invaded the bank at No. 3, as well as at No. 4 pit, and later for the air during 10 months when No. 3 pit was used as an upcast. What was meant by "heavy percussive effects" (paragraph 13, page 272)? As in all extensive explosions, damage was done on the roads traversed by the blast, timber was blown out, and falls of roof occurred in consequence. The bearing on the percussive theory of the position of the fires caused by the explosion was not apparent (paragraph 14, page 272).

With reference to the indications in the Lecœuvre gallery (which they had not been able to examine minutely), he did not

---

* *Trans. Inst. M. E.*, 1907, vol. xxxiv., page 271 ; and *Trans. M. G. M. S.*, 1907, vol. xxx., page 159.

† *Trans. Inst. M. E.*, 1906, vol. xxxii., page 439.

‡ *Report to H.M. Secretary of State for the Home Department on the Disaster which occurred at Courrières Mine, Pas de Calais, France*, on March 10th, 1906, by Messrs. H. Cunynghame and W. N. Atkinson, 1906 [3171].

§ *Trans. Inst. M. E.*, 1907, vol. xxxiv., page 272 ; and *Trans. M. G. M. S.*, 1907, vol. xxx., page 160.

think that it had been previously pointed out that all the fragments of the fourth air-pipe, marked on the enlarged plan of the gallery (plate xxv., volume xxxii.), were found out-bye of the original position of the pipe, except some pieces (No. 142) under a fall just opposite the position of the pipe. Fragments of clothing and leather hats were also shown scattered from the face outward.

It might appear to Mr. Ashworth to be "practically certain" that his "proved facts" (some of which were not facts) demonstrated where the explosion began and all about it, but this paper was not likely to convert to his views those who made personal investigations of the explosion.

Mr. W. L. Hobbs (Pendleton) said that it was immaterial to the discussion of the theory of air-percussion whether the ignition, which Mr. Ashworth agreed with Messrs. Atkinson and Henshaw took place in the Lecœuvre heading, occurred from the cutting out of a missed shot, as Messrs. Atkinson and Henshaw thought, or whether it occurred from an ignition of explosive at the fourth air-pipe, as Mr. Ashworth, Mr. Stokes,* and Mr. Simcock† contended. It was a very interesting point, and a good deal could be said for both sides; but it was sufficient for that discussion to agree that somehow or other an ignition did take place at that point.

Mr. Ashworth's argument, however, that another, and as he (Mr. Hobbs) understood him to contend, the primary explosion, originated at the western end of the recovery-drift from the Marie south level at 1,070 feet (326 metres) to the Joséphine seam, did not appear conclusive. Messrs. Atkinson and Henshaw stated‡ that the reason why the explosive force from the ignition in the Lecœuvre heading did not take the shortest routes to the bowettes was that the dust was not inflammable on those routes, in proof of which they gave analyses of the dust (Nos. 16 and 17). The explosion, they stated, went eastward along the Joséphine level to No. 2 pit, and part of the force turned through the recovery-drift before mentioned to the Marie seam, and so to the 326 metre bowette. He (Mr. Hobbs) assumed for the moment that the fuel for the explosion was coal-dust. When the force emerged from the recovery-drift, it seemed most probable that it would branch right and left into the the Marie workings, provided that

* *Trans. Inst. M. E.*, 1906, vol. xxxii., page 340.
† *Ibid.*, 1907, vol. xxxiv., page 151. ‡ *Ibid.*, 1906, vol. xxxii., page 468.

there was fuel for it to feed upon. This seemed to have been the case, as the whole of these workings were traversed by the explosion, and all the men perished. There did not appear, therefore, to be any necessity to suppose that there was a separate explosion at the end of the recovery-drift. All the effects would take place without any such separate explosion if the fuel were coal-dust.

With regard to the statement in Section 11 of Mr. Ashworth's paper, namely, that "the wet condition of the 1,070 feet (326 metres) north bowette did not restrict the flame of the explosion,"* he would point out that Messrs. Atkinson and Henshaw had said† that from the evidence they obtained, the bowette was probably not wet before the explosion, except about the bottom of the staple pit from the Joséphine level, and that it was probable that there was an ample supply of dust available at that point, as the coal was tipped down the staple pit from the Joséphine seam just above. But in the south 1,070 feet (326 metres) bowette there was a wet zone 200 feet long, which stopped the explosion at that point; and it was from workings beyond this point that a party of 13 men escaped twenty days after the explosion.

It was also remarkable that in numerous cases the explosive effects disappeared when the dust contained a high percentage of incombustible matter, or was absent.‡ Several instances of this had been given by Messrs. Atkinson and Henshaw in their detailed description of the effects in the various districts. If Mr. Ashworth's theory as to air-percussion were correct, why should the explosive force stop at these places? Why should not the air-pressure restart the burning and coking effects after passing over these barren zones? The workings were most complicated: branch roads and staple pits were abundant. The percussive effect of the air would have passed many junctions, including the area about the bottom of No. 3 shaft, where the roads must have been of large size, before reaching these barren spaces. If it had persisted so far, why should not the effect have been the same all over the workings until it arrived at the open shafts? Messrs. Atkinson and Henshaw had given many instances where the burning effects ended with the end of inflammable dust. Why should they have ended just at those points if the effects were caused by heavy air-pressure?

* *Trans. Inst. M. E.*, 1907, vol. xxxiv., page 272; and *Trans. M. G. M. S.*, 1907, vol. xxx., page 160.

† *Trans. Inst. M. E.*, 1906, vol. xxxii., page 467.   ‡ *Ibid.*, page 455.

With regard to accidents caused by air-blasts, a paper had been read on the subject by Mr. Thomas Adamson,* and an interesting discussion had ensued, to which Mr. Ashworth made an important contribution. Details were there given of a number of cases in which damage was done and men killed at a considerable distance from the fall. In all the instances, however, the cause was a huge fall of a considerable thickness of a strong roof, over extensive areas of goaf measuring as much as 100,000 and 200,000 square feet. Naturally, when the large quantity of air displaced by this huge fall was forced along the contracted cross-section of the roads, very high pressure would be developed, and the resulting damage was quite understandable. But they had heard of no such huge falls occurring at Courrières. On the contrary, it was evident from the system of timbering enforced at these collieries that the roofs were not strong enough to stand over a large area, even if no packing had been done. It was likewise impossible that a fall in the recovery-drift, which, being a cross-measure road, would not be of large cross-section, could develop an air-pressure anything like sufficient to cause burning and coking effects in roads with a cross-section at least as large as its own.

Again, Mr. Ashworth, in his contribution to the discussion of Mr. Adamson's paper, in discussing the Mount Kembla disaster,† had stated that the large fall of roof was admitted by all the witnesses. He proceeded to state that the effects were felt only in the straight main haulage way, and not in branch roads which were drier than the main road. Perhaps he would explain why if in an example which was admitted to be a case of air-blast the damage was confined to the straight outlet, the force at Courrières branched along all dusty roads, many of which led into districts which were cul-de-sacs. To his (Mr. Hobbs') mind the evidence seemed conclusive that the effects at Courrières were caused by the combustion of dust, and not by air-percussion.

Although it might be wandering a little from the strict subject-matter of the discussion, still as the Courrières explosion was largely dealt with in Mr. Ashworth's paper, he would like to refer to one aspect of the explosion. The ignition that took place in the Lecœuvre heading was either of gas or of dust. When the head-

---

* "Goaf-blasts in the Giridih Coal-field, Bengal, India," *Trans. Inst. M. E.*, 1905, vol. xxix., page 425.

† *Ibid.*, page 434.

ing was reached after the explosion, it had been unventilated for nearly three months, yet no trace of fire-damp could be detected.[*] Subsequently, bore-holes were put into the coal and also into the adjacent fault, but no fire-damp was found.[†] Yet if it was dust that carried forward the explosion, it was dust at the working-face. And it had so far been understood that it was the old dust back on the main roads, which had been deposited by the ventilating current, and which had absorbed additional oxygen, that was the chief danger.

The point to which he (Mr. Hobbs) wished to draw attention was illustrated in the "Experiments illustrative of the Inflammability of Mixtures of Coal-dust and Air,"[‡] by Dr. Bedson and Mr. Widdas. These experiments pointed to a theory that dust, on standing in contact with air, instead of becoming more inflammable, lost some of its inflammability. A specimen of freshly ground coal-dust ignited at a temperature of about 1,335° Fahr. After standing only 17 days in contact with air, it required about 1,490° Fahr., showing an increase of 155° Fahr. This raised the point as to whether coal-dust at the working face, where machines were working, especially punching-machines, might not be in some cases as dangerous as dust farther out on the main roads, or even more so where such main-road dust was largely diluted with shaly matter.

Mr. W. OLLERENSHAW (Denton) said that he was of the same opinion as Mr. James Ashworth, namely, that there had been in a large number of colliery explosions of the past a variety of effects that could not be satisfactorily explained, unless the theory of air-percussion was accepted. The indications of the direction of the force of the explosion in the Lecœuvre heading at Courrières could scarcely be explained by any other method of reasoning than that advocated by Mr. Ashworth, and this also applied to a large number of the great explosions of recent times.

Having had some experience of the effects of air-compression in the mine, he could understand that most of the effects described in Mr. Ashworth's paper might, with truth, be attributed to this cause. Some years ago, whilst working at the coal-face at the Dukinfield collieries, engaged in working out a pillar of coal on the lower side of the main level, for the purpose of lodge-room for

* *Trans. Inst. M. E.*, 1906, vol. xxxii., page 473.
† *Ibid.*, page 474.   ‡ *Ibid.*, 1907, vol. xxxiv., page 91.

water, a space measuring something like 600 by 60 feet was left behind; and, owing to the lack of support, the roof, which was of a very hard and compact rock, commenced to "weight," this going on for a considerable time, until a break took place. Hearing this break, he had warned his workmate, and they both began to run along the lower side of the pillar. They had got about 150 feet along this level, when the roof in the waste previously referred to fell with a tremendous crash. The compression of the air was so great as to blow them forward for a considerable distance. The workmen in other parts of the mine felt the concussion, and came to the conclusion that there had been an explosion in the pit. He (Mr. Ollerenshaw) believed that if the air compressed by the fall had not had several roads through which it could escape, and if they had been running along a level cut in the solid coal, they would have been thrown either against the face or sides of the level, and their lamps broken or damaged. The natural inference would have been that the accident was due to an explosion of gas liberated by the fall and ignited by the lamps, although as a matter of fact no gas was present, either before or after the fall took place. It was owing to this experience and to the unsatisfactory results of some of the investigations into the causes of colliery explosions in the past that he welcomed the paper of Mr. James Ashworth, and was pleased that the matter had been brought before the members. as probably same satisfactory explanations might be elicited.

Mr. JAMES ASHWORTH (Rushton Spencer), replying to Mr. W. N. Atkinson, said that he would, in the first place, ask Mr. Atkinson to give references to " the several previous occasions " when percussive effects had been discussed. He was not surprised that Mr. Atkinson and others did not agree with his deductions; in fact, he expected such to be the case, as stated in his paper on page 274. Unfortunately, Mr. Atkinson did not discuss the examples given in the paper as to "time," and therefore he would call his attention and that of other non-believers to the reports of the disaster which occurred in December, 1907, at Monongah, U.S.A..* where all the expert witnesses unanimously agreed that the effects of a blown-out shot were carried throughout two mines known as Nos. 6 and 8, by " simultaneous " explosions

* "The Monongah Mine Disaster," *Mines and Minerals*, 1908, vol. xxviii., pages 277, 327, and 394.

in various panels and districts of the two mines, and that these simultaneous centres of explosion phenomena were probably due to the ignition and explosion of gunpowder in canisters.   So effective were these "simultaneous" explosions in carrying the disaster, first throughout No. 8 mine, and then throughout No. 6 mine, that the fact of both mines being ventilated by separate fans, and the workings laid out with a view to the very highest point of safety, did not enable one single man to escape alive. Both mines were worked with open lights.

One point not noticed by Mr. W. N. Atkinson, but prominently referred to in the paper, namely, "time," had received most valuable support from the demonstration at Monongah. Thus the difference of time between the explosion making itself obvious on the surface at No. 8 mine and then at No. 6 mine was five seconds, and this accounted for a speed of about 3,000 feet per second.   In addition to this interesting fact, although there was tremendous force demonstrated at the mouth of No. 8 mine, there was practically none at the mouth of No. 6 entry.*   Inside this entry, about 200 feet, three men were killed in a tool shanty without a burn or bodily injury, and the cause of death was undoubtedly concussion of the brain from "percussive" effects on the air.

In all probability, no such complete disaster had previously occurred in any part of the world, and its demonstrations of force and effect had entirely upset several theories, such as the possibility of restricting the extension of an explosion in a mine or mines when divided into separate districts, and these districts again subdivided into separate panels.   Then, as to Courrières, if the time when the explosion effects appeared at the top of No. 3 pit had been taken, and compared with similar indications at the top of Nos. 4 to 11 pits, most valuable information would have been forthcoming.

The evidence of the upheaval of the floor in the Lecœuvre gallery, at the point where the air-pipe was smashed into fragments (fig. 1, plate xiv.), was a matter of fact, and not of theory, and would be found in the reports, and also by reference to the report of M. G. Léon, Chief Inspector of Mines.   Mr. Atkinson had said that the floor "was not" disturbed, without giving his authority for this denial of a fact which was demonstrated on

* There were similar demonstrations at the entry of No. 2 mine at the Fernie explosion ; see *Trans. Inst. M. E.*, 1902, vol xxiv., page 450.—J.A.

the official plan. Before Mr. Atkinson again insisted that an air-pipe made of sheet-iron or steel could be smashed into, say, 89 pieces by a so-called blown-out shot, he (Mr. Ashworth) would suggest that the Home Office should be asked to make experiments to prove its possibility, as at present its positive impossibility was proved by the fact that the coal tram which was being loaded by Regis was in the direct line of fire, and instead of being reduced to matchwood, was comparatively little damaged.

With regard to paragraph 14 of his paper, the bearing of this fact was that there was no proof of the flame of the explosion reaching the points named, and that these fires were therefore caused by the effects of percussion on mixtures of fire-damp and air or some other easily ignitable substances.

Referring to the subject of facts or no facts, he (Mr. Ashworth) could not reply to these where the " no facts " were not indicated, but he might add that the official plan of the Lecœuvre gallery showed that the pieces of the dismembered air-pipe were found under a fall of roof from just above where the pipe had originally been placed, and therefore it was clearly demonstrated that the pieces of pipe were in position before the roof fell. Mr. Atkinson possibly had not had the opportunity of noticing that the lower side rail of the waggonway was lifted considerably above (11 inches) the higher side rail (fig. 1, plate xiv.)

Referring to the plans of the Monongah mine, Mr. Ashworth said that the experts called in after the explosion, though not agreeing as to the point of origin, were unanimous in their opinion that the explosion was practically simultaneous at the several suggested points of origin. The day before, or on the morning of the explosion, the fire-bosses reported gas in many places in the workings, but it was instructive to note that after the explosion, only two places were found where it was possible to locate any small traces of gas. Besides the danger arising from fire-damp, there was an ever-present danger from a large number of overhead electric wires carried throughout the mines, and as these wires were blown down by the explosion, it was more than probable that some of them short-circuited, and therefore may have added to the effects by causing the ignition of mixtures of air and carbonic oxide. The explosion seemed to have been carried throughout the pit simultaneously, or at a speed of 3,000 feet per second, and he (Mr. Ashworth) suggested that this was by air-percussion. Reference had been made to "goaf-blasts," and in these there was a remarkable instance of the effects of air-percussion, namely, in the little injury sustained by the men who were near to the falls, whereas those farther away were killed. The Mount Kembla disaster differed from the Monongah disaster in several ways: there was no fire-damp, no blown-out shot, no means of ignition so far as was known, nothing but an exceedingly heavy fall of roof; the effects of that fall were almost entirely confined to one haulage road inbye and outbye, and the only explosion which took place was on the surface, where people were burned and materials set on fire, and not in the pit. The effects inside the mine were caused entirely by air-pressure and the mechanical force of the blast.

Replying to Mr. W. L. Hobbs as to Dr. Bedson's tests, Mr. Ashworth said that he did not consider old dust half as dangerous as new dust. New dust, as ascertained by Dr. Bedson in his experiments, commenced to give off gas immediately after separation from the solid coal. Under such conditions, he (Mr. Ashworth) considered that each particle of coal-dust floated along in its own balloon of gas, and, therefore, that it ought to be estimated as a gas and not as a solid, and that it was the greatest danger in a coal-mine where explosives were used.*

Mr. Hobbs did not seem to think that it was necessary to suppose that there was more than one explosion at Courrières. Therefore, as there was an undoubted explosion in the Lecœuvre gallery, which he (Mr. Ashworth) had shown conclusively could not have originated from a blown-out shot, it became necessary to find a possible cause, and he had found that there were two, and probably more, simultaneous explosions which could not now be traced for lack of precise information; and he had considered the place where the recovery-drift crossed the Marie level in No. 3 pit as the centre of the demonstrated force (see fig. 8, plate xxii., vol. xxxii.).† As to this point, the indications of force radiated away from it in every direction; thus the door which directed the intake air into the Marie deep workings was completely swept away, and no trace of it was found. The Marie was the only mine in No. 3 pit lighted by safety-lamps, and therefore

* See "An Ignition of Coal-dust at Middleton Colliery," by Mr. John Neal, *Trans. Inst. M. E.*, 1907, vol. xxxiv., page 221.

† *Trans. Inst. M. E.*

if this door were left open for a time, gas might have accumulated and have been carried on to one of the open lights in the stone drift, and ignited when the door was again closed. The ignition of explosives in many parts of the pits involved in the disaster was quite possible, but no investigation appeared to have been made to ascertain what became of the stocks of explosives. He believed that ignitions of explosives took place in other parts of the workings, and, as at Monongah, were main factors in extending the area affected by the disaster. Excess of dust was a more certain condition to bring any explosion to a termination than dirty dust; and certainly where fire-damp was a factor in an explosion, fine dirty dust was as dangerous as coal-dust.

He (Mr. Ashworth) did not dispute that the combustion or explosion of dust took place at Courrières, but the conclusion to which he had come was, that the explosion was extended by air-percussion : that was to say, if air-percussion had not been a factor, the explosion would have been confined to a smaller area, that the upheaval of the floor of about 11 inches (fig. 1, plate xiv.), the non-destruction of the tram in the end of the Lecœuvre gallery (fig. 2, plate xiv.), the indications at the top and bottom of the Joséphine drop pit, and the reported indications of force in the recoupage, all showed most distinctly that the effects from the originating cause of the explosion were not entirely due to the ignition of coal-dust, and that air-percussion was a principal factor, in which opinion he was pleased to find he had at least one supporter in Mr. Ollerenshaw.

On many occasions the possible pressure exerted by colliery explosions had been debated, and a few people had ventured to place a figure to their estimates, but quite recently Mr. J. T. Beard in his book on *Mine Gases and Explosions,* had placed it at 196 pounds per square inch*; and later still Dr. H. N. Payne, of the West Virginian University. when reporting on the Monongah disaster, estimated the pressure at 50 to 146 pounds per square inch,† according to the volume of air available for combustion. The writer therefore suggested that these suddenly exerted pressures were sufficient to extend any explosion to all parts of a mine, and also that there were at the present time no known means of neutralising them.

* Page 214.

† "The Monongah Mine Disaster," *Mines and Minerals,* 1908, vol. xxviii., pages 329 and 330.

DISCUSSION OF MR. GEORGE H. WINSTANLEY'S PAPER ON "ACCIDENTS IN WINDING, WITH SPECIAL REFERENCE TO ROPES, SAFETY-CAGES, AND CONTROLLING DEVICES FOR COLLIERY WINDING-ENGINES."*

Mr. G. H. WINSTANLEY, in replying to Mr. Henry Hall, said that he was sorry if any observations in his paper could be construed into an implication derogatory to the colliery manager. Nothing was farther from his intention. As the President, for the time being, of the Lancashire Branch of the National Association of Colliery Managers, he desired to make this quite clear, and to remark that no one was more ready at all times than himself to uphold and safeguard the reputation of the colliery manager.

With regard to cheap ropes, he thought that Mr. Hall would be willing to admit, as Mr. Ollerenshaw had indicated, that cheap ropes were manufactured, and sold, and that therefore there must be both purchasers and users. Whenever a purchaser had the choice of two similar articles, one of which was offered at a lower price than the other, there was a right and proper tendency at least to consider why the cheaper one should not be selected. If it proved to be undoubtedly inferior, then, of course, no sensible person would hesitate. But in the case of steel-wire ropes, it was not always possible to make a selection so easily. Two ropes might be offered, both of the same size and weight, both made from wire of the same size, both professing to possess the same degree of strength, and apparently, on a test, giving proof of this equality. How then was the purchaser to decide? Be he colliery manager or proprietor, he would most naturally feel disposed to purchase the cheaper rope, which apparently was as good as the more expensive one. The object of this portion of his paper had been to show that it was possible to produce two ropes, one much cheaper (in first cost) than the other, and yet each should be capable of giving the same tests of tensile strength, torsion, and bending. Even the analysis of the two samples of steel might be the same, and still in use the cheaper rope would have the shorter life and would accomplish less.

* *Trans. Inst. M. E.*, 1908, vol. xxxv., page 134; and *Trans. M. G. M. S.*, 1908, vol. xxx., page 240.

He had endeavoured to show how the cost of manufacture could be cheapened at the expense of the life of the rope. The valuable invention of Messrs. Vaughan and Epton had made it possible to apply a more practical and more satisfactory test, one which would, in a manner of speaking, crowd the working life of the rope into a few hours. By this means one would be enabled to determine which of two or more wires was of the better quality, and to ascertain relatively how much more work one rope would do than another under the same conditions and treatment.

He feared that Mr. Livesey had rather lost sight of this particular point in the paper, perhaps the most important point in that part of the paper relating to ropes. He thought that the "fatigue" test carried out by such an appliance as that invented by Messrs. Vaughan and Epton went a long way towards meeting Mr. Livesey's, as well as every other rope-user's, requirements.

In further reply to Mr. Leonard R. Fletcher, he would like to say, on the subject of the "factor of safety," that he foresaw this difficulty in the near future when very deep shafts became more common. The task of surmounting this difficulty, however, would, he thought, be undertaken very largely by the wire manufacturer and the steel expert. He had already raised the question with steel experts and wire manufacturers, and he was assured that when the time came, that was, when the demand arose, there would be found a means of meeting it without reducing more than to a very slight extent the present factor of safety. He was glad to see that Mr. John Gerrard supported him in this view. Finality had by no means been reached in the manufacture of steel wire, with a material having a strength of 130 tons per square inch of sectional area. Indeed, he believed that wires having a much higher degree of strength were already drawn and used for certain purposes. This, of course, would make it possible to use ropes of moderate dimensions and weight for heavy loads and great depths.

He agreed with Mr. Fletcher that the laws relating to the working and management of collieries were altogether too numerous and too involved. The Coal-mines Regulation Act, with its Amendments and Special Rules (to say nothing of other Acts of Parliament affecting colliery working), was to his mind far too cumbersome and unwieldy; and it was very difficult, even for the best-intentioned colliery manager, to avoid unconsciously trespassing. He would like to see these laws "boiled down" into a more compact and convenient, but not necessarily less comprehensive or less effective, form.

He was scarcely prepared to take up the subject introduced by Mr. Eckmann; it was rather outside the scope of his paper. He was quite prepared to admit that electric winding arrangements possessed many advantages; but his paper dealt rather with arrangements as they existed to-day, and he had little doubt that the steam winding-engine would, for many years to come, figure very largely in the equipment of British collieries.

Mr. Rushton's observations, with regard to the use of keps, would, he believed, be supported by many colliery managers. He was in favour of their use whilst raising and lowering men.

Mr. JOSEPH DICKINSON (Pendleton) said that Mr. Winstanley's paper was so nearly exhaustive as almost to preclude addition or comment; but a few remarks might be allowed on some breakages of ropes, and on safety-catches for cages.

As to rope breakage, whilst cordially approving of holding managers responsible for everything within the scope of their duties, it seemed perhaps too exacting to include the breakage of every winding-rope. Strain resulting in breakage might occur in various ways with heavy loads being brought into rapid motion, even with all the care so generally bestowed by the winders. Breakage had also occurred without sufficient sign of failure, from either bad material to begin with, or decay from improper storage, over which the manager had no control, and for which it would be unfair to hold him responsible.

As to safety-catches on winding-cages, they had saved life and also much damage to property; therefore as a safeguard in the event of breakage they might be looked upon otherwise than as a beginning at the wrong end. At any rate, some of the circumstances associated with their use might be mentioned.

Premising that they certainly added weight to the load, which was a disadvantage, and that they also required some care to keep in order, substantial guides were required to support the arrested load, which generally helped to keep all steady. They also afforded the comforting assurance that, even if they did not on every occasion secure immunity, they did at times avert danger. Like safety-boats for escape from shipwreck, they did not always succeed; but the sailors who were most exposed were said to like them. As an occasional assurance in winding, catches should not therefore be hastily parted with.

Formerly, single link-chains were used as guides in shafts, and also iron rods; then came wooden guides suitable for all depths, and with them the opportunity for the safety-catches; whilst now wire ropes, with fly ropes between to prevent collision of cages at meetings, were becoming common, but very much less suited for the catches.

The first automatic safety-cage known in this country was that devised by Mr. Edward N. Fourdrinier, as described in the Report from the 1849 Committee of the House of Lords. It was received with mixed hope and doubt, and, not proving satisfactory, disappeared. Others followed, some of which succeeded in running successfully at all speeds.

Confidence in them became such that at one time a proposal was made for their compulsory use, together with the use of the disconnecting apparatus with safety-hook as a protection against overwinding. Consequently, enquiry was made, from which it appeared that in 1879 in the Manchester and Ireland district, 189 safety-catches were in use, and 115 disconnecting appliances. Of the catches, 184 were the Owen, 4 the Broadbent, and 1 the Walker catch, a drawing of the Owen catch appearing in the official report for that year.* Of the disconnecting appliances, 86 were the Ormerod, 16 the Bryham, 9 the Walker, 2 the invention of another Mr. Walker, and 2 the Broadbent.

Ten years afterwards, in 1889,† in the same district, another count was made: the catches numbered 180, a diminution of 9; the disconnecting appliances 186, an increase of 71.

It might be that from the continually increasing use of wire rope as guides instead of wood, the number of safety-catches was diminishing.

Mr. GEORGE B. HARRISON (H.M. Inspector of Mines, Swinton) said that with regard to ropes, he was not quite so optimistic as Mr. Winstanley, because he knew some collieries where ropes had broken even recently, although everything apparently had been done to keep them in order, and he was afraid that more ropes were broken than was generally thought to be the case. When all

---

* *Report on the Inspection of Mines*, etc., 1879, by Mr. Joseph Dickinson, page 9.

† *Ibid.*, 1889, page 27.

was done that was possible, occasional breakages must
Mr. Winstanley said that he had expressed a rather ext
in order to elicit discussion.   He (Mr. Harrison) was
colliery managers were not lacking in courage in this
they wished to be assured that what was intended
accidents would not be likely to cause them.   He had ne
an accident through safety-appliances coming into opei
case had been recorded of an accident which caused
property, but no loss of life; but he knew personally se
where they had prevented damage from being done.   I
proved that even the old form of Owen catch would
good work when it was in proper order.   It was no gr
of time since, in the Manchester district, a cage was
going to the bottom.   He knew also of several other c
the safety-catch had acted, and also of cases where
broken where it had not acted; but in all the cases witl
was personally acquainted, only that had happened wl
have happened whether they had had a safety-hook or n
all his experience indicated that it had been a great
and a source of safety in pits where they had wooden g
had had no personal experience, however, of safety-
operation with wire-rope guides, although he had seer
models; but he was optimistic, and hoped that in the f.
would be safety-catches that would do as good work
engaging-hooks.   But exhaustive experiments would I
made, say in some disused shaft, to find out wherein the
lay.

The PRESIDENT said that the thanks of all the mei
due to the author of a paper so valuable in itself, and
led to so much interesting and instructive discussion.

Mr. GEORGE H. WINSTANLEY said that he would
press his thanks to Mr. Dickinson for bringing to bea:
extensive knowledge and interesting reminiscences on sc
a subject, and he was exceedingly sorry that Mr. Dicl:
not present on the occasion when the paper was read.
marks of Mr. Dickinson and Mr. Harrison, with regarc
cages, could be dealt with in the same breath.   The saf:
in use at the present time—those that had undoubtedl
again, saved a falling cage—were at old collieries wher

of winding was comparatively slow, where the load was comparatively small, and the conductors made of wood. They had, however, to face, not the conditions of the past, but those of the present and the immediate future. For every colliery with small shafts,
wooden conductors, and slow speed of winding, with comparatively
small loads, there were several collieries with deep shafts, high
speeds, and heavy loads ; and the problem of arresting the downward movement of a comparatively slow-moving cage, of moderate
weight, and that of arresting the movement of a cage weighing
10 or 12 tons, moving with terrific velocity, were widely different. He was afraid that it would be a long time before they could
get a safety-device—which he would be most eager to welcome—
that in the event of accident would take charge of the cage, not
attempting its sudden arrest, but controlling it till it came safely to
rest. To arrest a heavy body, moving at a high velocity, was a dangerous operation. Nearly all the safety-catches now in use tended
to stop a cage where the velocity was not so high as to produce destructive effects, and these might answer the purpose at the places
where they were used. The fact, however, should not be lost sight
of, that whatever experiments were undertaken they must, as in
those recorded in the Report of the Transvaal Commission, be
with full-sized appliances, under ordinary conditions, and moving
at ordinary winding speeds, and not with models. The Transvaal
report was on the whole unfavourable to safety-cages.   Of
four safety-devices experimented with, it gave a sort of qualified
approval.   All these were complicated, and likely to get out of
order, and to be out of order when they were expected to come
into operation. If a safety-catch could be found which would
gradually and safely arrest the cage, by all means let it
be adopted.

MANCHESTER GEOLOGICAL AND MINING SOCIETY.

ORDINARY MEETING,
HELD IN THE ROOMS OF THE SOCIETY, QUEEN'S CHAMBERS,
5, JOHN DALTON STREET, MANCHESTER,
MAY 12TH, 1908.

MR. JOHN ASHWORTH, PRESIDENT, IN THE CHAIR.

The following gentlemen were elected, having been previously nominated :—

MEMBERS—

Mr. ROBERT ARTHUR FORT, Mine Surveyor, Moss Hall Coal Company, Limited, Platt Bridge, near Wigan.

Mr. F. LLEWELLIN JACOB, Mining Engineer, Ferndale Colliery, Ferndale, Glamorganshire.

Mr. JOSEPH DICKINSON read the following paper on "Deviation of Bore-holes":—

DEVIATION OF BORE-HOLES.

By JOSEPH DICKINSON, F.G.S.

It is fairly well known that the usefulness of bore-holes in searching for minerals or water, tapping water or gas, and ventilating workings, is occasionally thwarted or rendered less helpful by deviation from the intended direction. Appliances have been invented with partial success for ascertaining the line of deviation, but something of a more practical nature is required; and as the want is becoming more apparent, improvement is likely to follow.

The idea of preparing these few notes on the subject was suggested by reading the careful description of a recent deep boring at Barlow, near Selby, which had proved the existence of coal about a dozen miles east of the present collieries in the Yorkshire coal-field.*

This bore-hole was 18 inches in diameter at the top, and diminished to a few inches at a depth of 2,371 feet. It took fully two years to bore, delays being occasioned by want of water and by serious deviations of the hole from the perpendicular, one of these deviations necessitating the re-boring of 80 feet of strata.

The angle or curve of deviation from the vertical was approximately measured by the setting of cement in a bottle enclosed in a case, which showed up to 15 degrees. But the direction of the deviation was not ascertained, it being assumed to be in the same direction as the dip of the strata. The result of the exploration was very much lessened in value by the deviation of the boring.

The comprehensive account sets forth that deviation to the dip is not held universally. Yet the assumption of that direction at the Barlow bore-hole seems to have been arrived at after consideration. The deviation is shown on a diagram (fig. 4, plate

---

* "Deep Boring at Barlow, near Selby," by Mr. H. St. John Durnford, *Trans. Inst. M. E.*, 1907, vol. xxxiv., page 426.

x., vol. xxxiv.),* and is said to have been the opinion of the
person who had charge of the boring. This opinion seems also
to have been accepted by the several eminent mining engineers
and geologists who took part in the discussion on the paper, and
being thus authoritatively introduced it may be left unques-
tioned; indeed, under such circumstances, it seems desirable to
add that the following observations are not intruded as controversial discussion, but as a distinct expression, lest the introduction on such high' authority may be misconstrued into an
axiom applicable generally to bore-holes in inclined strata.

Premising that fissured strata may divert the course of a bore-
hole into almost any direction, yet ordinarily, with regular dip,
force in boring seems likely to operate otherwise than when
acting on bodies moving in fluids or on slopes. In a plumb bore-
hole in inclined strata, the boring-cutter comes first on the rise
of the bedding; thus—the rise-side of the cutter takes the weight,
allows the dip-side to sink, the boring-rods above follow, bending to the dip-side with the cutter below thrusting to the rise;
and, as a natural consequence, the bore-hole follows the cutter
towards the rise.

As a practical illustration supporting this theory, the writer
may mention that sixty years ago, having occasion to sink a
second shaft to a seam of coal found previously by boring and
proved from a trial shaft, the second shaft was begun with the
bore-hole in the centre, in the hope that it would take the water,
which it did for a time. The strata had a moderate dip, and the
shaft was perpendicular. As the depth of the shaft increased, the
bore-hole diverged to the rise-side, and there passing out was
ultimately found in the coal-working some yards away on the
rise-side. Therefore, whatever definite information may hereafter be obtained as to the direction of the bore-hole at Barlow,
deviation to the dip cannot be accepted as an axiom applicable
generally to bore-holes in inclined strata.

Without going into details, which are well worth perusal, it
may be added that the Barlow boring was by the new Calyx
system, in which a rotary cylindrical steel-cutter is used for soft
strata and another cutter, with triangular notches and chilled
shot, for hard rock, along with combinations for regulating pressure and improving observation of the small outcome borings.

* *Trans. Inst. M. E.*

The other bore-hole, where the shaft proved the deviation to the rise, was bored with an ordinary chisel-cutter, spring-pole, and hand-turning.

Whether deviation of direction in either bore-hole was influenced by the system of boring is not noticed; nor are the positions of the Red Sandstone and Coal Measures in the geological series. Each of these items is an important but distinct factor; but what the writer has at present in view is merely to draw attention to the direction most likely to be taken by bore-holes in inclined strata.

Mr. HENRY BRAMALL (Pendlebury) said that he had much pleasure in moving a vote of thanks to Mr. Dickinson for his interesting paper. As to the theory now put forward by that gentleman respecting deviation in bore-holes, he thought there was a good deal in it. His own view was that the tendency would be for the bore-hole to deviate towards the rise on the tool coming upon a hard bed in the strata.

Col. GEORGE H. HOLLINGWORTH (Manchester), in seconding the motion, said that he had quite recently come across a concrete example of the truth of the argument put forward by Mr. Dickinson. He had consulted one of the most experienced borers as to the probable direction of a deviation, who stated that in his experience deviation not caused by a fault, or an obstruction in the bore-hole, took the direction of the rise of the measures. He was glad to have this opinion confirmed by Mr. Dickinson. With respect to the survey of bore-holes, there had been a considerable amount of this work done in the Transvaal and reported to have given satisfactory results, and a German inventor had perfected an apparatus which had been much used in Germany and France.

The motion was carried unanimously.

Mr. JOSEPH DICKINSON (Pendleton), in acknowledging the resolution, said that he was surprised that none of the experienced men present at the reading of Mr. Durnford's paper raised any doubt as to the direction of the deviation of the bore-hole at Barlow.

Col. GEORGE H. HOLLINGWORTH (Manchester) said that he thought that arose from the similarity of the strata shown in the boring cores over a considerable depth.   He had had a similar case where the bore had gone a long distance in ground of the same nature, and that pointed to its following the dip. But the expectation as to this bore was that it went to the rise, and Mr. Dickinson had given them particulars of a confirmatory example.

Mr. WILLIAM OLLERENSHAW (Denton) said he thought that the deviation of bore-holes depended very much on the character of the strata in which the bore-hole was driven.   He had had an experience of a bore-hole which, about 90 feet from the surface, had deviated about 7½ feet to the dip.   Most mining men were aware that there were sometimes ironstone-nodules present in a bore-hole, which would deflect it to the weaker side.

————

Mr. A. E. MILLWARD read the following paper on " Sinking and Tubbing a Well 15 feet in diameter, and Boring Two Holes 44 inches in diameter, through Gravel, Running-sand, Boulder-clay, and other Measures, at Altham Bridge Pumping-station, near Accrington " :—

SINKING AND TUBBING A WELL 15 FEET IN
DIAMETER, AND BORING TWO HOLES 44 INCHES
IN DIAMETER, THROUGH GRAVEL, RUNNING-
SAND, BOULDER-CLAY, AND OTHER MEASURES,
AT ALTHAM BRIDGE PUMPING-STATION, NEAR
ACCRINGTON.

By A. E. MILLWARD.

Introduction.--In the spring of 1906, the Accrington and
District Gas and Water Board obtained by Act of Parliament
powers to make new water-works at Altham.   For some years the
consumption of water had been periodically restricted throughout
the district supplied, and upon two occasions (in 1904 and 1905)
the reservoirs had only contained 15 and 12 days' supply
respectively, owing to the increased requirements of a growing
population and the exceptional drought of the previous two or
three years.   The whole of the water-supply, up to this period,
was obtained from the gathering-ground of the outlying hills,
but the supply from that source could not be increased.

    The neighbouring township of Altham gave positive indi-
cations of a vast supply of water, continuous at all seasons of
the year, and of excellent potable quality.

    In the *Memoirs of the Geological Survey** of that district, Prof.
Edward Hull, guided by the difficulties in attempting to work
one of the coal-seams from the outcrop, and by his knowledge
of the geology of the locality, spoke of the obstacles that would
have to be encountered in sinking to the deeper-lying coal-seams
owing to the presence of water.   This opinion was verified fully
at the sinking of the Calder pit by Messrs. George Hargreaves
and Company, who had to contend for a number of years against
two very heavy feeders of water in the shaft sunk about 150 feet
north of the site of the Altham pumping-station.   The first of

    * "The Geology of the Burnley Coal-field and of the Country around
Clitheroe, Blackburn, Preston, Chorley, Haslingden and Todmorden," *Memoirs
of the Geological Survey of England and Wales*, by Messrs. Edward Hull, J. R.
Dakyns, R. H. Tiddeman, J. C. Ward, W. Gunn and C. E. de Rance, with
a Table of Fossils by Mr. R. Etheridge, 1875, page 74.

these heavy feeders came from a compact bed of sandstone, lying 90 feet beneath the surface, and 30 feet below the position of the Arley Mine, which had been worked out in this neighbourhood many years before.

The Altham pumping-station lies nearly at the bottom of an extensive synclinal fold, the edges of the porous rocks of which, outcropping north and south, rapidly absorb the rainfall; the overflow, when the water is not drawn upon, gravitates into the river Calder.

It was decided to sink a well 15 feet in diameter into the " Blue Clay," and tub it with cast-iron plates, and then to bore two holes to the bottom of the water-bearing rock, a depth of nearly 180 feet from the surface. As will be seen by the section (fig. 1, plate xvii.), the ground sunk and bored through consists of a series of running-sand, gravel, clay, and boulders.

For winding out the sinking-stuff, a temporary headgear and a pair of engines, with cylinders 9 inches in diameter, were supported on pitchpine baulks, carried upon brick cross-pillars, so as to enable the tubbing to be raised a little above the original level of the surface. The baulks were so placed as to allow the guide-tubes of the bore-holes to be centred from them by lines hanging down the sides of the baulks. The well was centred by placing, before the circle was struck, two points on each of the four sides, from which the true centre could be checked at any time, and which was transferred to a cross-beam bounding the movement of the banking-wagon, when this was fixed.

The Sinking.—The ground was excavated to a diameter of 18 feet, and the first $5\frac{3}{4}$ feet, consisting mostly of loamy sand, was readily got out. At a depth of 8 feet, water was met with in the sump-hole leading in the centre of the shaft. At this point the sides were supported, from the top downward, with skeleton rings and battens. The rings of wrought-iron were 3 inches deep and $1\frac{1}{4}$ inches thick, there being eight segments to the circle, connected together by overlapping joints, with five holes in each end for regulating to the exact size required. These were placed every 3 feet apart. The battens were $1\frac{1}{2}$ inches thick and 8 inches broad, and usually overlapped vertically by 8 inches. These iron rings are, as a rule, only 1 inch thick, but, in this case, the greater thickness was chosen

under the expectation of some heavy work in driving forward the
battens at the back of the rings, which expectation was fully
verified, and thus rendered a stiff ring very necessary.

The first of these rings was suspended by hangers from eight
planks, 3 inches thick and 10 inches broad, placed, as shown in
fig. 3 (plate xvii.), upon the original surface, the first four being
sunk in the ground at four equidistant points, to the extent of
their own thickness tangential to the circle of the ground got out.
The other four planks lay evenly upon the first four, and the
ground between them, tangential to points of the same circle,
midway between the others. Very little weight could come upon
the planks when the other supports were fixed tightly.   The battens were forced well against the sides of the pit by wedges of hard
wood, 2 or 3 inches thick at the head, driven between them and the
wrought-iron rings.   Where the ground was of a loose and running nature, it was found necessary at periods to play upon
the top of the wedges to keep the battens perfectly tight.

After the ground had been cleared out to a depth of 8 feet,
a pulsometer pump was put in to deal with the water that
rapidly drained into the pit through the sandy ground. The
quantity of water increased as the sinking proceeded, and before
long some of the beds of sand commenced to work like barm,
and in other cases squirted through the joints of the battens.
This was quickly stopped by caulking those joints with hemp;
and it was very seldom that the water or sand broke out a
second time in the same place. At times the sand was so loose
and fluid in the pit-bottom that the sinkers had to stand on
broad planks crossed at the ends upon others, even though the
battens were driven as much as 2 feet in advance of and below the
ground that was being removed. A hole, 4 feet square, closely
timbered by means of 1½-inch boards, was kept in advance, near
the centre of the pit, for the pump to draw from.

For the convenience of supervision, the whole of the work
connected with the sinking was carried on with single shifts
of about 10 hours per day; but it was necessary to keep the
pump continually going, not only to be ready for the next
shift, but because the water, if allowed to rise into the ground
that had been already drained, caused a further disturbance
when once again lowered. It may be mentioned that, in lining
up the shaft with brickwork later on, after the bottom of

the well had been reached, even what had beei
ground, and a source of great trouble during sink
stood during the withdrawal of the battens, and di
slightest inconvenience; but here and there, v
had been purest, there were hollow places extend
horizontally. These were puddled up with loamy
rammed afterwards.

This method of work proceeded without hind
Blue Clay was reached, where a water-garland v
depth of 23¾ feet, to enable the lower work to be p
ring was 9 inches broad, with sheet-iron shroi
high, and was at first supported on flat wrought-i
long, driven into the clay; but, owing to the water
the failure of the pump, and the clay being of r
nature, the ring became damaged on the weal
shaft, and afterwards was supported by iron han
from the lowest iron skeleton ring.

The bed of the tubbing-curb was prepared o
of concrete, 3 feet thick vertically and 3 feet b
out a little further at the bottom, and brought
of the position of the curb, being made level witl
same and reinforced at this point, with one of th
skeleton rings bedded in the concrete. Before tl
built, a lining of brickwork, 9 inches thick, set in
was put in to form a solid backing; and to enable
set beneath and around the position of the wedgin
the latter might be effectually wedged without dela
of the concrete, the bottom of the brickwork was set
in fig. 2 (plate xvii.), and gradually brought into th
leaving here a clear space of 2 inches between the
the position of the flanges of the tubbing-plates. \
work had been brought up to within two courses
ring, four short lengths of 2-inch wrought-iron pi
horizontally at equidistant points in the brickwc
wards vertical pipes were connected to them and c
at the back of the brickwork, which pipes were
with ¾-inch holes, 3 inches apart, in order to drain
thus remove any pressure at the back until tl
set. The brickwork was backed with small bro
rammed.

This operation having been finished, the tubbing-curb was put together on the bed prepared for it; sheeting of yellow-pine, ¾-inch thick, and cut to the section of the curb, was placed between each segment; and "glutting," 2 inches thick, also of yellow-pine, with the grain of the wood placed vertically, just filled up the space at the back of the curb. Wedging was then commenced, and the curb was wedged tight in five shifts, six men working in each shift. In the process of wedging, at first yellow-pine wedges were employed and continued to be used so long as they could be got in without breaking; and these were followed by pitchpine wedges, until even the steel chisel could not be made to enter. Before placing the first ring of tubbing-plates, it was necessary to brick up with a single brick in each case from the walling, set further back at this point, for the purpose before mentioned, to within 2 inches of the position of each pair of vertical flanges, so as to enable this ring of plates to be securely and truly fixed in the usual manner by spear-wedges driven into position with the utmost force. These wedges, applied in pairs, and "wedded" as shown in fig. 2 (plate xvii.), were of pitchpine, 2 feet long, 4 inches broad, and from $1\frac{1}{2}$ to 2 inches thick at the head.

Between all the joints, both vertical and horizontal, yellow-pine sheathing, ¾ inch thick, was placed, the grain of the wood in each case running in the direction of the subsequent wedging. The whole of the plates were thus built up, each ring made even, plumb, and tight, and the space behind closely filled with small broken stone, so as to afford a solid backing and yet allow the water to escape and run without resistance through the central holes in the plates into the sump.

The tubbing was anchored at the top by building into the walling at the back two plates to each segment of the curb at a distance of 5 feet below the finished top of the tubbing. Through each hole in the end of these holding-down plates, immediately at the back of the flanges of the tubbing-plates, a bolt, $1\frac{1}{2}$ inches in diameter and 5 feet 2 inches long, was passed, and the upper ends of the bolts passed through holes prepared in the holding-down curb. This curb was surrounded by two of the skeleton rings used for the temporary support of the sides, the rings being placed one on the top of the other, and embedded in and surrounded with 2 feet of concrete. The wedging of the

top three rings of plates and the curb was left unti
to allow of the concrete setting.

The wedging began at the bottom, and the vert
almost completed before the horizontal joints w
practice not always observed—but the object air
method was the prevention of the lifting of the
forcing them well against the sides and thus
thickening of the lower horizontal joints at the
upper horizontal joints; otherwise, it might have b
get a chisel in and to make the joint watertight, a
there would have been serious risk of fracture of t]

As a rule, the tubbing was twice wedged wi
wedges and then twice wedged with pitchpine wec
of comparison, pitchpine wedging alone was tried
places, no wedges of yellow-pine being used a
points; but, in such places, a greater difficulty w
in finally stopping the percolation of water throug]

Almost the whole of the wedges used wer
machine-sawn pieces, pointed with a clogger's knif
and entirely hand-shaped wedge is undoubtedly
the two for driving, but is not considered worth
pense, the cost of making them being fully thr
of the sawn wedge; and there is very little diffe
between the waste of timber arising from extra
sawn wedges, and that incurred in the operation of
making the other kind. It may be pointed o
care was taken to use only seasoned timber of the
straight-grained, and free from knots and defects.

In the bottom of the well, cast-iron guide-pipe:
holes, of 4 feet inside diameter, were fixed on t]
their centres being 8 feet apart, and were surro
depth of 4 feet of concrete. The Blue Clay was f
a little water, and to meet this, and to enable t]
be got into position and to set, without being inter
the water, the two cast-iron guide-pipes were conn
at the bottom with a 3-inch pipe, the pump keep
down " on the snore " in one of the 4-foot pipes.
drains were also arranged at the bottom leading
tubes, so that the water coming from every direc
successfully dealt with.

The Boring.—Before starting to bore inside the 4-foot in diameter guide-pipes, steel lining-tubes were lowered to the bottom of the hole, and these were kept well driven forward as the boring proceeded through the alluvial drift. These tubes were 44 inches in diameter and $\frac{1}{2}$ inch thick, formed of two plates, each $\frac{1}{4}$ inch thick; and were in lengths of 4 feet, each plate forming a complete ring of itself, being butt-jointed, and placed horizontally quarter-circle on with its companion plate, and overlapping 1 foot vertically for riveting the lengths together.

The American method of boring was adopted, which appears to be the most rapid, and ensures a very true and straight hole. The derrick, built of 2-inch planks, 8 inches broad, was about 60 feet high. The boring-tool first employed consisted of a wrought-iron head, 40 inches in diameter and 9 inches deep, studded with twenty-six cutters attached to a boring-bar, 15 feet long and 5 inches in diameter; the whole weighing about 2 tons, suspended and worked from a manilla rope 10 inches in circumference.

The percussive motion was imparted to the tool by connecting one end of a " walking-beam " to an engine-driven crank, and the other end of the beam to the boring-rope; the intermediate connection of the latter being a strong stirrup and temper-screw, for gradually lowering the tool as it cut its way into the strata, with cross-head attached, and a clamp for gripping the boring-rope. The stirrup was attached to the walking-beam by means of a strong chain, and the rope-clamp was connected to the lower end of the temper-screw by two large loose rings. About a dozen yards of rope were paid out from the reel, and one of the operators with a lever, which was readily attached to the boring-rope by means of a doubled piece of rope twisted round it, walked round and round in a backwards-and-forwards direction, and thus imparted a steady rotary motion to the cutting-tool.

From one spot on the boring-stage the chargeman could start the engine or reverse it; could throw the drum containing the sludger-rope into or out of gear, or apply a brake to the same; and could bring the reel containing the boring-rope into gear with the engine, as required, to lower or withdraw the boring-tool.

Actual boring operations commenced on June 4th, 1907, in

clay and stone at a depth of 39 feet from the bo
working single shifts of 12 hours a day, the alluv
through on June 26th at a depth of 68½ feet
datum-level.   The Boulder-clay presented ma
The Arley Mine, which was thought to be too n
to be workable, was expected to be reached at a d
but the seam had been worked out; and the shale a
this operation, bored unevenly and slowly.   The
tubes was driven tightly down to a depth of 88 fe
a smaller boring-tool was used, and the bore-hole
a depth of 121 feet before a further set of steel tub
Then the hole was continued to a depth of 124½
depth the top of the water-bearing rock was reac
14th, 1907, and the 40-inch steel tubes were
down into it.   On August 19th, 1907, in orc
matters, a second shift was started, and on
the black shale at the bottom of the water-be
found at a depth of 173½ feet, the bore-hole bei
depth of 174 feet.

After thoroughly sludging the hole, 6 hu
cement was lowered in a specially-prepared bag,
flattened out upon the bottom, to receive the risi
pumps   After this cement had been allowed to
bearing rock was lined with steel tubes, 36 inc
diameter and ⅜-inch thick, formed in the same
others, and perforated with holes ¾ inch in dia
inches pitch, so as to allow free entry of the wate
and yet to prevent detached pieces of rock from f
pipe-column and making difficult the withdrawal
case of need.

The second bore-hole was put down and finish
manner.

Engines are being put in at the present time ca
1½ million gallons of water per day under a total h
It is expected that the feeder will average 1 mill
day if exhausted to a minimum, which event, h
occur with certainly less than a year's continu
the maximum speed of the engines employed.

On the motion of Mr. G. B. HARRISON (H.M. Inspector of Mines, Swinton), seconded by the Honorary Secretary (Mr. SYDNEY A. SMITH, Manchester), the thanks of the meeting were accorded to Mr. Millward for his paper.

The PRESIDENT (Mr. John Ashworth) remarked that this was one of those practical papers that the Society was always glad to receive, and he was sure that they would be very pleased to have a discussion upon it. ·

Mr. A. E. MILLWARD (Accrington), in reply to Mr. Charles Pilkington, said that the thickness of the tubbing-plates was ⅞ inch ;· and they had three vertical and three horizontal ribs; he was of opinion that more depended upon the strength of the ribs of tubbing-plates than on the thickness of the plates.

The HONORARY SECRETARY (Mr. Sydney A. Smith, Manchester) said that he would like to ask a question on the following statement in the paper, namely, that " the wedging began at the bottom, and the vertical joints were almost completed before the horizontal joints were started " . . . in order to prevent " the lifting of the plates, by first forcing them well against the sides and thus avoiding the thickening of the lower horizontal joints, etc."[*]   Did that mean that the wedges were put in from the inside, in order to swell out the tubbing to fit the shaft? Would it not have been very much better to have wedged the tubbing outside between it and the strata or side of the shaft, and fix tight the segments in correct position, and then wedge the joints? The method of wedging described must have disturbed the tubbing to a considerable extent. ·

Mr. A. E. MILLWARD said that the first process in building up the plates was to make them as tight as possible. The spear-wedges fixing them must be driven to the utmost extent with heavy hammers.

Mr. SYDNEY A. SMITH said that this fact was not stated in the paper. It appeared from the description given that the joints were wedged from the inside, in order to swell out the tubbing to fit the shaft, and that the vertical joints were

---

[*] *Trans. Inst. M. E.*, 1908, vol. xxxv., page 405 ; and *Trans. M. G. M. S.*, 1908, vol. xxx., page 299.

completed before the horizontal joints were ˙
(Mr. Smith) was of opinion that Mr. Millwar‹
wedged the tubbing first of all on the outside, a
position that when he came to wedge inside to
water-tight it would not have disturbed the tu

Mr. A. E. MILLWARD said that the pressure
wedging was exceedingly great.   An acquaint
said: "We had such a job when we came to ˅
horizontal joints; we could hardly get the chi‹
case, not having the holding-down curb suff
anchored, he could not risk any lifting of the pla¹
that it was inadvisable to wedge alternately vertic
joints, and so he had every vertical joint wedg
possible before proceeding with the horizontal j

Mr. SYDNEY A. SMITH said that the point ˅
to make clear was this:—Mr. Millward had
vertical joints were wedged first of all, the
being the prevention of the lifting of the plates
them well against the sides.   His (Mr. Smith'˷
Mr. Millward ought to have wedged the ring
tight in the shaft—from the back—that the w‹
inside would only be to make the joints water

Mr. A. E. MILLWARD did not think that tha
He did not think that one could wedge the tubbin
at the back without the subsequent wedging ɴ
plates.   The plates did not stir to any obvious
vertical and horizontal joints were wedged from
the tubbing was now perfectly water-tight.

In reply to the President, Mr. Millward s˙
was a very good supply of water, and there ha
as 60,000 gallons an hour flowing about a for
occurrence of heavy rain-storms.   The pumps
were a pair of buckets, 18 inches in diameter, ˙
The engines were of a local make, and the pressɪ
under steam was 150 pounds per square inch.
very free from sand, and a little hard.

Mr. JOSEPH DICKINSON (Pendleton) said that
that " between all the joints, both vertical and ho

pine sheathing, ⅜ inch thick, was placed, the grain of the wood in each case running in the direction of the subsequent wedging."* As to the grain of the wood, he had heard some difference of opinion expressed as to what was the best way of placing sheathing between the tubbing. Which grain did the writer mean?

Mr. A. E. MILLWARD said that he meant the vertical grain. He did not think that any other grain would matter, so long as the grain was in such a direction that it would easily cleave and take the wedges. That was the only point to notice, for unless that were done the wedges could not be driven in.

Mr. WILLIAM PICKSTONE (Kersal) enquired whether the rainfall very rapidly found its way to the hole.

Mr. A. E. MILLWARD replied, moderately so. Roughly speaking, it took about a fortnight to get through. He had a similar experience when engaged at a neighbouring sinking. In about a fortnight after a continuous rain-storm they would find the water increasing while the pumps were doing their usual work.

Mr. A. E. MILLWARD, in reply to Mr. Dickinson, said that he did not think that workings in the Mountain Mine (900 feet below) would disturb the water, as there were several thick layers of shale that were bound to keep back the water in its native rock. His experience of other workings gave him assurance that there was no danger to be feared of the water going. If it did, they would have to stop the workings, as they would not be able to pump the water.

In answer to further questions, Mr. Millward said that there was no impervious cover on the water-bearing rocks ; water flowed in a continuous stream from an old pit-shaft about ¾ mile from the bore-holes. There was no Boulder-clay there, the Boulder-clay through which they had to sink ran from about 300 to 600 feet on either side of the river Calder, the ancient river-bed. In places, there was nothing but the soil on the top of the water-bearing rock. They got varying layers of strata at the surface.

* *Trans. Inst. M. E.*, 1908, vol. xxxv., page 404 ; and *Trans. M. G. M. S.*, 1908, vol. xxx., page 298.

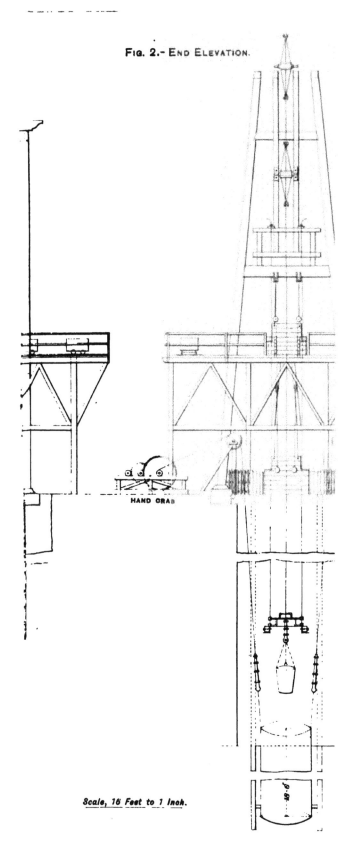

FIG. 2.- END ELEVATION.

HAND CRAB

Scale, 16 Feet to 1 Inch.

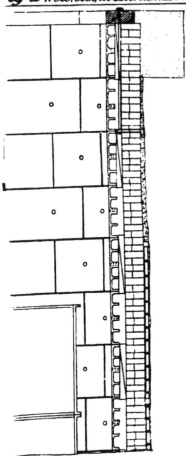

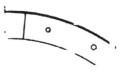

ARRANGE BRIDGE PUMPING-STAT

-PLAN.

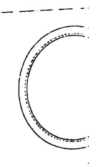

Mr. SYDNEY A. SMITH asked what was the total area of the gathering-ground from which the water flowed to the bore-hole.

Mr. A. E. MILLWARD replied that, speaking roughly, it would not be more than 5 or 6 square miles; and on either side there were hills that must bring down water, and when the ground was well drained a great quantity of this must also be absorbed. There was a large layer of sandstone rock that would rapidly take the whole of the overflow of water; he simply took the ground over the rocks which could be measured in many ways.

———

MANCHESTER GEOLOGICAL AND MINING SOCIETY.

ORDINARY MEETING,

HELD IN THE ROOMS OF THE SOCIETY, QUEEN'S CHAMBERS,
5, JOHN DALTON STREET, MANCHESTER,
JUNE 16TH, 1908.

MR. JOHN ASHWORTH, PRESIDENT, IN THE CHAIR.

The following gentlemen were elected, having been previously nominated : —

MEMBERS—

Mr. JOHN WILLIAM JOBLING, Mining Engineer, Clifton Colliery, Burnley.

Mr. ARTHUR MOORE LAMB, Mining Engineer, Eskdale, Birkdale, near Southport.

Mr. JOSEPH CRESSWELL ROSCAMP, H.M. Assistant Inspector of Mines, Prestwich, Manchester.

Mr. FRANK G. L. SAINT, Colliery Manager, Glen Tarn, Hindley Green, near Wigan.

.

The following paper by Mr. A. GHOSE on "The Mode of Occurrence of Manganite in the Manganese-ore Deposits of the Sandur State, Bellary, Madras, India," was read in his absence : —

THE MODE OF OCCURRENCE OF MANGANITE IN THE MANGANESE-ORE DEPOSITS OF THE SANDUR STATE, BELLARY, MADRAS, INDIA.

By A. GHOSE, F.C.S.

The deposits of manganese-ore found by the writer in the hill-ranges of the Sandur State, Madras Presidency, India, are not only remarkable for the extraordinary development of the is general in the series of deposits, and their mode of occurrence of some minerals of manganese in finely crystallized forms. Of these, manganite, which is comparatively rare in other parts of India, is so frequently met with in the manganese-ore deposits situated on the Sandur Hills, that no deposit is known in that area where it has not been observed. Although the crystals of manganite are not always found in the state of development which is noticeable in some specimens, yet their distribution is general in the series of deposits, and their mode of occurrence is identical with rare exceptions. They are invariably found lining cavities or infilling narrow fissures formed in the beds of the ore, either consisting of psilomelane or of psilomelane and wad, and more rarely of an intimate mixture of psilomelane and braunite. A typical specimen of manganite from the Ramandrug mine, which was the first to be identified, has been described by Mr. L. L. Fermor,* Deputy Superintendent of the Geological Survey of India.

The cavities assume various irregular shapes, and are frequently angular and very rarely ellipsoidal. They vary in size from the fraction of an inch to a few inches across. Their presence inside the ore-mass cannot always be detected by examination of the outward appearance of the ore, and geodes lined with crystals have been found inside solid-looking ore, although their occurrence is more frequent in ore which shows an apparently brecciated structure. Sometimes the exterior surface of the ore-mass containing the geodal cavities shows a

---

* *Records of the Geological Survey of India*, 1906, vol. xxxiii., pages 229-232.

peculiar " pitted " structure, resembling " thumb-marks." These depressions were caused by the removal of crusts concentrically deposited in hollows. The drusy cavities, in rare instances, have been found immediately on fracture to contain minute traces of a liquid which has hitherto afforded no opportunity for an examination. Where the concentric lining of the cavity consists of compact psilomelane without any capillary fissures, the acicular crystals of manganite are found to retain a shimmering brilliancy almost approaching irridescence. On the other

Fig. 1.—Ramandrug Mine, showing Main Working and Ore-stacks.

hand, subsequent infiltration, in the absence of a strongly coherent envelope, has left the crystals with ferruginous or siliceous encrustations.

The mass of ore containing the cavity frequently shows numerous ramifications of compact psilomelane filling narrow fissures, which, crossing each other, give a brecciated appearance to the ore. The inside of the cavity is lined with concentric layers of compact fine-grained psilomelane showing successive deposition, which is sometimes marked by a thin separating ferruginous film due to subsequent percolation or deposition

of suspended impurities. Delicate needle-like pi
of manganite grouped in bundles, and sometim
pendicularly to the walls of the cavity, showing bru
tions and more frequently exhibiting radiated fi
project towards the central druse.   The comp
and the crystals of manganite are symmetrical :
Pseudomorphous crystals of pyrolusite exactly 1
acicular form of manganite are very frequentl
the cavities, and are probably the alteration-
dehydration of original crystals of manganite.   It
in many cases the crystals of manganite have ui
tion, and are no longer true manganite.   In s(
fissures alternate deposition of compact psilom
crystalline mineral with silvery lustre (probal
has been noticed. The layer of this hitherto unid
is covered by another crust of compact psilom(
surface of which crystals of manganite spring
metric succession of minerals is an evidence of
pitation, and the oldest crust always rests directly
ore.

Of the various hypotheses propounded from
account for the origin of the manganese ore-i
Sandur State, two theories appear to be based oi
sound deductions.   One of these advocates a c(
sedimentary origin, and the other maintains th
are essentially the result of replacement of the o
rock by mineralizing solutions.   On the assun
ore-deposits are of contemporaneous origin witl
Dharwar rocks, it is admissible to suppose that tl
of later origin, and were formed and filled up su
folding and uplifting of the ore-beds. Whether t
of the fissures and the formation of the cavities, t
of psilomelane, and the crystallization of m
the resultants of uprising magmatic waters
igneous intrusions is problematical. The presen(
mangan-magnetite, and specularite in the ore appe
support to such a theory. On the other hand, that t

* A new manganese-mineral, discovered by Mr. L. L.
F.G.S., and described by him in his paper on "Manganese in]
of the Mining and Geological Institute of India, 1906, vol. i., pa

formed and the manganese-minerals were deposited by ascending
and descending solutions is easily inferred.   The immense stress
to which the rock-formation in the Sandur area was subjected
in the pre-Cambrian Era (which resulted in the strata being
compressed and uplifted locally into great synclinal folds),
and the intrusion of the younger granites (which are exposed
in the surrounding regions by the denudation of the overlying

FIG. 2.—PORTION OF MAIN BED AT RAMANDRUG MINE.

Dharwars), crumpled the rocks and produced numerous narrow
fissures and tension-joints which are so conspicuous in the Sandur
ore-beds.   Circulating waters produced enlargement of these
fissures by solution of the more soluble portion of the walls, the
effect being greatest where the channels crossed parts of the
deposit containing lenticles of wad formed in the original ore-

sediment during deposition, which were partl
moved, leaving cavities of proportionate dimen
containing varying proportions of manganese w
On relief of pressure and temperature in the u
deposit, a portion of the manganese in the s
was precipitated on the walls of the cavities
psilomelane of compact texture. The fissures
came plugged by the gradual deposition of
further circulation was obstructed, and
manganese-charged solution, being confined wi
deposited crystals of manganite on the walls
defined fine structure of the crystals of ma
indicates periods of quiescence of the mangan
tion, which facilitated slow crystallization.

To reconcile the hypothesis according to
deposits have been formed by the replacemen
rock by manganese-bearing solutions derived f
rocks, the origin of the cavities may be expla
assumption of pre-existing fissures. The orig
bearing rock of sedimentary origin contained
erals as mechanically-deposited disseminations
rock-mass. The manganese or the manganifero
not evenly distributed throughout the mass, :
there was considerable variation in the saturatic
ing waters which carried away the manganes
the degree of concentration depended on the lea
or barren portions of the rock. The manganifer
deposited its burden in another part of the rock
the mineralizing solution was not always of u
it may be inferred that there was considerable ·
rate of dissolution of rock and deposition of 1
which took place simultaneously. At times, the :
a quantity of mineral equal to the amount of
undergoing replacement, dissolved and carried :
the supply of the manganese-mineral held in sc
there was a corresponding diminution in the pr
ore, which was unable to fill up the space create
of the rock by dissolution. Thus, when diss
precipitation, the inability of the mineral-bea
deposit the ore in sufficient quantity to fill 1

dissolution resulted in the formation of narrow open spaces and
cavities, which mark periods of cessation or interruption of the
process of deposition. These spaces of dissolution subsequently
served as loci for the deposition of compact psilomelane and
manganite.

Fig. 1 is a view of the Ramandrug mine, where geodes lined
with manganite are abundant, showing the main working and
ore-stacks at a distance. The protruding mass cresting the hill
on the left of the view is entirely composed of manganese-ore.
The light railway conveys the ore to the loading station of the

FIG. 3.—PORTION OF A GEODE OF PSILOMELANE FROM THE RAMANDRUG
DEPOSIT, CONTAINING CRYSTALS OF MANGANITE.

aerial tramway (2,600 feet long) which connects the deposits,
situated at a height of 3,150 feet above the sea-level, with the
railway-terminus in the plain.

Fig. 2 is a view of a portion of the main bed at Ramandrug.
The ore-body originally formed a fine cliff 30 feet high, extending over 400 feet in length. With the progress of mining operations, it has been exposed to a further depth of over 20 feet, and
forms a conspicuous deposit, showing bedded structure. This is
one of the numerous extensive manganese-ore deposits (number-
ing over sixty) found in the Sandur State.

Fig. 3 is a portion of a geode of psilomelane from the Raman-drug deposit, containing fine, fibrous, divergent crystals of manganite.

———

On the motion of Mr. J. S. BURROWS (Atherton), the thanks of the meeting were tendered to Mr. Ghose for his paper, the motion being seconded by the HONORARY SECRETARY (Mr. Sydney A. Smith), who said that the paper was a most interesting one, dealing, as it did, with one of the less common minerals.

———

The following paper, by Mr. LEONARD R. FLETCHER, was read, on "The Patent Keps under the Cages at Chanters Pit, Atherton Collieries":—

THE PATENT KEPS UNDER THE CAGES AT CHANTERS PIT, ATHERTON COLLIERIES.

By LEONARD R. FLETCHER.

For many years the ordinary fang-keps, with which all colliery-managers and engineers are more or less familiar, were in use at Chanters No. 1 pit. A few years ago, however, the attention of the management was called to the various forms of improved keps on the market; and, after consideration, it was decided to give a trial at this pit to the Beiens mechanical kep-arrangement. The chief advantages claimed for these keps are:—

(1) The allowing of the withdrawal of the keps from under the cage with the full load of the cage upon them, so that the engineman has no need to reverse his engines for banking purposes.

(2) A consequent saving of labour to the engineman in manipulating the winding-engines.

(3) A consequent saving of time and steam.

(4) A consequent reduction of wear-and-tear on the winding-appliances.

After three years' experience with a set of these keps at this pit the writer can say with confidence that these claims have been justified in actual practice.

Fig. 1 (plate xxxii.)* shows the general arrangement of the keps; whilst figs. 2 and 3 show the construction of one catch with casing, B, twin-lever, C, and shaft, D, and the three different positions, f, A, and g, which the catch takes up when brought into action.

The essential parts of the keps are:—A, the supporting bar on which the cage rests; B, the casing; and C, the twin or forked lever (figs. 2 and 3).

The twin-lever, C, is keyed on the same shaft, D, as the banksman's lever, and makes with it a circular movement when operated

* *Trans. M. G. M. S.*, vol. xxx., plate x.

by the banksman. The connexion of this tw:
supporting bar, A, is effected through a pin, a, wh
b, and is caused to move in a circular slot, c, i
bar, whereby this bar executes a backward or fo

The winding-cage rests on a level surface o
bar, A. The pressure exerted by the winding-c
taken by the surfaces of the casing at the
and e (fig. 3). Both of these surfaces form in t]
tion a part of the circumference of two concer
are so arranged that the pressure on the pin
slot is reduced to a minimum.

The supporting bar, A, when withdrawn (fig
only a backward, but also a downward moven
of the concentric surfaces of the bar and the ca:
the winding-cage receives a slow downward mov

In case it should happen that the supportin
be pushed forward too soon, before the winding-
the proper height, the ascending cage will tur:
bar round the pin and push it out of the w
ascending cage can pass freely through (fig. 3, f).
has passed the bar, the latter again falls into its
by reason of its own weight, and is ready for
on (figs. 2 and 3, A).

For comparatively light loads these supportin
with a broad flat surface, and the frame of
directly upon them; but for heavy loads thes:
with a sloping face, and a counter-shoe with
face is riveted firmly on the underframe of
Chanters pit these bars have the flat face, but f:
a shoe with a flat face is riveted on the cage.
works in a notched quadrant, and is easily ma:
banksman. The makers of these keps claim th:
suitable than any others for heavy loads, and
already been made for loads of 27 tons, and a:
satisfactorily.

Since these patent keps have been put in a:
pit, three-deck cages have been substituted f
cages formerly in use, and the advantage gain:
these keps has been most pronounced. Apart
of time, labour, and steam, there is, of cou:

accidents from overwinding due to the engineman starting the wrong way.   As soon as the cage is resting on the catches, the reversing lever can be moved over ready for the next winding, the keps being withdrawn from under each deck of the cage without any help from the engineman.   Thus one source of accidents due to overwinding is minimized, and the winding-engineman is relieved of much unnecessary work and worry. Objection is sometimes raised to the catches from the mistaken theory that slack chain may be resting on the cage, which will suffer a severe shock and strain when these keps are withdrawn. As a matter of fact, this can never happen in actual practice, as the weight of the opposite cage-rope hanging in the pit will instantly take up any slack rope or chain that there may be when the top cage comes to rest upon the keps.

In conclusion, the author would say that this short paper is not written with the object of raising the question whether or not keps should be used under pit-cages when winding coal or men, but to bring before the members an ingenious and simple contrivance which has all the advantages and none of the dis-advantages of the old form of fang-keps.

———

Mr. JOHN GERRARD (H.M. Inspector of Mines, Worsley) proposed a vote of thanks to Mr. Leonard R. Fletcher for his admirable paper, which was seconded by Mr. W. SCOTT BARRETT (Liverpool).

The resolution was passed.

Mr. L. R. FLETCHER (Atherton) thanked the members for their vote.

———

Mr. T. H. WORDSWORTH read the following paper on " The Lee Safety-appliance for Cages " : —

THE LEE SAFETY-APPLIANCE FOR

By T. H. WORDSWORTH.

The following is a description of an applian
vented to prevent a cage from falling to the botto
in the case of a winding-rope breaking where wire-
are in use.

The principle of the invention is a cam to gri
ductors, actuated in the first instance by springs, a
weight of the cage. The chains which keep these
action during the winding would also act as safety
of any of the four corner-chains breaking.

Figs. 1 and 2 (plate xxxiii.)* are side and end
double-decked colliery cage fitted with the applianc
4 are enlarged views of the gripping device.

Each side of the cage, a, is fitted with a vert
upper ends of the bars being connected together by
and to the cage suspension-chains, d, which in to
to the winding-rope.

The bottom ends of each bar are secured to a
under the cage-bottom, f, and two or more coile
are disposed between the bar, e, and the cage-b
right angles to each vertical bar, b, is a horizonta
slides freely in a vertical direction in suitable gui
i, connected to the cage, a. The guides, i, a
clearly in the end view (fig. 5, plate xxxiii.),* anc
plates, i,i, spaced apart to receive the horizontal ba
of the horizontal bars, h, are extended around an
side-conductors or guide-ropes, j, of the cage, for
for the guide-ropes. This is more clearly shown
plan (fig. 4, plate xxxiii.).* The side bars, b, are
bent portion, b^1, as shown in the side view in fig.
to each vertical side-bar, b. At this point are a
levers, k, the outer ends of which are also pivoted t
bars, h, at h^1, the cranked ends, k^1, of the levers, k,

* *Trans. M. G. M. S.*, vol. xxx., plate xl.

within the trough. The levers, k, are loosely pivoted to brake-shoes, m, so formed in conjunction with the troughs on the ends of the horizontal bars, h, that as when the levers, k, are pulled downwards with the vertical bars, b, the levers, k, turn on their fulcra, and, by means of the brake-shoes, securely grip the conducting ropes, j.

The action of the apparatus is as follows;—When the cage is in use, the weight of the cage is transmitted through the middle suspension-chains to the vertical side-bars, b, which are raised, and thus compress the springs, g, g, interposed between the bottom cross-bar, e, and the cage-bottom, f. The angle-levers, k, are also lifted centrally, removing the brake-shoes, m, from contact with the guide-ropes, j, the cage running freely on the guide-ropes. In case of breakage of the winding-rope, the lifting strain on the vertical bars, b, will be released and, the compressed springs, g, will force the bars, b, downwards. The angle-levers are thus turned on their fulcra, as shown in dotted lines in fig. 1, (plate xxxiii.),* and the brake-shoes, m, brought into action to grip the guide-ropes, j, thus securely binding the horizontal bars, h, and the angle-levers, k, to the guide-ropes, and immediately arresting their descent.

The descending cage compresses the springs, g, until the whole weight of the cage, acting through b, is brought to bear upon the angle-levers, k, and exerts such a powerful gripping action upon the guide-ropes, j, that the fall of the cage is arrested.

Latches or detents, n, are pivoted to one of the guide-plates, i, to act as abutments to prevent the premature lifting of the horizontal bars, h, the latches being removed by a stop, o, attached to the pivoted angle-levers, k, when the apparatus comes into action, as will be seen from fig. 5 (plate xxxiii.).*

The writer does not suggest that this appliance is absolutely perfect, in case the rope should break on the empty side when the cage is going down during a fast wind: as it is doubtful whether rope conductors would under that condition be capable of withstanding the strain. But, even under such conditions, the damage to property would be no greater than if the appliance were not in use. In all cases of the full-side rope breaking the appliance would hold the cage. There are also many instances on record of overwinding where, when the detaching-hook has acted properly, the chains, bell, or some other

* *Trans. M. G. M. S.*, vol. xxx., plate xi.

portion of the tackle has given way, and allowed the cage to fall to the bottom of the shaft. In such a case this appliance would hold the cage, and prevent an overwind from developing into a serious accident.

An objection to catches of this nature has been that they come into operation when the cage is at rest on the keps, or at the bottom of the pit. In this arrangement, however, when the cage is at rest on the landing at the pit-bottom, the springs are kept compressed, so that the cams do not grip the conductor; and an amplification can be made which causes the springs to be compressed when the cage is at rest on the keps.

The following " Description of a New Patent Appliance for arresting the descent of Cages in Shafts, in the Event of the Winding-Rope Breaking," by Messrs. Joseph Hindley and John Stoney, was read by Mr. JOSEPH HINDLEY : —

DESCRIPTION OF A NEW PATENT APPLIANCE FOR ARRESTING THE DESCENT OF CAGES IN SHAFTS, IN THE EVENT OF THE WINDING-ROPE BREAKING.

By JOSEPH HINDLEY and JOHN STONEY.

The appliance has for its object the arresting of a cage in a mine-shaft in the event of the winding-rope breaking, or becoming detached from any cause; and is shown in detail in figs. 1 to 9 (plate xxxiv.).* In the first place, if the rope breaks a considerable distance above the cap, the lower part of the rope must receive an acceleration downwards relative to the cage. The springs, a (fig. 1), have sufficient tension to pull the rope down even in an extreme case (that is, the longest length of rope), within a reasonable time.

The braking force is obtained by means of wedges and forked levers (one of each to each guide-rope). The wedge, b (fig. 8), moves in a suitable box, c (fig. 6), with the thin edge of the wedge uppermost. The brake is applicable to cases where two, three, or four guide-ropes are used, and can be applied to existing cages, the above-mentioned boxes, c (figs. 1, 2, and 3), being placed in the position formerly occupied by the guides or thimbles. The wedge is lifted in position by means of the lever, d (figs. 4 and 5), which is provided with a jaw-end; the latter, when raised, approaches the guide-rope and grips the same without bending it, thus giving an additional pull to the wedge.

The braking force is dependent upon the angle of the wedge. The angle adopted by the authors is sufficient to cause the cage to stop dead, if the rope breaks on the ascent. If the descending cage has a considerable velocity when the rope breaks, then the arrest of the cage will be gradual, the braking distance being proportional to the speed prior to the rupture. An important feature of the appliance is that no parts of the brake are under stress (except the springs) during ordinary winding conditions. The mechanism is extremely simple, and no movable parts

* *Trans. M. G. M. S.*, vol. xxx., plate xii.

project beyond the guide-ropes, so that there is no liability of these parts coming into action through meeting any obstruction in the pit-shaft. The boxes (figs. 4, 5, and 6) are provided with guides or thimbles at both top and bottom, which are easily and cheaply renewed when worn.

The face of the wedge, b (fig. 8). and the corresponding part of the box is grooved in semi-circular form, so that there is no cutting, kinking, or alteration in the shape of guide-ropes.

Mr. GEORGE H. WINSTANLEY (Manchester) said that it was very refreshing to find a safety-cage for which the inventor did not claim that it was capable of doing everything. Mr. Wordsworth had, for the inventor, frankly admitted a doubt as to the action of the Lee appliance in the case of a rope breaking with a descending cage. That it would act in the event of a rope breaking when the cage was ascending was obvious, and there were many such appliances that would act under these conditions. It was interesting to note the important circumstance in this particular appliance, that its use would not prevent the employment of "decking platforms." With regard to Messrs. Hindley and Stoney's appliance, it was pleasing to note that the inventors had recognized and taken into consideration what many inventors of safety-cages had refused absolutely to admit, namely, the existence of kinetic energy, which was a tremendous force to be reckoned with in the case of a heavy cage descending at a high speed. Messrs. Hindley and Stoney had apparently recognized the necessity of stopping the descending cage gradually, with a sort of braking effect, and avoiding anything like a sudden stop. Any attempt to stop suddenly a rapidly moving cage was simply courting disaster. He moved a vote of thanks to the three gentlemen.

Mr. H. STANLEY ATHERTON (Bolton) seconded the motion, which was adopted.

Mr. GEORGE H. WINSTANLEY asked whether Mr. Hindley or Mr. Stoney could explain what would happen in the case of commencing a winding where the acceleration of speed was very quick. One could conceive that there would be for a moment an apparent reduction in the weight of the descending cage, due to acceleration. Would the springs then come into operation? He would like to be assured that this was not likely to happen. It was possible for the springs in some appliances to come into operation, on account of a temporary reduction in the weight of the cage.

Mr. JOHN STONEY (Manchester) said that they had made an indicator, and obtained diagrams from the same which showed the acceleration of the cage, and the tension on the winding-rope at any part of the pit-shaft. This appliance he exhibited to the members.

Mr. JOSEPH HINDLEY (Tyldesley) exhibited diagrams shew-ing the different effects produced by varied conditions, which demonstrated that the tension on a winding-rope varied from half the weight of the cage upwards. This gave a factor of safety of 4, if the tension of the springs be equivalent to one-eighth of the weight of the empty cage.

In reply to another question as to whether the greatest tension of the winding-rope had been found by this indicator, Mr. Hindley said that such information would be of no service to them in their work, but it was possible to get it easily by a slight alteration in the indicator. The catches had been tested up to a weight of 6 hundredweights. They had not been tried, as yet, under ordinary working conditions.

Mr. G. H. WINSTANLEY suggested that Messrs. Hindley and Stoney should make further experiments, and give the results to the Society in a future paper. He regarded the instrument for gauging the tension on the rope as of very great value.

Mr. HINDLEY said that he and his friend would be pleased to read such a paper before the Society. He assured the meeting that the guide-ropes, which were an essential part of the safety-appliance, would stand the force that was likely to be thrown upon them at any time during action.

Mr. JOHN S. BURROWS (Atherton) asked what would happen if a large portion of the broken rope fell on the cage. He apprehended that the rope would wrap itself round the chains and be most difficult to get away.

Mr. HINDLEY replied that the rope falling
a gradual force applied to the cage; if it droppe
(which it would do or miss the cage), it would c
tighter.

Mr. JOHN S. BURROWS said that he had kn
itself so much round a cage that the rope had
in pieces. Would not 1,200 feet of rope falling
cage in the shaft upset the safety-arrangement.

Mr. HINDLEY replied that it could not possi
wedges; in fact, it would tighten them. The)
account the falling winding-rope in their cal
strengths of various parts.

The Institution of Mining Engineers
Transactions. 1907-1908.

Vol. XXXV, Plate.

To illustrate Mr Leonard R. Fletcher's Paper on "The Patent Keps under the Cages at Chanters Pit, Atherton Collieries."

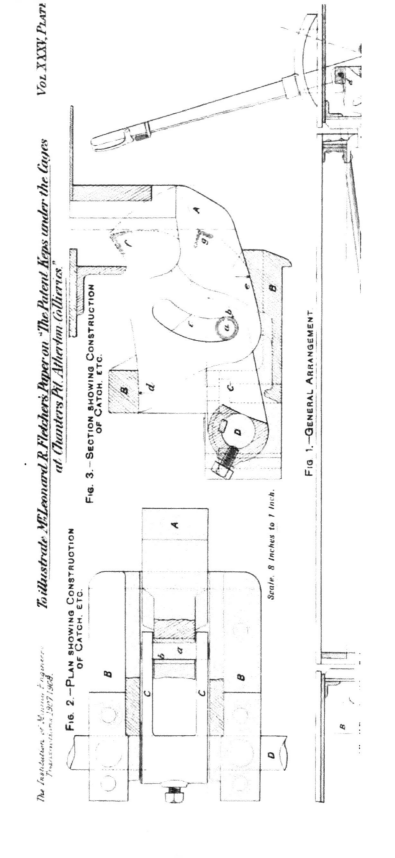

Fig. 2.—Plan showing Construction of Catch. etc.

Fig. 3.—Section showing Construction of Catch. etc.

Scale, 8 Inches to 1 Inch.

Fig 1.—General Arrangement

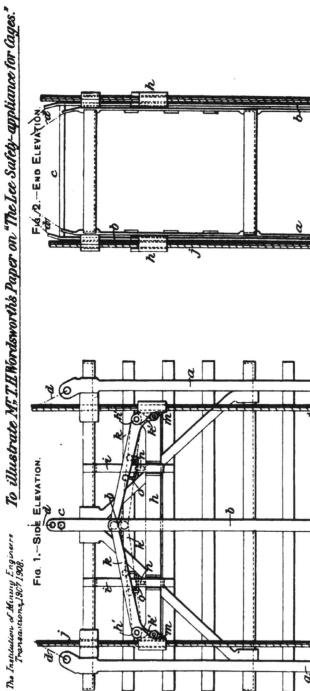

The Institution of Mining Engineers
Transactions, 1907-1908.

To illustrate Mr. T. H. Wordsworth's Paper on "The Lee Safety-appliance for Cages."

FIG. 1.—SIDE ELEVATION.

FIG. 2.—END ELEVATION.

MANCHESTER GEOLOGICAL AND MINING SOCIETY.

ORDINARY MEETING,

HELD AT THE UNIVERSITY, MANCHESTER, JULY 3RD, 1908.

PROF. W. BOYD DAWKINS, VICE-PRESIDENT, IN THE CHAIR.

The following gentlemen were elected, having been previously nominated : —

MEMBERS—

Mr. THOMAS JOHNSON, Mining Engineer, Dover Castle Hotel, Dover ; and P.O. Box No. 1056, Johannesburg, Transvaal, South Africa.

Mr. FRANCIS A. LINTON, Colliery Manager, Bickershaw Lane, Bickershaw, near Wigan, Lancashire.

ASSOCIATE MEMBER—

Mr. ROBERT CARTWRIGHT, Colliery Under Manager, Strangeways Hall Colliery, near Wigan, Lancashire.

Mr. HENRY HALL read the following paper on "Ignition-points of Wood and Coal" : —

IGNITION-POINTS OF WOOD AND COAL.

By HENRY HALL, I.S.O., H.M. Inspector of Mines.

There has always been a certain amount of mystery in connexion with spontaneous fires underground and on board ship,
and the scientific explanation of the phenomenon does not always
fit in with the circumstances which present themselves when
these outbreaks occur. Formerly, it was the fashion to assign
them to pyrites disseminated either in the coal-seam itself, or
in the strata immediately overlying or underlying the coal.
Nowadays we are told that the coal is the culprit in the entire
absence of pyrites, and that its hunger for oxygen causes heat
to be given off to such an extent that, unless there is free radiation, it may become ignited, and cause an underground fire with
all its attendant dangers. This latter view, if correct, would
appear to bring all coal-seams into the same category, but experience teaches that only a very few seams are liable to ignite
spontaneously: generally, in any mining area only one or two
earn this reputation. The same is the case when coal is put into
stock in large quantities, or is shipped for long voyages: that is,
only certain seams are implicated; but, curiously enough, coal
from seams which have never been known to fire underground in
the mine frequently do so when stocked or shipped. This may
probably be explained by the fact of the much larger bulk in
the one case than in the other.

In almost every instance it is found that the outbreak starts
from the point where the coal is most dense and subjected to
crush or pressure: on board ship, immediately under the hatches
through which it is loaded; in the mine, the "waste" seems to
be the dangerous point, although we hear sometimes of fires
starting on the side of pillars, and said to be due to the heat
given off through friction. With the exception of these latter
cases, of which the writer has no personal experience, all such
fires start at a point where the material is densest and radiation
most impeded.

This subject of spontaneous combustion has
means thoroughly thrashed out, and a pap
chemist would be especially welcomed by the
of the Society. It is very difficult, almost in
adequate precautions against a danger of the o1
development of which so little is known. E2
doubt, taught many how to recognize some of
indicate that a fire is imminent, but as to its w
one's knowledge is small; and when the fire ac1
it comes as a surprise, whereas the "scientif.
with a smile that what had happened was a
under the circumstances.

It is well known that if a chemical action
developed is constantly going on in a mine,
methods in operation are such as to hinder that
freely radiated as it is generated, there will in t
of " something " catching fire, and that " somet
material within the heated area which ignites'a
perature. But still there is a lack of informatio1
nature of the physical or chemical property of
seam which renders it liable to spontaneous c
rapid rise of temperature which takes place v
prevented can be simply shown by covering an
light bulb with a material such as small coal
found that in the course of an hour or so the he1
that the glass of the lamp will melt and collaps

The writer was led to make the tests which h
consequence of what he saw at an undergrou
which arose in an underground boiler-house,
which were several beams of timber, put i1
strength. These had in most cases entirely s
and although both the cannel-seam and some of
fire, it looked as if the timber had begun the m
of this timber (it had been built into the wall:
and of the cannel were tested, and it was foun
ignited at a temperature about one-third lowe
needed in the case of the cannel.

* See in this respect "An Outbreak of Fire, and its
Colliery," by M. F. Holliday, *Trans. Inst. M. E.*, 1905,
vol. xxx., page 167 ; and 1906, vol. xxxi., page 2.

This experiment was followed by more careful and elaborate tests of various kinds of wood and coal from different seams, which were carried out in the laboratory at Kirkless through the courtesy of the Wigan Coal and Iron Company, Limited, and with the assistance of Mr. T. H. Byrom, the chemist to the company. The results of these experiments are contained in the following table:—

RESULTS OF EXPERIMENTS TO TEST THE EFFECT OF HIGH TEMPERATURE ON TIMBER AND COAL.

Sample Tested.	Description of Sample.	Results of Testing in a small Fire-clay Muffle.
Swedish timber	Old sleeper from Wigan Four Feet mine; approximate age, 6 years.	With various tests this sample smouldered at 392°, 356°, 410·, 428° and 428° Fahr. (200·, 180°, 210°, 220°, and 220° Cent.). Average temperature, 403° Fahr. (206° Cent.).
Welsh larch ...	Part of 11-foot bar from same mine as above, and about the same age.	Test No. 1: Smouldered at 680° Fahr. (360° Cent.); temperature increased to 752° Fahr. (400° Cent.), at which point it had not burst into flame. Test No. 2: Smouldered at 680° Fahr. (360° Cent.); temperature increased to 752° Fahr. (400° Cent.), at which point it had not burst into flame.
American pitch-pine	From a baulk in use for 5 years in haulage engine-house, Wigan Four Feet mine.	Smouldered at 599° Fahr. (315° Cent.).
Norway fir ...	Old prop, out of goaf in Wigan Four Feet mine.	Test No. 1: Smouldered at 520° Fahr. (260° Cent.) (this test was carried out the previous day with a specially dried piece of wood). Test No. 2: smoked at 536° Fahr. (280° Cent.), and smouldered at 680° Fahr. (360° Cent.). Test No. 3: smoked at 520° Fahr. (260° Cent.), and smouldered at 671° Fahr. (355° Cent.). Average for tests Nos. 2 and 3: smoked at 518° Fahr. (270° Cent.), and smouldered at 675° Fahr. (357° Cent.).
Old timber ...	From seat of fire in Pemberton Four Feet mine; rotten, but fairly solid. Do. do.	Test No. 1: smouldered at 446° Fahr. (230° Cent.) (this test was carried out in a copper air-bath, the sample being placed 2 inches from the bottom). Test No. 2: smouldered at 446° Fahr. (230° Cent.) (this test was carried out in a copper air-bath, the sample being placed 2 inches from the bottom).
	Do., and charred.	Test No. 3: smouldered at 266° Fahr. (130° Cent.) (this test was carried out in a copper air-bath, the sample being placed on bottom of bath).
	Do., and softer than in the first test.	Test No. 4: smouldered at 320° Fahr. (160° Cent.) (this test was carried out in a fire-clay muffle).
	Decayed (not out of a mine).	Test No. 5: smouldered at 392° Fahr. (200° Cent.) (this test was carried out in a fire-clay muffle).

Sample Tested.	Description of Sample.	Results of Testing in a
Arley coal ...	From a colliery near Wigan.	Smoked at 392°] coal caked and density to 752° when the inflamr
Cannel	From Wigan Cannel-mine.	Smoked at 356°] great amount of s fired about 752°]
Pemberton Five Feet coal	From Pemberton Five Feet mine.	Smoked at 284°] started smoulde (360° Cent.).
Black bass ...	From Wigan, about 12 feet below the Five Feet mine. The fire was chiefly in this metal.	Smoked at 338°] Smoke increase(ignition-point o (400° Cent.).
Six Inch coal ...	From 18 inches above the Pemberton Five Feet mine. This coal is thrown into the goaf.	Smoked at 320° smoke increased ignition-point o (400° Cent.).

The general results of these tests show that t
in an atmosphere at a lower temperature than co
that partially-decayed timber, such as old pit-pr
readily of all.

It is possible that fires underground would n
if care were taken to clear out all old or used t
seams where heating takes place. The actio
timber seems to make it inflammable at quite l

———

Mr. T. H. Byrom (Wigan Coal and Iron Cc
said that it had been a pleasure to him to make
ulated in the table, in regard to the ignition-poi
specified. As the members would probably unde
a very easy matter to determine the exact ignit
substances, because so much depended upon th
prevailed. It was almost impossible to reprodu
the conditions which existed in mining operati
peratures given might be accepted, as he believ
within a few degrees of the exact ignition-poi
time they were not, of course, strictly com
results that might be proved in occurrences on
was only natural to suppose that the conditi
ignition would take place would vary conside

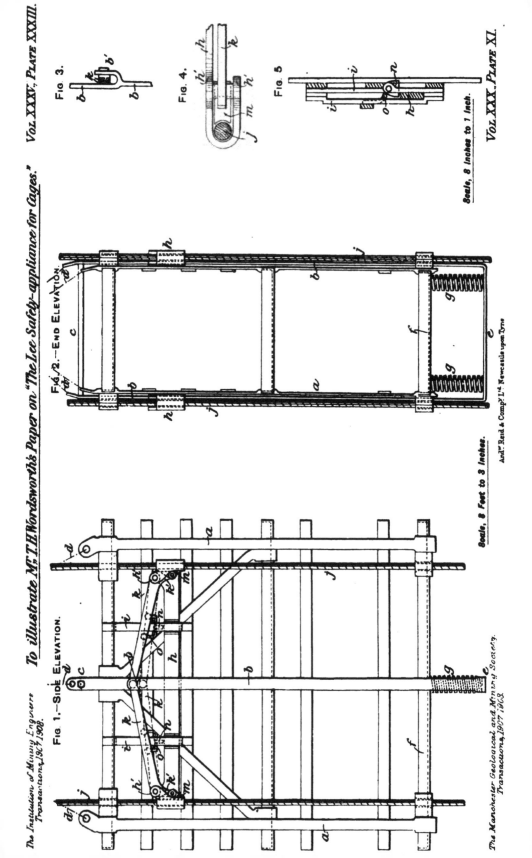

The Institution of Mining Engineers.
Transactions, 1907-1908.

To illustrate Mr. T. H. Wordsworth's Paper on "The Lee Safety-appliance for Cages."

VOL. XXXV., PLATE XXXIII.

FIG. 3.

FIG. 4.

FIG. 5.

VOL. XXX., PLATE XI.

Scale, 8 inches to 1 inch.

FIG. 2.—END ELEVATION.

FIG. 1.—SIDE ELEVATION.

Scale, 8 Feet to 3 Inches.

Andw Reid & Compy Ltd. Newcastle upon Tyne

The Manchester Geological and Mining Society.
Transactions, 1907-1908.

FUSING OF ELECTRIC CABLES.

Mr. JOHN GERRARD (H.M. Inspector of Mines, Worsley) exhibited a portion of a thick electric cable which had been fused at one end. It was part of an insulated and armoured cable, made on the three-phase system. It had been in a fire, and the question was how it came to be fused so that the copper, iron, and lead had been melted; because the electricians had told him (Mr. Gerrard) that the moment a cable was broken the protecting fuse blew out and the current was stopped. When this cable was in order, it was subjected to a voltage of about 440, the pressure on the surface some distance away being 465. The cable was in the return airway, which for about 360 feet had been on fire, including timber and coal. He had hoped that some of the electrical members would be at the meeting, and able to throw some light on the question of how the end of the broken cable had come to be fused in the manner shown; for, so far as be was concerned, it was very difficult to understand how a fire could melt iron, copper, and lead, and fuse the metals together, as was done in this case. No other part of the cable was so fused. It was between two brick walls, each $2\frac{1}{2}$ feet thick, and across these walls were iron girders with a covering of wood; behind the walls were coal-seams. Electricians said that if anything occurred to a cable and it got short-circuited, a fuse would instantly blow in the power-house, giving warning that there was something wrong, and the fuse blowing out would stop the current from passing along the cable. If that were so, would the iron, copper, and lead be fused as they were in this case? It was difficult for him to conceive how coal from the seam behind the brick walls could operate upon a cable armoured as this one was, and for no other part of the cable to show such a result. To him it had the appearance of electric fusion. His opinion was that the cable was crushed by the edge of one of the iron girders falling upon it, and that then there had been an electric flash, which would produce considerable flame. The whole of the return airway was lined with the finest possible coal-dust, thickly coated, and that might have played a part in initiating the fire.

Mr. VINCENT BRAMALL (Pendlebury) suggested that if the melting were due to external heat, the lead would have gone first.

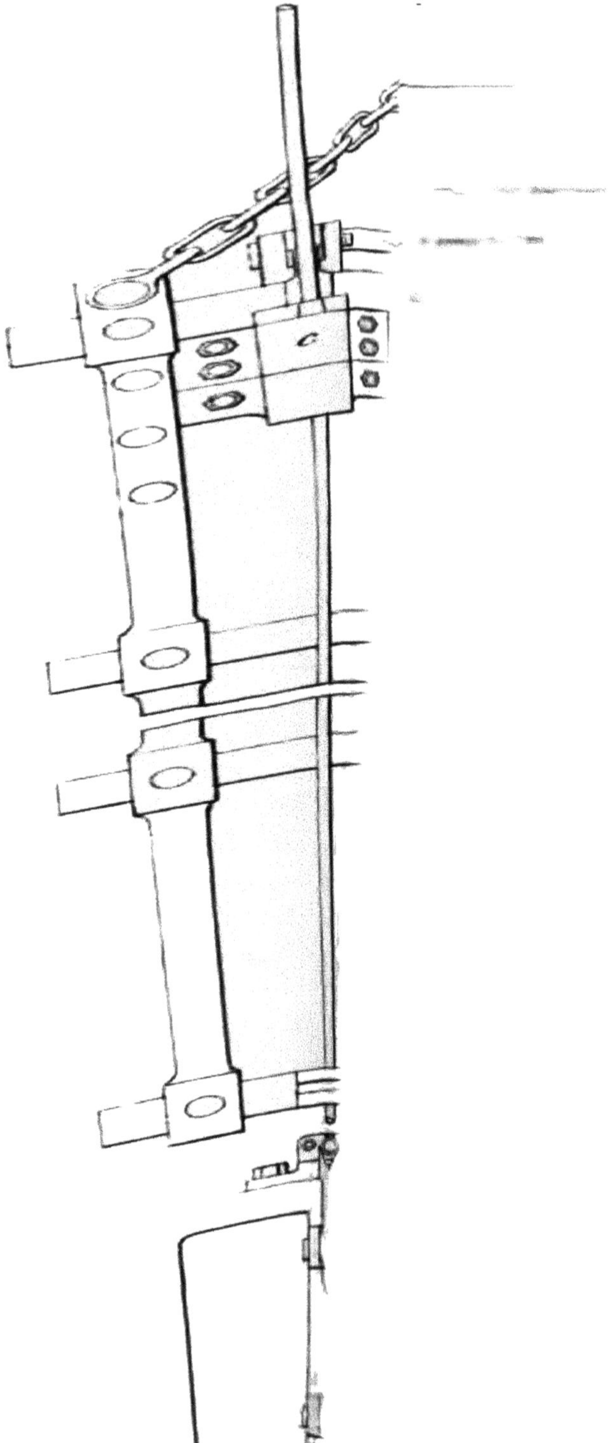

CHINA-CLAY: ITS NATURE AND ORIGIN.

By GEORGE HICKLING, B.Sc., Lecturer in Geology at the
University of Manchester.

Introduction.—The following paper contains the chief results
of an investigation into this matter, made at the request of Prof.
W. Boyd Dawkins, in view of the action brought by the Great
Western Railway Company against the Carpalla China Clay
Company. The work was carried out in the Geological Labor-
atories of the University of Manchester.

The material which is worked in the china-clay pits of
Cornwall is of great commercial value and deep scientific
interest. Hence it is a matter for surprise and regret
that no adequate description of it has ever been published, the
more so as a great deal of speculation has been indulged in
regarding both its nature and origin. The general nature of the
deposits, their relation to the granite, and the method by which
they are worked, were clearly described, three-quarters of a
century ago, by Sir Henry De la Beche.[*] He refers to the
deposit simply as "decomposed granite," clearly meaning granite
decomposed by ordinary weathering. But the idea was early
introduced that this modification of granite was due to the
ascent of hot acid vapours from below through the fissures in the
rock,[†] and during recent years the theory has been revived,
being now very popular under the title of "pneumatolytic
action." The author trusts that the facts here brought forward
will effectively dispose of the idea that such a process has had
any connection with the formation of china-clay-rock.

The china-clay, or "kaolin" of commerce, is prepared by
directing on to the "rock" in the quarry a head of water, which
sweeps the material away to a series of pools or catch-pits. In

---

[*] *Report on the Geology of Cornwall, Devon, and West Somerset*, by H. T. De
La Beche, 1839, pages 449 to 453 and 509 to 513.

[†] *Mineralogisches Taschenbuch*, by von Buch, 1824 ; and "Sur le Gisement,
la Constitution, et l'Origine des Amas de Minerai d'Etain," by Daubrée, *Annales
des Mines*, 1841, series 3, vol. xx., pages 65 to 112.

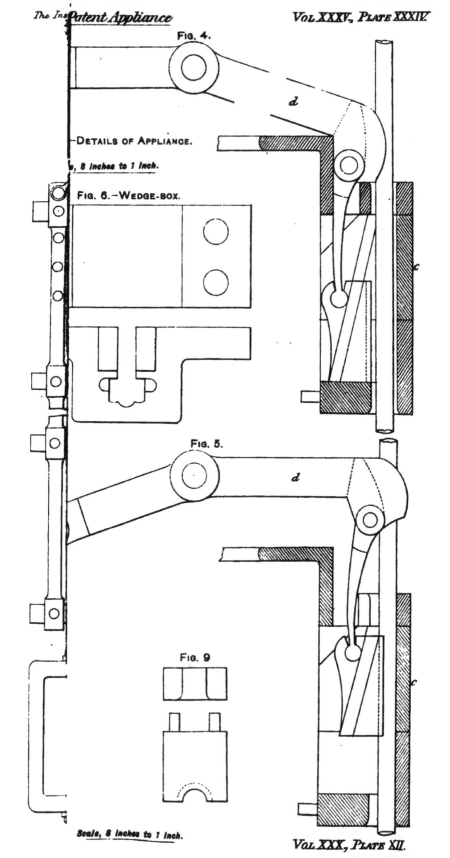

FIG. 4.

DETAILS OF APPLIANCE.

*, 8 inches to 1 inch.

FIG. 6.—WEDGE-BOX.

d

c

FIG. 5.

d

c

FIG. 9

Scale, 8 inches to 1 inch.

up the great bulk (90 per cent., or more) of the material, while tourmaline forms an insignificant proportion. The amount of quartz is difficult to estimate, as it is very readily missed; while the relative abundance of kaolinite and mica is difficult to judge for reasons which will be obvious when they have been described. These proportions will be discussed later. The size of the various fragments varies from about 0·0079 inch (0·02 millimetre) down to 0·00079 inch (0·002 millimetre) in diameter. Consequently, magnifying powers of 500 to 2,000 diameters must be used for effective work.

The quartz shows its usual characters, generally exhibiting glassy fracture with sharp edges, evidently the result of recent mechanical disintegration of larger grains.

The tourmaline is usually in fragments of needles or prisms, such as make up the well-known " tourmaline-suns " of some Cornish granites. More rarely, it is in irregular grains; and, like the quartz, it shows evidence of recent fracture. Though not abundant, the fragments are easily recognized, on account of their colour and characteristic pleochroism.

The mica appears in all cases to be some form of muscovite. Two classes, however, are readily recognized. The comparatively large irregular flakes are, no doubt, derived from the primary micas of the granite. They are quite distinct from the smaller, more regularly shaped plates or short prismatic aggregates (seldom perfectly hexagonal) which correspond exactly with the secondary muscovite formed in the weathered felspars of the partly decomposed granites (Figs. 6 and 7, Plate I.). Both types commonly show " crenulated " edges which strongly suggest chemical corrosion (Fig. 5, Plate I.).

The kaolinite will be treated by itself. No other minerals, so far as the author has observed, occur as normal constituents of the clay. An odd crystal of something else may be seen now and then, but it must be looked upon as quite accidental. The author would also emphasize the fact that there is no trace of any amorphous material. It is all crystalline.

The Kaolinite of China-clay.—This constituent occurs typically in the form of irregular hexagonal prisms, with rough faces, which show strong transverse striations corresponding to the basal cleavage (Fig. 1, Plate I.). These prisms are invariably curved, sometimes quite vermiculiform. The shorter prisms commonly present a fan-like arrangement (Fig. 2, Plate I.) and exactly resemble the similar forms of mica, from which they can only be distinguished by their lower interference-tint. This point will be referred to later. Isolated plates or very short prisms usually lie, of course, on the basal faces; and this is the condition of the great majority of the fragments, both of kaolinite and of muscovite. Even with convergent polarized light, it is not easy to judge the amount of birefringence under such circumstances, and, consequently, to decide to which mineral a given fragment belongs; hence the difficulty of estimating their relative proportions. Both kinds show the same irregular form (due, probably, to development within the decaying felspars) and the same evidence of corrosion on the edges (Figs. 3 and 4, Plate I.).

The low interference-tint and low index of refraction definitely distinguish these crystals from mica. The identification with kaolinite rests on the following evidence:—

(a) The index of birefringence is distinctly low, about that of quartz, but variable.

(b) The index of refraction (by immersion in oils) is very near to 1·560. This is a little higher than that given by Dick for the Anglesey kaolinite (1·54). Through the kindness of Dr. F. H. Hatch, however, the author has been able to examine a small sample of Dick's material, and finds that it also appears to have an index of 1·56.

(c) The prismatic crystals extinguish parallel to the basal plane.

(d) Basal flakes show a biaxial interference-figure.

(e) The author has not been able to determine accurately the specific gravity. By using only the coarsest particles of "mica-clay" he was able to effect a partial separation of the kaolinite and mica in a mixture of bromoform and chloroform. In a diffusion-column of these liquids, almost pure mica was obtained at the bottom, while kaolinite with a little quartz and entangled mica collected around a quartz-crystal used as an indicator, some a little higher, some slightly below. It is perhaps noteworthy that scattered particles floated at various lower levels, and that the main mass of the kaolinite would not come to rest at a uniform level. Thus, while the fineness of the particles seriously interferes with the precision of this determination, it is clear that the specific gravity is very near to that of quartz (2·66), which is certainly as near to the figure given by Dick for kaolinite (2·62) as the conditions of the determination will allow one to approach.

(f) The crystals are completely soluble in superheated strong sulphuric acid.

The above facts appear to the author to leave no doubt as to the identity of the mineral; they are all in accord with the known characters of kaolinite, and the close agreement in specific gravity and complete agreement in refractive index seem specially significant. It would have been interesting to get a chemical analysis to confirm the identification; but, after several trials, the author gave up the attempt to isolate the mineral in quantity. Only the coarsest particles can be used for separation by the usual specific-gravity method, and it is a very tedious matter to obtain such material even in small quantity; even if this were obtained, the separation of the mica would be still more tedious, while the quartz could not be got rid of. But, in the absence of a separate analysis, fairly definite evidence may be obtained from the bulk-analysis of the clay; and it will be useful to discuss this evidence somewhat fully for various reasons.

Evidence of Chemical Composition of the China-clay on its Mineral Constitution.—It has been seen that there are four minerals making up the clay: three of these are well-known and readily identifiable; the fourth has been identified as kaolinite. Knowing the composition of all these minerals, we can test the accuracy of the determinations by seeing whether a mixture of them can be obtained which will correspond with the bulk-analysis of the clay. This may be done more readily, as one constituent (tourmaline) may be left out of consideration as being far too rare (certainly much below 1 per cent.) to affect the bulk-analysis appreciably. Of the other three, quartz is not liable to any variation in composition, while muscovite is fairly constant. The kaolinite should be pure hydrated silicate of alumina. These facts enable us to test the analysis with much greater precision and delicacy than would otherwise be the case.

TABLE I.—ANALYSIS OF CHINA-CLAY NO.

Chemical Symbol.	Actual Analysis. Per cent.	Theoretical Constitution.		
		Muscovite. 32·2. Per cent.	Kaolinite. 55·3. Per cent.	Quartz. 12·5. Per cent.
SiO₂	51·55	14·06	25·70	12·5
Al₂O₃	34·17	12·31	21·86	—
Other bases (K₂O)...	3·87	3·87	—	
H₂O	10·52	1·44	7·74	
	100·11			

TABLE II.—ANALYSIS OF CHINA-CLAY NO.

Chemical Symbol.	Actual Analysis. Per cent.	Theoretical Constitution.		
		Muscovite. 28·2. Per cent.	Kaolinite. 66·0. Per cent.	Quartz. 5·8. Per cent.
SiO₂	49·12	12·74	30·59	5·8
Al₂O₃	36·94	10·85	26·09	—
Other bases (K₂O) ..	3·38	3·38	—	
H₂O	10·50	1·26	9·24	
	99·94			

TABLE III.—ANALYSIS OF CHINA-CLAY NO.

Chemical Symbol.	Actual Analysis. Per cent.	Theoretical Constitution.		
		Muscovite. 34·3. Per cent.	Kaolinite. 58·2. Per cent.	Quartz. 7·5. Per cent.
SiO₂	50·29	15·45	27·08	7·5
Al₂O₃ ...	36·17	13·16	23·01	—
Other bases (K₂O) ..	4·12	4·12	—	
H₂O	10·33	1·54	8·15	
	100·91			

TABLE IVA.—ANALYSIS OF KAOLINIZED GRANITE, GEO
BREAGE, CORNWALL.§

Chemical Symbol.	Actual Analysis. Per cent.	Theoretical Constitution.		
		Muscovite. 37·1. Per cent.	Kaolinite. 13·0. Per cent.	Quartz. 50·02. Per cent.
SiO₂	71·15	16·76	6·04	50·02
Al₂O₃	19·41	14·28	5·13	—
Other bases (K₂O)...	4·45	4·45	—	
H₂O	5·09	1·66	1·82	
	100·10			

* "Analyses of Samples of China-clay (Kaolinite), C
Macadam, *Mineralogical Magazine*, 1885.1887, vol. vii., page
　† *Ibid.*, page 76.　　　　‡ *Ibid.*
　§ "The Geology of the Land's End District," by Messrs. C
anal. Pollard, *Memoir of the Geological Survey of England a
tion of Sheets Nos. 351 and 358*, 1907, page 59.

TABLE IVb.—SAME ANALYSIS AS IN TABLE 4A, WITH TOURMALINE CALCULATED FROM B_2O_3.

Chemical Symbol.	Actual Analysis. Per cent.	Theoretical Constitution.				Theoretical Total. Per cent.	Total with Collins Kaolinite. Per cent.
		Muscovite. 32·7 Per cent.	Tourmaline. 3·3. Per cent.	Kaolinite. 14·6. Per cent.	Quartz. 49·52. Per cent.		
SiO_2... ...	71·15	14·78	1·20	6·79	49·52	72·29	72·82
Al_2O_3	19·41	12·59	1·15	5·77	—	19·41	19·41
Other bases ...	4·47	3·92	0·55	—	—	4·47	4·47
H_2O	5·09	1·47	0·12	2·04	—	3·63	3·10
	100·12					99·80	99·80

TABLE V.—ANALYSIS OF KAOLIN FROM PIEUX (MANCHE), FRANCE.[*]

Chemical Symbol.	Actual Analysis. Per cent.	Theoretical Constitution.			Theoretical Total. Per cent.	Total with Collins' Kaolinite. Per cent.
		Muscovite. 7·92. Per cent.	Kaolinite. 90·56. Per cent.	Quartz. 1·45. Per cent.		
SiO_2	46·71	3·58	42·12	1·45	47·15	50·45
Al_2O_3	38·74	2·96	35·78	—	38·74	33·74
Other bases (K_2O)...	0·95	0·95	—	—	0·95	0·96
H_2O	13·50	0·35	12·68	—	13·03	9·73
	99·90				99·87	99·87

TABLE VI.—ANALYSIS OF KAOLIN FROM MANCHE, FRANCE.[†]

Chemical Symbol.	Actual Analysis. Per cent.	Theoretical Constitution.			Theoretica Total. Per cent.	Total with Collins' Kaolinite. Per cent.
		Muscovite. 4·08 Per cent.	Kaolinite. 87·09 Per cent.	Quartz. 0 Per cent.		
SiO_2	46·28	1·84	45·14	—	46·98	49·66
Al_2O_3	39·92	1·57	38·35	—	39·92	39·92
Other bases (K_2O) ..	0·49	0·49	—	—	0·49	0·49
H_2O	13·73	0·18	13 59	—	13·77	10·23
	100·42				101·16	100·30

We are justified in assuming that all the bases other than alumina are present in the muscovite. In the following analyses, therefore, these bases have been reduced to their equivalents of K_2O, and from this the percentage of muscovite present has been estimated; the quantities of silica, alumina, and water due to it have then been calculated. Obviously, the remainder of the alumina is due to the kaolinite, and from it the percentage of that mineral may be estimated, and the quantity of silica and water due to it calculated.

The percentages of muscovite and kaolinite having been found, the remaining material may be taken as quartz. Hence, adding

* "Sur la Constitution des Argiles," by Th. Schlœsing, Comptes-rendus de l'Académie des Sciences, 1874, vol lxxix., page 474.

† Ibid., page 475.

its percentage of silica to that estimated for th
kaolinite, we should get the total silica in the (
and the closeness with which we approach th;
a very delicate test of the validity of all our
the percentages of muscovite or kaolinite are w
will be correspondingly affected, and every perc
the quartz is increased or diminished will affec
by over $\frac{1}{2}$ per cent.

Kaolinite has been assumed in these calculat
composition $Al_2O_3 . 2SiO_2 . 2H_2O$, or $Al_2O_3 = ($
$H_2O = 14$. Collins's suggestion[*] that it has a diffe
is dealt with below, the last column in the
referring to this point.

The first three analyses were selected from
ber given by Macadam, because their water-pe
most closely with those of the clays examined
who determined the water in the latter by ignit
analysis, that of the decomposed granite, was 1
tourmaline taken into account (estimated from
of B_2O_3).[†]

It will be seen that the theoretical totals of
in the foregoing tables are in close agreement wi
in the actual analyses. This makes very signifi(
the small differences which do exist are consta
Nearly without exception the silica is a little too
a little too low, the average silica-excess being 0·
water-deficiency 0·61 per cent. (Analysis No. ·
this consideration.). It is clear that there mu
explanation for these differences. There is no
why the silica should be too high. Two causes n
for the water being too low, namely :--

(1) The ratio of muscovite to kaolinite may b(
substitution of kaolinite for about 6·4 per cent. (
(that is, 6·4 per cent. of the whole clay; about 20

---

[*] "On the Nature and Origin of Clays : the Compositi(
J. H. Collins, *Mineralogical Magazine*, 1885-1887, vol. vii., pa

[†] Since writing the above, the author has learned the
concluded that china-clay consists of kaolinite, muscovite,
purely chemical examination, and has calculated their prop
manner. (See papers by Vogt and Lavezard, in the *Men*
Société pour l'Encouragement de l'Industrie Nationale, Paris, 1

muscovite itself) would raise the average total water to the
required figure. But, at the same time, this would actually raise
the silica slightly (0·08 per cent.), making its error so much
greater. Furthermore, about one-fifth of the K_2O would now be
left unaccounted for.

(2) The muscovite may be hydrated. If this be the case, it
will account both for the deficiency of water and for the excess of
silica. The average water-deficiency is 2·62 per cent. of the muscovite present. If the percentage-composition of the muscovite is re-
calculated with that amount of extra water, and the amount of
muscovite in the clay then re-estimated in accordance with the
new K_2O ratio, its percentage will be increased by 0·65. Nevertheless, of course, the amount of silica and alumina contained
in it will remain as before, and the kaolinite, therefore, will
also be unchanged. The difference of 0·65 per cent. must come
off the quartz present, and the total silica will therefore be
diminished by that amount.

. Making this correction, then, on the average of the six
analyses, we shall have the alumina, potash, and water of the
theoretical totals in perfect agreement with the actual analyses,
while the theoretical silica will be 0·19 per cent. too low. This
close correspondence appears to the author to form the strongest
testimony to the correctness of the determined mineralogical constitution of the clays.

In support of the view that hydration of the muscovite is
responsible for the small discrepancies originally found, it may
be noted (1) that they are especially marked (both excess of silica
and deficiency of water) in analysis No 4, in which the percentage of muscovite is comparatively high, and that of kaolinite
very low; (2) that the high percentage of silica and low percentage of water vary together; (3) that the muscovite in the decomposed granite has a pearly lustre, which, as remarked by Mr.
Collins, is not improbably a sign of hydration; and (4) that
Johnstone has shown that muscovite immersed, even in pure
water, does become hydrated.* It may be noted that the water
added (2·62 per cent.) is very nearly equal to the addition of one
molecule of H_2O (2·25 per cent.), but it may be doubted whether
this is more than a mere coincidence.

* "On the Action of Pure Water and of Water Saturated with Carbonic
Acid Gas on Minerals of the Mica Family," by A. Johnstone, *Quarterly Journal
of the Geological Society of London*, 1889, vol. xlv., pages 363 to 368.

The above evidence appears to the author amply to justify the conclusion that the mineral referred to as kaolinite really has the composition Al_2O_3, $2SiO_2$, $2H_2O$, and that it forms about 50 to 70 per cent. of ordinary china-clays. Should anyone remain sceptical, he believes that the argument may be strengthened to demonstration by reference to the china-clay of Aue, which was described by Ehrenberg[*] in 1836 as consisting almost entirely ("Fast die ganze Substanz") of crystals which, from his description and figures, are obviously the same as those considered by the author as kaolinite, while the clay itself has the composition of true kaolinite, as shown by the following analyses,[†] that by Forchammer being in almost perfect agreement with theory:—

COMPOSITION OF KAOLIN OF AUE, NEAR SCHNEEBERG.

		Forchammer. Per cent.	Wolff. Per cent.
SiO_2	... ...	46·53	48·49
Al_2O_3	... ...	39·47	37·88
H_2O	... ...	13·97	13·58
$CaCO_3$	... ...	0·31	0·18

Reference has been made to Mr. Collins' attempt to show that kaolinite, at least of china-clay, has not the composition generally attributed to it.[‡] He gives a number of analyses (Table II.) of Cornish china-clays which agree roughly with the theoretical percentages, namely, silica, 48; alumina, 41; and water, 10·8; answering to a formula $Al_2 H_6O_6 SiO_2 Al_2O_3$, $3SiO_2$. The average water-percentage in these analyses is rather over 2 per cent. lower than that in nearly all the analyses of his Table I.; and the author cannot avoid a suspicion that in specially drying these clays before the analyses of Table II. some constituent water was in some way removed, thus accounting for the disagreement between these and practically all other analyses of china-clays. However that may be, it is obvious that the whole bulk-analysis of the clay will not give the composition of kaolinite or anything else, seeing that several minerals certainly enter into its composition in an important degree. As some

[*] *Ueber Microscopische neue Charaktere der erdigen und derben Mineralien*, by C. G. Ehrenberg, *Poggendorf's Annalen der Physik und Chemie*, 1836, vol. xxxix., page 104, fig. 1.

[†] Rammelsberg, *Handbuch der Mineralchemie*, 1875, second edition, Leipzig.

[‡] "On the Nature and Origin of Clays: the Composition of Kaolinite," by J. H. Collins, *Mineralogical Magazine*, 1885-1887, vol. vii., page 209.

weight has been attached to this paper of Mr. Collins', the author
thought it worth while to test how far his formula for kaolinite
would work if substituted for the generally accepted one in the
partial analyses worked out above, and to give the results by the
side. It will be seen that the differences are about five times
greater than with the usual formula, so that we may confidently
say that the kaolinite present has not that composition.

Source of the Kaolinite.—It is, of course, a familiar statement that kaolinite is one of the decomposition-products of felspar, the turbidity of which as seen in most sections is commonly
attributed to the development of this mineral. The author's
own observations lead him to doubt this. A careful search through
the felspars of a number of decomposed granites with high powers
has failed to show him any trace of a decomposition-product
which he could regard as kaolinite. Muscovite is abundant, as
also calcite, epidote, and very minute flakes and specks of iron-ore
(? micaceous ilmenite). It is evident that kaolinite might be
difficult to detect, owing to the approximation of its refractive
index and birefringence to that of the felspar: but the author
does not think that this can account for his failure to find it. The
only other mineral in the parent-granite of the china-clay which
could conceivably give rise to kaolinite is the muscovite. But it
is equally clear that the primary muscovite-crystals show no trace
of any such decomposition-product: indeed, they are usually
quite fresh. In a word, no sign of kaolinite can be found in
the still coherent granite: the crystals are only to be seen in the
fine powder which gives rise to the clay: hence it seems evident
that the development of kaolinite takes place in the clay itself.
Yet it is as true of the fragments of muscovite and felspar (if any
remain) in the clay as of those in the granite itself, that they
show no trace of kaolinite as a decomposition-product. Two
possibilities therefore remain: the kaolinite may represent
material which passes into solution and recrystallizes, or muscovite may be directly converted into kaolinite. The author has
little doubt now that the latter alternative is the correct one.
The thoroughly micaceous habit of the crystals has already been
noted, as likewise the fact that there is no visible difference in
form between the kaolinites and the smaller micas. The curious
forms alone are almost sufficient proof that the crystals have not

been deposited freely from solution. They stand in marked contrast with the beautifully regular crystals from Anglesey, which are probably much more characteristic of the proper habit of the mineral. It has been seen, too, that both the micas and kaolinites show signs of chemical corrosion, the action being more marked in kaolinite. This fact would be very surprising if the mineral were crystallizing out in the clay, but is obviously natural if the mica is undergoing a partial solution by which it is converted into kaolinite.

The complete similarity of the kaolinite and the mica seems to the author the strongest argument in favour of the former being merely a derivative of the latter. There is, however, one partial distinction which is significant: the *long* vermicular prisms are almost invariably kaolinite; the shorter ones include a much larger proportion of muscovite. It seems to the author probable, as a matter of inference, that the longer prisms would be more likely to form in the later stages of the decay of the felspars, when there would naturally be more room for development, than in the earlier. This is confirmed by the rarity of any such long prisms actually in the decayed felspars, so far as the author has been able to observe. This *may* indicate that in the final stages of decay there is a deficiency of alkali, and kaolinite may consequently be formed directly, new material probably being added to already existing crystals.

Lastly, the author may adduce, with some hesitation, a final observation supporting the theory that mica is converted directly into kaolinite. The prisms may be seen showing every variation of birefringence from that of muscovite down to *nil*. The writer fully realizes the difficulty of being sure of the optical orientation and of the thickness of such minute fragments, but he feels confident that, when these disturbing factors are eliminated, there still remains a wide range of birefringence which is no doubt correlated with a variable chemical constitution. Prisms may be seen which appear to be muscovite at one end and kaolinite at the other.

A paper by A. Johnstone,[*] already referred to, has a direct bearing on this point. He exposed muscovite to the action of

* "On the Action of Pure Water and of Water Saturated with Carbonic Acid Gas on Minerals of the Mica Family," by A. Johnstone, *Quarterly Journal of the Geological Society*, 1889, vol. xlv., pages 363 to 368.

pure water, and also of water saturated with carbonic acid, for twelve months. At the end of this period, he found that both specimens had taken water into combination, being converted into hydro-muscovites. Biotite was acted upon in the same way; it was also superficially bleached, and lost a certain amount of iron and magnesium by solution in the carbonic-acid water. These changes certainly appear to be in the direction of a conversion into kaolinite. It has been seen above, also, that the muscovite in the china-clay appears to be hydrated.

Schlœsing, in his paper "Sur la Constitution des Argiles Kaolins,"* from which the above analyses, Nos. 5 and 6 were taken, found, after allowing water in which kaolin had been stirred to stand for a month, that there still remained in suspension [? or partly in solution] some "argile colloïdale" which was notably rich in potash and magnesia, containing as much as 4·25 per cent. of the former, and 1·29 of the latter. If, as would seem likely, some of this material was in solution, the presence of the large quantity of potash would seem to indicate again the removal of this material from the muscovite. That there is material in solution the author determined by putting a little raw china-clay into distilled water, carefully filtering, concentrating the solution, and evaporating a spot of concentrated material on a glass slip. A beautiful finely-crystalline deposit was obtained, of minute cubes, rhombic plates, and feathery stars. With a flame-test, the solution showed clearly the presence of potassium.

The question will naturally arise whether these changes can be observed in the larger primary micas of the granite. Their pearly lustre to the naked eye has already been mentioned. Under the microscope they usually shew more or less corroded edges. No further change is generally apparent in the specimens in the china-clay. With the micas in the Coal-measure clays, examined by the author, the case is different. In a grey underclay, for example, from Glascote, he found the bulk of the material to consist of minute, more or less hexagonal plates, or aggregates of these. These "aggregates" he took, at first, to be merely accidental, until he found examples which themselves had definitely [illegible] hexagonal forms. Further observa- [illegible] with no doubt that the "aggregates" were simply

* [illegible] 1874, vol. [illegible], page [illegible]

the result of the complete decomposition of larger mica-crystals, every stage of the process being observable. Incidentally, he may add that many of the less decomposed micas were distinctly tinged brown, and contained numerous secondary rutile-needles —evidently altered biotite-crystals. Rutile was abundantly strewn free through the clay. Of course, one wants to know the nature of the small hexagonal plates into which these micas are resolved. So far as the writer has been able to observe, they show no characters incompatible with their being kaolinite, although one hesitates to identify them as such. The double refraction appears to be weak.

Schmid,* however, describes what is evidently the same phenomenon. Describing the principal constituents of the kaolin of Eisenberg he says of, the larger flakes ("glimmerartige" Blätter):—

"Their outline commonly is completely rounded, often at the same time indented, very seldom partly straight, and rarely regularly six-sided. They are clear to dusty and turbid. With this turbidity, the indentations show further towards the middle, and allow the commencement of a falling into smaller plates."

The other constituents of the clay are quartz-grains and the "microvermiculite" which we have identified as kaolinite, a slight admixture of needles of rutile, and some other mineral (? tourmaline) completing the list. In view of the facts that these "micaceous" plates seem to be the most abundant constitutent of the material, that there seems little doubt that they were originally mica, that the whole material is now a good "kaolin," and that the analysis shows that nearly 74 per cent. of it is soluble in superheated sulphuric acid and has a composition very closely agreeing with kaolinite, while the remaining 26 per cent. appears to be nearly pure quartz, it appears to the author that this decomposed micaceous material must approximate closely to kaolinite in composition, and that the conversion of mica into kaolinite is here seen taking place.

It would appear highly probable, therefore, that while the minute secondary muscovites from decomposed felspars are converted directly into kaolinite-pseudomorphs, a similar conversion may take place (of course, much more slowly) in the larger primary micas, and is accompanied in the latter instance by the splitting-up of the crystal.

* "Die Kaoline des Thüringischen Buntsandsteins," by E. E. Schmid, *Zeitschrift der Deutschen Geologischen Gesellschaft*, 1876, vol. xxviii., page 92.

Agencies concerned in the Formation of Kaolinite.—In view of what has already been said, there does not require much to be added under this head. It would be unnecessary to say anything at all, were it not for the recent revival of the idea of kaolinization by pneumatolysis. The following observations should be sufficient to show that such a hypothesis is both unnecessary and untenable.

Muscovite, which represents the first stage of the transformation, is a perfectly normal decomposition-product, formed whenever potash-felspar is exposed to the influence of atmospheric moisture. Seeing, therefore, that ordinary weathering is able to accomplish the first stage of the change, there seems no reason why it should not complete it.

The china-clay rock is essentially a surface-sheet covering the irregularly corroded surface of the solid granite. For the accuracy of this statement the accompanying section (Fig. 10, Plate I.) is sufficient evidence. It might be contended that the difference between the superficial material and the solid rock below was due merely to mechanical disintegration. But the author has shown that it is only in the soft powdery material that typical kaolinite is to be found—it is *not* present in any of the solid material which contains unaltered felspar. It is self-evident that this surface-layer of rubbly material is due to atmospheric weathering, and it therefore follows that the conversion of muscovite into kaolinite, as the formation of the muscovite from felspar, is a part of the same process. The downward limit of the decomposition is naturally extremely irregular, as is well seen in the irregular surfaces left where the soft material has been swept away (Fig. 9, Plate I.). Occasionally, sinking along fissures, the water has caused decomposition under extensive masses of solid granite, as in other rocks it may give rise to caves.

Again, there is evidence to show that the muscovite in clay is transformed *in situ* into kaolinite; and no one, it is imagined, will suggest pneumatolytic action in that case.[*]

The following figures may be added, giving the estimated loss per cent. of each of the given constituents of a biotite-gneiss

[*] See footnote, page 28, on the similar decomposition of felspar in millstone grit.

from North Garden, Albemarle County, Virginia, U.S.A., as the result of complete atmospheric weathering[*] :—

	Lost per cent.			Lost per cent.
SiO₂	52·45	K₂O		83·52
Al₂O₃	0·00	Na₂O		95·03
Fe₂O₃	14·35	H₂O increased from 0·62		
CaO	100·00	to 13·75 per cent. of		
MgO	74·70	whole mass.		

The correspondence of this with the changes which must occur in the kaolinization of granite is obvious.

The arguments in favour of kaolinization by pneumatolytic action[†] are the depth to which the kaolinized material extends ; its improved quality from the deeper parts; the occurrence of schorl, topaz, gilbertite, fluor, etc., in association with it; and Mr. Collins' statement that felspar artificially treated with hydrofluoric acid is reduced to a material chemically and microscopically resembling exactly china-clay.

The first argument has already been disposed of. The second points, in the author's opinion, to the extraction of free oxygen from the water in the more superficial parts, so that the biotites in the rock below have not been oxydized and decomposed, and the clay is therefore a more pure white.   The third does not appear to prove anything beyond the undoubted fact that schorl-veins were formed in the granite before its kaolinization; there is no obvious connection between the production of the veins and the subsequent kaolinization, as shown by the fact that the veins are equally characteristic of the undecomposed granite. The last piece of evidence has been examined by the author, by repeating the experiment. Clear, glassy, rather pink orthoclase was immersed for 52 hours in hydrofluoric acid at the natural temperature. The fragment soon lost its colour and fell to pieces, breaking along cleavage-planes. A coarse white powder remained after removing the acid, washing, and drying. Under the microscope the fragments, mounted in oil, were still sharply angular and glassy, but *absolutely isotropic.* They had not the remotest resemblance to china-clay.   The same experiment was tried by

* *Rocks, Rock-weathering, and Soils,* by Dr. G. P. Merrill, 1897, pages 213 to 215.

† "On the Nature and Origin of Clays : the Composition of Kaolinite," by J. H. Collins, *Mineralogical Magazine,* 1885-1887, vol. vii., pages 211 and 212 ; and "Geology of the Land's End District," by Messrs. Pollard, C. Reid, and J. S. Flett, *Memoir of the Geological Survey of England and Wales : Explanation of Sheets Nos. 351 and 358,* 1907, pages 58 to 60.

putting the acid on grey Irish granite, and brushing off the white powder left on the surface on drying. The felspar was affected in precisely the same way. The arguments advanced, therefore, scarcely appear to the author to justify the statement that " This method of decomposition [kaolinization] has undoubtedly originated from heated corrosive vapours ascending from below through certain broken or fissured parts of the granite."*

Previously Recorded Occurrences of Crystalline Kaolinite in Clays.—The statement that kaolinite is a common (or essential) constituent of clays is very general.   Nevertheless, Ries has justly remarked that, when the evidence is sifted, the foundation for the statement is very slight. The difficulty of identification of small particles of the mineral readily explains this state of affairs, but does not justify the replacement of evidence by mere assumption. Hence it may be well to record a number of cases in which it appears to have been definitely observed, if not always recognized.

(1) *Porcelain-earth of Aue.*—The observation by Ehrenberg† that this deposit is almost entirely composed of kaolinite-crystals has already been noted.

(2) " *Bunzlau Clay.*"—In the same paper,‡ he figures exactly similar crystals in this deposit.

(3) *Kaolin of Eisenberg, Thuringia.*—From this china-clay, E. E. Schmid§ figures perfectly typical kaolinite-crystals, under the merely descriptive appellation of "microvermiculite," and describes them as "richly strewn" among the other materials. His figures and descriptions leave no doubt as to the identity of the mineral. He further records the mineral in similar deposits from Osterfeld, Weissenfels, and Gleina.

(4) *Ball-clay of Edgar, Florida.*—Ries‖ gives a rough sketch of a preparation of this clay, which shows it to consist largely

---

* "The Geology of the Land's End District," by Messrs. C. Reid and J. S. Flett, *Memoirs of the Geological Survey of England and Wales: Explanation of Sheets Nos. 351 and 358*, 1907, page 58.

† *Ueber Microscopische neue Charaktere der erdigen und derben Mineralien*, by C. G. Ehrenberg, *Poggendorf's Annalen der Physik und Chemie*, 1836, vol. xxxix., page 104, fig. 1.

‡ *Ibid.*

§ "Die Kaoline des Thüringischen Buntsandsteins," by E. E. Schmid, *Zeitschrift der Deutschen Geologischen Gesellschaft*, 1876, vol. xxviii., page 92.

‖ *Clays: their Occurrence, Properties, and Uses*, 1906, page 119.

of the typical crystals, while in another place* he gives an analysis which shows its composition to be remarkably close to pure kaolinite.

(5) Johnson and Blake,† in their well-known paper on kaolinite and pholerite, record many substances having the composition of kaolinite as made up of hexagonal plates and "prismatic aggregates" of such plates. In many cases there seems to be little doubt that these "aggregates" are identical with the kaolinite-crystals of our own china-clay, for example, those of Richmond (Virginia) and Summit Hill (Pennsylvania), etc. Unfortunately, the geological nature of the materials examined is usually not clearly stated, and in some cases, at least, they appear to be mere deposits in fissures or similar positions. Nevertheless, the following statement is important:—

"We have examined microscopically twenty specimens of kaolin, pipe-, and fire-clay . . . . In them all is found a greater or less proportion of transparent plates, and in most of them these plates are abundant, evidently constituting the bulk of the substance."‡

(6) According to Ries,§ similar crystals have been observed by Cook in the clays of New Jersey, U.S.A.

This list might be augmented almost indefinitely, if cases were added where "kaolinite" or "kaolin" has "been identified" in clays A single example may show what these "identifications" are worth.   Regarding the sedimentary kaolin of Hersey (Wisconsin), Buckley states that:

"Under the microsc pe the crude kaolin was observed to consist mainly of kaolinite-grains, although numerous small grains of quartz were observed. The largest grains of kaolin were not over 0·0025 mm. in diameter." ‖

In other words, the largest grains of kaolinite were just so small as to be beyond the range of identification. The same tendency to describe as "kaolin," "kaolinite," "clay-substance" or in some such way any material too fine for proper examina-

---

* *Clays: their Occurrence, Properties, and Uses,* 1906, page 169, table 12.

† "On Kaolinite and Pholerite," by Messrs. S. W. Johnson and J. M. Blake, *American Journal of Science,* 1867, second series, vol. xliii., pages 351 to 361.

‡ *Ibid.,* page 356.

§ *Clays: their Occurrence, Properties, and Uses,* 1906, page 98.

‖ "The Clays and Clay-industries of Wisconsin," by E. R. Buckley, *Wisconsin Geological and Natural History Survey: Bulletin No. vii., Part 1, 1901,* page 237.

tion, has prevailed again and again, and is almost entirely responsible for the reputed occurrences. The author has himself seen kaolinite only in a single stratified clay, referred to below.*

Relation of China-clay to the Stratified Clays.—Our present knowledge of the nature and history of common clays is still unsatisfactory. The investigations which have been made leave some important points unsettled, while the accounts of the subject in most text-books contain statements which are actually wrong or misleading. In particular, the idea is very widespread that a more or less definite hydrous silicate of alumina, either in the form of crystalline kaolinite, or in some amorphous condition, constitutes the real pure clay, in which other substances that may be present are to be regarded as impurities. Another firmly-rooted idea is that clays are essentially the residue of the "felspathic" element of the igneous rocks.

The most successful observations hitherto made are those of Mr. Hutchings[+] on fire-clays, the results of which are almost completely in agreement with the few examinations the author has been able to make. As Mr. Hutchings remarks, microscopic examination at once demonstrates the uselessness of trying to evolve formulæ out of clay-analyses. It would seem to be a safe statement that no sedimentary clay exists which is so homogeneous that its bulk-analysis can afford, by itself, any trustworthy guide to its mineral constitution. Now, such analyses appear to form practically the whole evidence on which the existence of a "clay-base" of kaolinite, halloysite, or similar mineral rests. This evidence may therefore be put on one side as practically valueless. Beyond this, nothing seems forthcoming further than the statement that "plates" or "scales," which are sometimes taken for kaolinite, occur in clays. Mr. Hutchings has described such fine flakes as forming the "paste" of clays, but has referred the mineral to some form of mica.[‡] The author confirms the

* Since the above was written, the author has examined the completely-weathered felspathic millstone grit, which is quarried at Mow Cop, North Staffordshire, as a source of kaolin. It contains kaolinite in the usual curved prisms, and also in beautifully-crystallised hexagonal plates, exactly similar to the material described by Dick, but much finer. It would seem that this latter form of the mineral usually occurs where it has probably been freely exposed to percolating water.

+ "Notes on the Probable Origin of some Slates," by W. M. Hutchings, *Geological Magazine*, 1890, vol. vii., pages 264 to 273 and 316 to 322.

‡ *Ibid.*, page 269.

general occurrence of such a "paste," and can also state that in some cases it certainly is mainly micaceous. In other cases he is more doubtful of its nature.

The only case in which he has observed genuine kaolinite in stratified clay is that of the fire-clay of Heathfield, Bovey Tracey. The well-known Bovey Tracey clays may have been formed by the washing of the adjacent kaolinized granites, which hypothesis seems to be supported by their microscopical characters. They differ only from artificial china-clay in the less complete separation of the coarser particles, the absence of more than an occasional curved prism of kaolinite, and the presence of countless minute grains and rods of rutile. Mica is present, but it is impossible to say how much of the fine flaky material which forms the basis of the clays is that mineral, and how much is kaolinite.

The rarity of *prisms* of kaolinite is not surprising, in view of the readiness with which those prisms must break up. The rutile is a difficulty. It appears to be an almost universal constituent of clays,* and in normal cases it seems probable, as suggested by Mr. Hutchings, that it arises as a decomposition-product of biotite; at least, what appear to have been biotite-crystals may be found reduced to aggregates of minute colourless plates, with minute rutiles scattered between them. But if the Bovey Tracey clays are to be compared with china-clay, some other source must be found, as biotite is never present; tourmaline would seem to be the only possible source, if the rutile of these clays has been derived from the material indicated, but it is far more probable that they have been partly derived from the coarser particles of the china-clay rock, among which biotite is abundant.

However that may be, there are stronger reasons for supposing that most clays have not been derived from material exactly comparable to china-clay-rock. It would seem very significant that recognizable felspar is commonly to be found in them†; while it is entirely absent in china-clay-rock. The obvious conclusion is that the ordinary clays are the residue of rock which has undergone less complete chemical decomposition.

* "On the Occurrence of Rutile-needles in Clays," by Dr. J. J. H. Teall Mineralogical Magazine, 1885-1887, vol. vii., pages 201 to 204.

† "Notes on the Probable Origin of some Slates," by W. M. Hutchings, Geological Magazine, 1890, vol. vii., page 267.

The absence of typical prisms of kaolinite confirms this ; were they at all common, they must have been observed. The typically coloured biotite tends to the same conclusion. Hence the author concludes that ordinary clays are the result of more rapid weathering, more mechanical and less chemical in character, than that which has produced the china-clay-rock : the latter substance being produced only when the weathering is greatly prolonged, in consequence of the non-removal of the decomposed material.

The common clays appear to be formed, moreover, very largely of the micas of the parent-rocks. The more minute flakes may include secondary micas from the decomposition of felspar, either in the parent-rock or in the clay, but the larger fragments certainly are mostly original. The great bulk of many clays would appear to consist mainly of micaceous material. Yet chemical analyses show far too little alkali to allow of the main mass being muscovite, which species the absence of colour in the flakes usually suggests. The explanation, the author believes, is afforded by what has already been suggested regarding the origin of kaolinite in china-clay, and especially of the Thuringian clays described by Schmid. It seems at least probable that a similar alteration of the micas may take place in all clays, alkali being removed and water taken up, even though the process may never convert the mica entirely into kaolinite.

SUMMARY.

(1) China-clay is the product of the artificial extraction by washing of the finest particles from decomposed granite. It consists of crystals of kaolinite, muscovite, quartz, and tourmaline, the first two constituents making 90 per cent. or more of good clays.

(2) The kaolinite occurs in the form of irregular curved hexagonal prisms, showing strong transverse cleavages, or in isolated plates. The index of refraction and that of double refraction agree with those of the Anglesey kaolinite. The crystals are biaxial, and extinguish straight. The specific gravity agrees closely with that given by Dick. They are soluble in superheated sulphuric acid.

(3) It is shown that the Cornish china-clays agree very closely indeed in their chemical composition with a mixture of kaolinite,

muscovite, and a little quartz (the tourmaline is negligible). As the presence of the muscovite and quartz is undoubted, it must be admitted that the composition of the remaining mineral approaches very closely indeed to Al_2O_3, $2SiO_2$, $2H_2O$ (kaolinite). It appears that the muscovite is hydrated (hydromuscovite).

(4) Confirmatory evidence is adduced, showing that the almost pure kaolinite of Aue consists of similar crystals.

(5) It is shown that the composition of china-clay cannot be explained on the assumption that the kaolinite has the composition suggested by Mr. Collins.

(6) The turbidity of felspar seen in section, and commonly attributed to "kaolinization," is not due to the formation of kaolinite.   Muscovite is abundantly developed, and various other minerals help to cause the turbidity, but no trace of kaolinite has been found in any fragment of felspar, however decomposed.

It is very probable that the secondary muscovite is directly converted into kaolinite. The two minerals as they occur in the china-clay are closely similar in form, and there is some evidence of intermediate stages. It has been shown by Mr. Johnstone that water with carbonic acid hydrates muscovite or biotite, and also bleaches the latter, extracting iron and magnesium.   Pure water dissolves a minute part of raw china-clay, the solution containing potash.

(7) It is shown that the micas in common clays decompose, and that the product of decomposition is probably kaolinite.

(8) Atmospheric weathering is the cause of the "kaolinization" of the granite.   Secondary muscovite represents the first stage of the process, and is a normal product of the weathering of felspars.   The kaolinized granite is shown by borings to be in the form of an irregular surface-sheet. Muscovite appears to be changed into kaolinite in sedimentary clays.   "Kaolinization" has taken place subsequently to "tourmalinization," and is in no way comparable with that process. Felspar is decomposed by hydrofluoric acid, but the product presents no resemblance to china-clay.

(9) Kaolinite of the form here described has been recorded previously from several "kaolins," but no definite record is known from common clays.

(10) The common clays are always very heterogeneous. The fine "paste" generally consists of minute flakes of mica or kaolinite (these cannot be distinguished in such fragments). The decomposition of mica is probably responsible for the abundant rutile.

(11) China-clay-rock is the result of the complete weathering of igneous masses *in sitû*. Most common sediments are probably the result of less complete weathering, in which the material is removed so soon as decomposition has proceeded far enough to set it loose.

APPENDIX I.—BIBLIOGRAPHY.

BUCKLEY, E. R., "The Clays and Clay-industries of Wisconsin," *Wisconsin Geological and Natural History Survey: Bulletin* No. vii., part 1, 1901.

COLLINS, J. H., "On the Nature and Origin of Clays: the Composition of Kaolinite," *Mineralogical Magazine*, 1885-1887, vol. vii., pages 205 to 214.

DE LA BECHE, H. T., *Report on the Geology of Cornwall, Devon, and West Somerset*, 1839, pages 449 to 453 and 509 to 513, fig. 2 and Plate XII.

VON BUCH, L., *Mineralogisches Taschenbuch*, 1824.

DAUBRÉE, A., "Sur le Gisement, la Constitution, et l'Origine des Amas de Minerai d'Etain," *Annales des Mines*, 1841, series 3, vol. xx., pages 65 to 112.

DICK, A., "On Kaolinite," *Mineralogical Magazine*, 1888-1889, vol. viii., pages 15 to 27.

EHRENBERG, C. G., "Ueber Microscopische neue Charaktere der erdigen und derben Mineralien," *Poggendorf's Annalen der Physik und Chemie*, 1836, vol. xxxix., pages 101 to 106.

HUTCHINGS, W. M., "Notes on the Probable Origin of Some Slates," *Geological Magazine*, 1890, vol. vii., pages 264 to 273 and 316 to 322.

JOHNSON, S. W., and BLAKE, J. M., "On Kaolinite and Pholerite," *American Journal of Science*, 1867, second series, vol. xliii., pages 351 to 361.

JOHNSTONE, A., *Quarterly Journal of the Geological Society*, 1889, vol. xlv., pages 363 to 368.

MACADAM, W. J., "Analyses of Samples of China-clay (Kaolinite), Cornwall," *Mineralogical Magazine*, 1885-1887, vol. vii., page 76.

MERRILL, G. P., *Rocks, Rock-weathering and Soils*, New York, 1897.

RAMMELSBERG, C. F., *Handbuch der Mineralchemie*, 1875, second edition, Leipzig.

REID, C., and FLETT, J. S., "The Geology of the Land's End District," *Memoirs of the Geological Survey of England and Wales, Explanation of Sheets* Nos. 351 and 358, 1907.

RIES, H., *Clays, their Occurrence, Properties, and Uses*, London, 1906.

SCHMID, E. E., "Die Kaoline des thüringischen Buntsandsteins," *Zeitschrift der Deutschen Geologischen Gesellschaft*, 1876, vol. xxviii., pages 87 to 110.

SCHLÆSING, TH., "Sur la Constitution des Argiles," *Comptes-rendus de l'Académie des Sciences*, 1874, vol. lxxix., pages 376 to 380, and 473 to 477.

TEALL, J. J. H., "On the Occurrence of Rutile-needles in Clays," *Mineralogical Magazine*, 1885-1887, vol. vii., pages 201 to 204.

Vogt and Lavezard, *Mémoires publiés par la Société d'Encouragement pour l'Industrie Nationale*, Paris, 1906; Lavezard, E., *Contribution à l'Etude des Argiles de France*, pages 113 to 192; Vogt, G., *De la Composition des Argiles*, pages 193 to 218.

APPENDIX II.—Explanation of Plate I

All the micro-figures are *camera-lucida* sketches, except No. 8, which is drawn from a photograph.

Fig. 1.—Vermiform crystal of kaolinite from "mica"-clay of Carpalla works, St. Austell, Cornwall. Magnified 220 diameters.

Fig. 2.—Crystal of kaolinite in china-clay, Carpalla. Magnified 920 diameters.

Fig. 3.—Basal view of short crystal of kaolinite, showing irregular form, from "mica"-clay, Carpalla. Magnified 220 diameters.

Fig. 4.—Basal view of short crystal of kaolinite, showing irregular form and marked corrosion, from "mica"-clay, Carpalla. Magnified 220 diameters.

Fig. 5.—Basal view of crystal of muscovite (probably primary) showing marked corrosion. Magnified 220 diameters.

Fig. 6.—Two fan-shaped crystals of secondary muscovite in a decomposed felspar-crystal from the decomposed granite. Tre Rice works, St. Austell, Cornwall. Magnified 920 diameters.

Fig. 7.—Fan-shaped crystal of secondary muscovite in china-clay, Carpalla. *Cf.* fig. 2. Magnified 920 diameters.

Fig. 8.—Preparation of fine "mica"-clay, Carpalla. Magnified 200 diameters. Drawn from a photograph. In the photograph, the fragments of quartz cannot be distinguished from the flakes of kaolinite and muscovite, which form the bulk of the clay. Tourmaline is here proportionately more abundant than in the finer china-clay.

Fig. 9.—Granite tors at Roche Rocks, Cornwall, illustrating the irregular surface of solid granite left when the decomposed material (= "china-clay rock ") has been naturally washed away (after De la Beche).

Fig. 10.—Section along trench for Sheepstor Embankment, Burrator Reservoir, Plymouth Waterworks, Dartmoor (after Sandeman). Here the solid granite is still covered by the surface-sheet of decomposed material ("china-clay rock "), to a maximum depth of 100 feet. Note (1) irregular surface of the solid granite ; (2) undecomposed cores of granite in the china-clay rock ; and (3) tourmaline-veins passing through from solid granite to china-clay rock.

The Chairman (Prof. W. Boyd Dawkins) said that Mr. Hickling's paper was full of new matter and of great interest. It arose out of an enquiry which had been carried on for over six months with regard to the physical history and structure of china-clay, and clays in general. Mr. Hickling had been carrying on a series of investigations, largely under his (Prof. Boyd Dawkins's) direction, in the Geological Laboratory of the Manchester University, with the best and latest microscopical instruments. The paper, he hoped, was only the first of a series of original researches, the results of which would be brought by Mr. Hickling before the Manchester Geological and Mining Society.

Mr. John Gerrard (H.M. Inspector of Mines, Worsley), in moving a vote of thanks to Mr. Hickling, said that the paper brought to his mind the immense advantage which the Society had in its close association with the University of Manchester. They had again and again received advantages which the members present so fully realized that evening. It was much to be desired that the connexion between the two bodies would be closer, and would continue for many years to come. It was a little departure from the subject, but he was sure that it would be of interest to members to know that it had been announced that one of the students of the mining department of the University had qualified and would receive a degree. He was sure that they all congratulated the Mining Department of the University on that success.

Mr. Alexander Reid (Chester) seconded the resolution, which was passed.

The Chairman (Prof. W. Boyd Dawkins) said that he would like, before the meeting ended, to call attention to the fact that one reason why this elaborate enquiry had been possible was the possession of a new and superior microscope. They had found that the ordinary English microscope was not satisfactory, and, thanks to Mr. Hickling, the instrument obtained united the excellences of two of the best German microscopes. He hoped that the method of this enquiry, begun in the geological laboratory of the Manchester University, would be extended to other departments of geological research.

The discussion on the paper was postponed.

———

On the motion of Mr. George B. Harrison (H.M. Inspector of Mines, Swinton), seconded by Mr. William Pickstone (Manchester), the meeting tendered its heartiest thanks to the Vice-Chancellor (Dr. Alfred Hopkinson, K.C.) and Council of the University of Manchester, for permitting the meeting to be held in the University buildings.

A vote of thanks was also passed to Prof.   :
his services in the chair.

The CHAIRMAN, in acknowledging the vo   ı
that it was a pleasure always to receive ı
University. The University did not forget ı   ı
Society in the establishment of their Mu   :
organization of the teaching of mining.

———

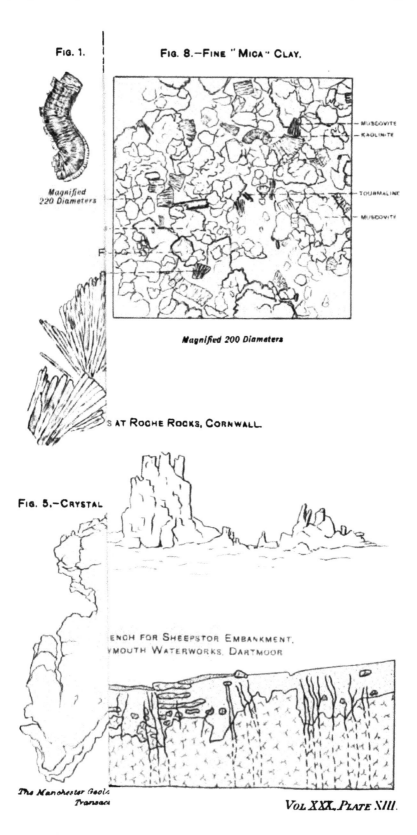

FIG. 1.

*Magnified
220 Diameters*

FIG. 8.—FINE "MICA" CLAY.

MUSCOVITE
KAOLINITE
TOURMALINE
MUSCOVITE

Magnified 200 Diameters

S AT ROCHE ROCKS, CORNWALL.

FIG. 5.—CRYSTAL

ENCH FOR SHEEPSTOR EMBANKMENT,
YMOUTH WATERWORKS, DARTMOOR

INDEX TO VOLUME X

INDEX.

THE

. MANCHESTE

GEOLOGICAL & MININ(

(Federated with the Institution of Mini

— — --- - —

LIST OF COUNCIL & (

SESSION 1908-1&

. —— —

LIST OF MEMI

YEAR 1907-1908.

PUBLISHED AT THE ROOMS OF THE
QUEEN'S CHAMBERS, 5, JOHN DALTON STREF

PRINTED BY J. ROBERTS & SONS, SAI

PAST PRESIDENTS OF THE SOCIETY.

Presidents Deceased are signified by *Italics*.

Year of Election.	
1838-39-40	*Egerton, The Rt. Hon. Francis, M.P.*
1841-2-3, 49-50-1	*Heywood, James, F.R.S., F.G.S.*
1843-4-5, 55-6-7	*Egerton, Sir Philip de Malpas Grey, Bart., M.P.*
1845-46-47	*Moseley, Sir Oswald, Bart.*
1847-48-49	*Thickness, Ralph, M P., Wigan.*
1851-52-53	*Black, James, M.D., F.G.S.*
1853-54-55	*Ormerod, G. W., M.A., F.G.S.*
1857-8-9, 65-6-7	*Binney, E. W., F.R.S., F.G.S.*
1859-60-61	*Kay-Shuttleworth, Sir, J.P., Bart., M.P.*
1861-3, 77-8, 87-8	Dickinson, Joseph, F.G.S.
1863-64-65	*Knowles, Andrew.*
1867-8-9, 84-5	*Greenwell, G. C., F.G.S., M.Inst.C.E.*
1869-70-1, 82-3	*Aitken, John, F.G.S.*
1871-2-3, 88-9	*Knowles, John, M.Inst.C.E., J.P.*
1873-74	*Knowles, Thomas, M.P.*
1874-5, 6-7, 86-7	Dawkins, Prof. W. Boyd, M.A., D.Sc., F.R.S.
1875-76	Smith, R. Clifford, F.G.S.
1878-79	*Forbes, John Edward, F.G.S.*
1879-80	Earl of Crawford and Balcarres, P.C., F.R.S.
1880-81	Shuttleworth, Lord, P.C.
1881-82	*Gilroy, George, M.Inst.C.E.*
1883-84	Pilkington, Edward, J.P.
1885-86	*Ormerod, Henry Mere, F.G.S.*
1889-90, 1902-3	Hall, Henry, I.S.O., H.M.I.M.
1890-91	Burrows, John S., F.G.S.
1891-92	*Tonge, James, F.G S., Assoc.M.Inst.C.E.*
1892-93	*Peace, Maskell William, F.G.S.*
1893-94	Saint, William, H.M.I.M.
1894-95	Watts, William, F.G.S.
1895-96	Winstanley, Robert.
1896-97	*Stirrup, Mark, F.G.S.*
1897-98	*Ridyard, John, F.G.S.*
1898-99	*Livesey, Clegg, J.P.*
1899-1900	Barrett, W. S., J.P.
1900-01	Greenwell, G. C., F.G.S., M.Inst.C.E.
1901-02	Barnes, J., F.G.S.
1903-04	Hollingworth, Col. G. H., F.G.S.
1904-05	Gerrard, John, H.M.I.M.
1905-06	Bramall, Henry, M.Inst.C.E.
1906-07	Pilkington, Charles, J.P.
1907-08	Ashworth, John, C.E.

LIST OF MEMBERS,

YEAR 1907-1908.

Year of Election.	Names and Addresses.
	HONORARY MEMBERS.
1874	Agassiz, Alexander, Cambridge, Massachussetts, U.S.A.
1891	Bell, Thomas, J.P., 40, Esplanade Road, Scarborough.
1904	Bolton, Herbert, F.R.S.E., F.G.S., F.L.S., The Museum, Bristol.
1877	Broeck, Ernest Van den, Place de l'Industrie 39, Brussels.
1874	Geikie, Sir Archibald, D.Sc., LL.D., F.R.S., 3, Sloane Court, London, S.W.
1874	Hull, Professor Edward, M.A., F.R.S., 14, Stanley Gardens, Notting Hill, London, W.
1879	Kinahan, G. H., M.R.I.A., Woodlands, Fairview, Co. Dublin.
1873	Martin, Joseph S., I.S.O., H.M. Inspector of Mines, The Vikings, 16, Durdham Park, Clifton, Bristol.
1905	Saint, William, H.M. Inspector of Mines, Cromer House, 58, Cathedral Road, Cardiff. *Past-President.*
1890	Whitaker, William, B.A., F.R.S., F.G.S., 3, Campden Road, Croydon.
	MEMBERS.—(FEDERATED.†)
	M.INST.M.E.
1906	Allott, Henry Newmarch, M.Inst.C.E., 46, Brown Street, Manchester.
1893	Ashworth, John, C.E., 8, King Street, Manchester, *President.*
1878	Ashworth, Thomas, Kenmal Wood, Chiselhurst, Kent.
1895	Atherton, H. Stanley, Heath Cottage, Sharples, Bolton. *Member of Council.*
1877	Atherton, James, 13, Mawdsley Street, Bolton.
1897	Baker, Godfrey, 457, Bolton Road, Pendlebury, near Manchester.
1901	Baldwin, Walter, F.G.S., 5, St. Albans Street, Rochdale.

† For List of Non-Federated Members see page xiv.

Year of Election.	Names and Addresses.
1875	Barrett, W. S., J.P., 64, The Albany, Old Hall Street, Liverpool. *Past-President.*
1904	Bastow, S. E., Messrs. Bruce Peebles & Co., Ltd., East Picton, Edinburgh.
1907	Beales Henry Batson, 64, Cross Street, Manchester.
1903	Bentley, George, Bradford Colliery, Manchester.
1889	Bolton, Edgar O., Burnley Colleries, Burnley.
1880	Bolton, H. H., High Brake, Accrington.
1902	Bouchier, C. F., Strangeways House, Platt Bridge, Wigan. *Member of Council.*
1904	Bradshaw, Hubert, Stoneclough, near Manchester.
1886	Bramall, Henry, M.Inst C.E., Pendlebury Collieries, Pendlebury, Manchester. *Past-President.*
1904	Bramall, Vincent, Pendlebury Colleries, Pendlebury. *Member of Council. Hon. Auditor.*
1902	Brewerton, Joseph, 72, Bridge Street, Manchester.
1906	Brown, Francis Verrill, Guardian Buildings, 3, Cross Street, Manchester.
1904	Buckley, Charles Arthur, Miramar, Upper Arthog Road, Hale, Cheshire.
1907	Burr, Malcolm, B.A., F.G.S., Eastry, S.O., Kent.
1878	Burrows, John S., F.G.S., Green Hall, Atherton, near Manchester. *Past-President.*
1904	Carter, J., Rainford Colleries, Rainford, St. Helens.
1904	Cass, J., 568, Liverpool Road, Platt Bridge, Wigan.
1908	Chambers, Sydney A., 96, Gresham House, London, E.C.
1901	Chapman, C. H., 293, Liverpool Road, Salford.
1906	Christoper, George Alfred, The Mount, Park View, Wigan.
1898	Clark, Robert F., Bickershaw Collieries, Leigh, Lancashire.
1903	Clark, William, Garswood Coal and Iron Co., Park Lane, Wigan.
1894	Clarke, Robert, 42, Deansgate, Manchester.
1901	Coleman, W. H., 18, Egerton Road, Fallowfield, Manchester.
1904	Constantine, E. G., Stirling Boiler Co., Ltd., 25, Victoria Street, Westminster, London, S.W.
1901	Cotterill, H. W. B., Waterworks Engineer's Office, City Hall, Cardiff.
1905	Coulston, P. Barrett, 5, Cross Street, Manchester.

Year of Election.	Names and Addresses.
1878	Cowburn, Henry, 253, Westleigh Lane, Westleigh, Manchester.
1904	Crankshaw, Hugh M., 11, Ironmonger Lane, London.
1883	Crankshaw, Joseph, F.G.S., 11, Ironmonger Lane, London.
1907	Cross, T. Oliver, 77, King Street, Manchester.
1908	Davies, Edward T., Wynnstay Colliery, Ruabon.
1905	Dawes, Alfred, 4, The Square, Blaenau Festiniog.
1869	Dawkins, Professor W. Boyd, M.A., D.Sc.. F.R.S., F.G.S., F.S.A., Assoc.Inst.C.E., Fallowfield House, Fallowfield, Manchester. *Past-President.*
1903	Dickson, James, Westhoughton Collieries, near Bolton.
1903	Dixon, Walter, Birkacre Collieries, Coppull, near Chorley.
1904	Dixon, William R., Coalbrook Vale Colliery, Nantyglo, R.S.O., Mon.
1889	Dobbs, Joseph, Coolbawn House, Castlecomer, Co. Kilkenny.
1898	Douglas, Ernest, Pemberton Collieries, Wigan.
1898	Eagle, George, 37, Brown Street, Manchester.
1905	Eames, Cecil W., Moss Hall Collieries, Platt Bridge, Wigan.
1903	Edmondson, R. H., Garswood Hall Collieries, Wigan.
1884	Elce, George, Rock Mount, Altham, Accrington.
1878	Ellesmere, The Right Hon. The Earl of, Worsley, near Manchester.
1895	Ellis, Thomas Ratcliffe, King Street, Wigan.
1897	Evans, Walter, J.P., Royton, Oldham.
1880	Fairclough, William, F.S.I., Leigh, near Manchester.
1907	Files James, 572, Manchester Road, Swinton, near Manchester.
1891	Finch, John, 51, 52, Exchange Buildings, Birmingham.
1906	Fletcher, Clement, the Hindles, Atherton, near Manchester.
1896	Fletcher, Leonard R., Atherton Collieries, Atherton, near Manchester. *Member of Council.*
1908	Fort, Robert Arthur, Moss Hall Colliery, Platt Bridge, near Wigan.
1873	Garforth, W. E., M Inst.C.E., F.G.S., Snydale Hall, Pontefract. *Vice-President.*

Year of Election.	Names and Addresses.
1893	Garside, Edward, B.Sc., Assoc. M.Inst.C.E., Town Hall Chambers, Ashton-under-Lyne.
1902	Garton, Walter T., F.G.S., Brookfield, Wigan Road, Ashton-in-Makerfield.
1892	Gerrard, John, F.G.S., H.M. Inspector of Mines, Worsley, near Manchester. *Past-President.*
1908	Ghose, A., F.C.S., 42, Shambazar Street, Calcutta.
1898	Glover, J. W., Cyprus Government Railway, Locomotive Department, Famagusta, Cyprus.
1903	Glover, Robert B., c/o Glover Bros., Mossley, near Manchester.
1903	Graham, George, 15, Montague Road, Sale, Cheshire.
1905	Grave, J. U. Roger, c/o Messrs. J. P. Bissett & Co., Foo-Chou Road, Shanghai, China, *via* Siberia.
1882	Greenhalgh, Robert, 167, Bolton Road, Atherton, near Manchester.
1874	Greenwell, G. C., M.Inst.C.E., F G.S., Beechfield, Poynton, Cheshire. *Past-President.*
1879	Greenwood, John, 1, Marsden Street, Manchester.
1877	Grundy, H. Taylor, Blackburn Street, Radcliffe.
1888	Grundy, James, F.G.S., Walthew House Farm, (Crooke Delivery), Wigan.
1908	Hackett, George Booker, P.O. Box 3117, Johannesburg.
1877	Harris, George E., Assam Railway and Trading Co., Limited, Margerita, Assam, India.
1904	Harris, H. P., Brynn Dedwydd. Wrexham.
1894	Harrison, George B., H.M. Inspector of Mines, Worsley Road, Swinton, Manchester. *Vice-President.*
1906	Hart-Davies, Capt. Henry Vaughan, Wardley Hall, Worsley, near Manchester.
1905	Harvey, R. H., Messrs. Dick Kerr, & Co., Praça, Castro Alves, No. 55, Bahia, Brazil.
1889	Higson, Charles H., The Chestnuts, Helsby, via Warrington.
1904	Higson, Peter, 18, Booth Street, Manchester.
1906	Hobbs, William Lowbridge, Dyserth, S.O., Flints.
1904	Hodge, W. Guy, 35, Bryn Road, Swansea.
1904	Hollingworth, F. H., The Oak, Hollinwood, Oldham.
1878	Hollingworth, Col. George H., F.G.S., 37, Cross Street, Manchester. *Past-President. Hon. Treasurer.*

Year of Election.	Names and Addresses.
1903	Hollingworth, Henry, Ellerbeck Colliery, Coppull, Chorley.
1904	Hooghwinkel, Gerald H. J., Dacre House, Victoria Street, London, S.W.
1906	Hopkinson, Austin, A.M.I.Mech.E., A.M.I.E.E., 86, Cross Street, Manchester.
1905	Horrobin, William, Bedford Collieries, Leigh, Lancashire.
1906	Houghton, Henry, Oak Mount, Ormskirk Road, Skelmersdale, Lancashire.
1894	Hughes, Owen, Hardman House, Hollinwood, Oldham.
1903	Humphris, Henry, Blaenau Festiniog, North Wales.
1907	Hunter, Sherwood, 20, Mount Street, Manchester.
1908	Jacob, F. Llewellin, Ferndale Colliery, Ferndale, Glam.
1884	Jobling, John, Cliviger Collieries, Burnley.
1908	Jobling, John William, Clifton Colliery, Burnley.
1908	Johnson, Thomas, c/o Miss B. Johnson, Box 370, Johannesburg, Transvaal.
1903	Johnson, W. H., B.Sc., Woodleigh, Altrincham, Cheshire.
1907	Jones, John T., Foggs House, Little Lever, near Manchester.
1903	Jones, O. R., H.M. Inspector of Mines, 5, Spring Gardens, Chester.
1905	Kay, Joseph, Agecroft Colliery, near Manchester.
1893	Kenrick, John P., Honan, China (communications to Messrs. Grindlay & Co., 54, Parliament Street, London, S.W.)
1905	Kneebone, C. Maitland, Cerro Muriano Mines, Ltd., Estacion de Cerro Muriano, Prova de Cordoba, Spain.
1889	Knowles, John, Brynn Mount, Westwood, Lower Ince, near Wigan.
1887	*Knowles, Lt.-Col. Sir Lees, Bart., M.A., LL.M., F.G.S., F.Z.S., D.L., Westwood, Pendlebury. *Trustee.*
1908	Lamb, Arthur Moore, Eskdale, Birkdale, Southport.
1904	Landless, J. E., Habergham Colliery, Burnley.
1898	Landless, Richard, Colliery Offices, Bank Parade, Burnley.
1890	Law, J. Illingworth, Willow House, Waterfoot, Manchester.
1897	Lees, Frederick, J.P., The Rookery, Ashford, Bakewell.
1899	Leigh, Oswald B., North Lincoln House, Frodingham, Doncaster.
1908	Linton, Francis A., Bickershaw Lane, Bickershaw, nr. Wigan.

Year of Election.	Names and Addresses.
1898	Livesey, John, Rose Hill Colliery, Bolton.
1908	Lomas, J. E. H., 32, Gt. St. Helens, London, E.C.
1905	Lord, Chadwick, Jubilee Colliery, Crompton, Oldham.
1898	Lowe, Henry, Chisnall Hall Colliery, Coppull.
1904	Macalpine, G. L., M.Sc., Altham and Great Harwood Collieries, Accrington.
1902	Machin, Thomas H., 7, Mottram Street, Hoverley, Hyde, Cheshire.
1901	McKay, William, Fern Dell, Stockport Road, Bredbury, Cheshire.
1903	Marshall, Eustace A., 37, Queen's Road, Southport.
1892	Mathews, D. H. F., H.M. Inspector of Mines, Hoole, Chester. *Member of Council.*
1893	Matthews, E. L., Imperial Iron Works, Pendleton, Manchester.
1896	Matthews, Thomas, Imperial Iron Works, Pendleton, Manchester.
1892	Miller, Arthur, Bredbury Colliery, near Stockport.
1906	Millward, Albert Edward, Manchester Road, Accrington.
1895	Mitton, A. Dury, Assoc.M.Inst.C.E., The Old Vicarage, Malvern Wells.
1904	Morrison, Daniel, 41, John Dalton Street, Manchester.
1904	Mountain, M. B., Elton Grange, Bury.
1904	Munroe, Martin, 5, Fairlie Place, Calcutta.
1904	Nicholson, J., Jun., Prudential Buildings, Nelson Square, Bolton.
1907	Oldfield, William, West View, Minsterley, Shropshire.
1900	Ollerenshaw, W., Denton Colliery, Denton, Manchester. *Member of Council.*
1883	Peace, George Henry, M.Inst.C.E., Monton Grange, Eccles. *Member of Council.*
1905	Percy, Frank, Monument Cottage, Wigan Lane, Wigan.
1900	Pickstone, William, 5, Moor Lane, Kersal, Manchester. *Vice-President.*
1906	Pickup, William, Elmwood, Rishton, near Blackburn.

Year of Election.	Names and Addresses.
1879	Pilkington, Charles, J.P., The Headlands, Prestwich, Manchester. *Past-President.*
1873	Pilkington, Edward, J.P., Clifton, Manchester. *Past-President.*
1899	Pilkington, Lawrence, Firwood, Alderley Edge, Cheshire.
1903	Pilkington, Lionel E., Haydock, St. Helens. *Member of Council.*
1877	Place, W. H., Hoddleston Collieries, Darwen.
1904	Pope, P. C., 196, Deansgate, Manchester.
1904	Preece, G. G. L., 30, Great Western Street, Moss Side, Manchester.
1891	Prestwich, Joseph, 72, Eccles Old Road, Pendleton, Manchester.
1907	Raby, Gregory, Lota Alto, Lota, Chile, South America. Trans. : c/o Neal Miller, West Ridge, Prenton Road, West, Birkenhead.
1904	Ramsden, Cecil Sydney, 42, Deansgate, Manchester.
1897	Reid, Alexander, M Inst.C E., Witton Lodge, Hoole Road, Chester.
1897	Richardson, Isaiah, Blainscough Collieries, Coppull, near Chorley.
1900	Ridyard, G. J., Shakerley Collieries, Tyldesley.
1860	(d) Ridyard, John, F.G.S., Hilton Bank, Little Hulton, Bolton. *Past-President.*
1904	Rigby Harold, Greville Lodge, Winsford, Cheshire.
1893	Rigby, John, Greville Lodge, Winsford, Cheshire.
1897	Ritson, W. A., 4, Booth Avenue, Withington, Manchester.
1904	Robinson, Fred. J., The Gables, Newton-le Willows.
1889	Robinson, John, Haydock Collieries, St. Helens.
1908	Roscamp, Joseph Cresswell, H.M. Assistant Inspector of Mines, Prestwich, near Manchester.
1892	Roscoe, George, Peel Hall Collieries, Little Hulton, Bolton.
1905	Ross, A., Moston Colliery, near Manchester.
1903	Rushton, A., Maypole Colliery, Abram, near Wigan.
1905	Saike, Yoshima, Jagawa Colliery, Buzen, Japan.
1908	Saint, Frank G. L., The Marshes, Atherton Road, Hindley Green, near Wigan.

Year of Election.	Names and Addresses.
1887	Saint, William, H.M. Inspector of Mines, Cromer House, 58, Cathedral Road, Cardiff. *Past-President.*
1905	Scholes, T., Oswaldtwistle Collieries, near Accrington.
1889	Scott, William B., Eversley Cottage, Middleton, near Manchester.
1891	Scowcroft, Thomas, J.P., Redtborpe, Bromley Cross, near Bolton.
1904	Settle, William, Prestwich, Manchester.
1905	Smith, John, Bower Colliery, Hollinwood, Oldham.
1888	Smith, Sydney A., Assoc.M.Inst.C.E., 1, Princess Street, Albert Square, Manchester. *Hon. Secretary.*
1903	Spackman, Charles, Rosehaugh, Clitheroe.
1905	Speakman, F., Church Street, Leigh, Lancashire.
1886	Speakman, Harry, Bedford Collieries, Leigh, Lancashire.
1904	Spencer, R. S., Princess Royal Colliery Co., Ltd., Whitecroft, near Lydney, Glos.
1905	Stewart, J. E., c/o Pekin Syndicate, Ltd., Tientsin, North China
1907	Stone, Thomas, M.A., The Park Collieries, Garswood, near Wigan.
1882	Stopford, T. R., Woodley, Radcliffe, near Manchester.
1903	Sutcliffe, W. H., F.G S., Shore Cottage, Littleborough, Rochdale.
1907	Taite, Charles Davis, 196, Deansgate, Manchester.
1905	Takagi, Kiichiro, The Mitsui Mining Co., Tagawa, Buzen, Japan.
1899	Tansley, A. E., Springfield House, Coppull, Chorley.
1907	Taylor, Hugh Frank, Sandycroft Foundry Co. Ltd., Sandycroft, near Chester.
1905	Thompson, Fred J , Osborne Bank Esplanade, Fleetwood.
1892	Thompson, James, Westhoughton Road, Westhoughton, Bolton.
1904	Tong, Fred N., Spring Bank, Astley Bridge, near Bolton.
1891	Tonge, Alfred J., Hulton Collieries, near Bolton. *Member of Council.*
1898	Tonge, James, F.G.S., Westhoughton, near Bolton.
1907	Toplis, William Sherman, Novara, Rowan Avenue, Brooklands, near Manchester.

Year of Election.	Names and Addresses.
1895	Travers, T. W., Spring Bank, Broad Oak Park, Worsley, near Manchester.
1904	Turner, Charles, Irlam, near Manchester.
1876	Unsworth, John, J.P., Scot Lane Collieries, Blackrod, near Chorley.
1907	Wainewright, Wilfrid B., 534, Mason Buildings, Los Angeles, California, U.S.A.
1904	Walker, Howard J., Bank Chambers, Wigan.
1882	Wall, Henry, Tower Buildings, Wallgate, Wigan.
1893	Wallwork, Jesse, Bridgewater Collieries, Walkden, near Manchester. *Member of Council.*
1905	Walshaw, J., Astley and Tyldesley Collieries, Tyldesley, near Manchester.
1897	Walton, Thomas, Bank Hall Collieries, Burnley.
1905	Waterworth, Joseph, Westleigh Collieries, Leigh.
1907	Watson, Percy Houston Swann, 12, Cowper Street, Chapeltown Road, Leeds.
1880	Watts, William, M.Inst.C.E., F.G.S., Kenmore, Wilmslow, Cheshire. *Past-President.*
·1905	Whitworth, C. S., 13, Edmund Street, Rochdale.
1905	Wilkinson, H. Tatlock, Chloride Electrical Storage Co., Ltd., Clifton Junction, near Manchester.
1907	Williams, Thomas, Oakwood, Hexham.
1897	Winstanley, George H., F.G.S., 42, Deansgate, Manchester. *Member of Council. Hon. Auditor.*
1876	Winstanley, Robert, 42, Deansgate, Manchester. *Past-President.*
1897	Wood, John, Barley Brook Foundry, Wigan.
1897	Wood, Percy Lee, Clifton and Kersley Coal Co., Ltd., Clifton, Manchester. *Member of Council.*
1904	Wood, Thomas, Barley Brook Foundry, Wigan.
1895	Wordsworth, T. H., New Moss Colliery, Audenshaw, near Manchester. *Vice-President.*
1906	Young, William, c/o Thomas Fletcher and Sons, Ltd., Stopes Colliery, Radcliffe.

Year of Election.	Names and Addresses.
	ASSOCIATE MEMBERS.—(Federated.)
	Assoc.M.Inst.M.E.
1904	(*d*) Blackwell, G. G., The Albany, Old Hall Street, Liverpool.
1908	Cartwright, Robert, Strangeways Hall Colliery, near Wigan.
1906	Cunliffe, James, 81, Moor Road, Chorley.
1906	Dubois, Marcel, 6, Rue Gounod, Paris (XVII).
1907	Lomax, George Edward, Fern Hill, Huyton, Liverpool.
1905	Mellor, Edward Thomas, F.G.S., The Geological Survey Office, Box 387, Pretoria, Transvaal.
1905	Preston, S. C., Bolton Hey, Roby, Liverpool.
1906	Rogers, James Taylor, Calder Cottage, Littleboro.
1905	Ross, Arthur, 1, Glengall Road, Old Kent Road, London, S.E.
1907	Schember, Friedrich, Wiederhofergasse, 6, Vienna, IX/2.
	ASSOCIATES.—(Federated.)
	Assoc.Inst.M.E.
1898	Dickinson, Archibald, 283, Colne Road, Burnley.
1907	Galliford, John, 479, Edge Lane, Droylsden, Manchester.
1902	Jobes, R. A., 98, Carley Road, Sunderland.
1907	Nuttall, Theodore Hodson, Poste Restante, Johannesburg.
1905	Woodward, W., 83, Wolverhampton Road, Stafford.
1907	Wynne, George R., Bog Mines Ltd., Minsterley, Salop.
	STUDENTS.—(Federated.)
	Stud.Inst.M.E.
1907	Bolton, H. Hargreaves, junr., High Brake, Accrington.
1905	Cross, Charles Oliver, Snowdown Sinking, Nonington, near Dover.
1908	Eccleshall, George B., 34, Bradford Terrace, Worsley Road, Farnworth, near Manchester.
1905	Entwisle, George, 176, Swinton Hall Road, Swinton.
1908	Evans, Thomas Emrys, 5, Kelnerdeyne Terrace, Rochdale.
1904	Hark, J. R., 92, Market Street, Hindley, Wigan.
1907	Hampson, Ralph, Shotton Cottage, Shotton, Flints.

Year of Election.	Names and Addresses.
1904	Ormond, Percy, Myrtle Villa, Standish, near Wigan.
1908	Pilkington, Edward Fielden, B.A., The Headlands, Prestwich.
1906	Spencer, John, 212, Blackburn Road, Accrington.

MEMBERS—(NON-FEDERATED).

1907	Ackroyd, Alfred, Ellerslie, Victoria Crescent, Eccles.
1895	Barnes, J., F.G.S., 301, Great Clowes Street, Higher Broughton. *Past-President.*
1881	Black, W. G., F.R.C.S.Ed., F.G.S., 2, George's Square, Edinburgh.
1900	Brancker, Richard, 11, Old Hall Street, Liverpool.
1908	Clifford, William, Jeannette, Pa., U.S.A.
1894	Cole, Robert H., Endon, Stoke-on-Trent.
1879	Crawford and Balcarres, The Right Hon. The Earl of, Haigh Hall, Wigan. *Past-President.*
1856	Dickinson, Joseph, F.G.S., Hon.M.I.M.E., South Bank, Sandy Lane, Pendleton. *Past-President. Trustee.*
1907	Dobson, Benjamin Palin, South Bank, Heaton, Bolton.
1903	Edmondson, J. H., Garswood Hall Collieries, Wigan.
1903	Gillott, J. W., Lancaster Works, Barnsley.
1881	Hall, Henry, I.S.O., H.M. Inspector of Mines, Rainhill, Lancashire. *Past-President.*
1898	*Hall, Levi J., Morland House, Birch Vale, near Stockport.
1900	Henshaw, A. M., M.Inst.C.E., F.G.S., Talk-o'th'-Hill Collieries, Staffordshire.
1902	Hewitt, J., Eccleston Hall Colliery, Prescot.

Year of Election.	Names and Addresses.
1876	Higson, John, M.Inst.C.E., F.G.S., 18, Booth Street, Manchester.
1900	Hinnell, H. Leonard, M.Inst.C.E., 41, Corporation Street, Manchester.
1897	Hobson, Bernard, M.Sc., F.G.S., Thornton, Didsbury, Manchester.
1897	Howsin, Evelin G., Isles House, Padiham, Lancashire.
1897	Keen, James, Hindley Green, near Wigan.
1901	Knight, Henry, Rose Bridge and Ince Hall Collieries, Wigan.
1884	Leech, A. H., King Street, Wigan.
1887	Lord, James, Hill House, Rochdale.
1881	Macalpine, G. W., Broad Oak, Accrington.
1897	Noar, L. Lamb, c/o Mrs. Lomax, Stoneleigh, North Promenade, St. Annes-on-the-Sea.
1903	Owen, Richard, Pearson & Knowles' Collieries, Wigan.
1897	*Pickup, P. W. D., Rishton Collieries, Rishton, near Blackburn.
1887	Platt, Samuel Sydney, M.Inst.C.E., Morredge, Sudden, Rochdale.
1903	Ramsbottom, James, Church Road, New Mills, Stockport.
1899	Selby, John B., Atherton Hall, Leigh, Lancashire.
1882	Settle, Joel, The Hill, Alsagar, Stoke-on-Trent.
1877	Shuttleworth, The Right Hon. Lord, of Gawthorpe, Burnley (Lord Lieutenant of the County of Lancaster). *Past-President.*
1884	*Simpson, W. W. Winkley, near Whalley, Blackburn.

Year of Election.	Names and Addresses.
1864	*Smethurst, William, F.G.S.
1881	Smith, John, Bickershaw Collieries, Westleigh, near Manchester.
1873	*Smith, R. Clifford, F.G.S., Ashford Hall, Bakewell. *Past-President*.
1891	Sutcliffe, Richard, Horbury, near Wakefield.
1889	Taylor, William, 51, Park Road, Darwen.
1897	Tickle, James, Bamfurlong Collieries, Bamfurlong, Wigan.
1893	Timmins, Arthur, Assoc.M.Inst.C.E., F.G.S., Argyll Lodge, Higher Runcorn.
1886	*Trafford, Sir Humphrey Francis de, Bart., F.G.S. Charles Street, Berkeley Square, London, W.
1887	Walkden, Richard, 26, Watery Lane Terrace, Springvale, Darwen.
1882	Walker, T. A., Pagefield Iron Works, Wigan.
1888	*Walmesley, Oswald, 2, Stone Buildings, Lincoln's Inn, London, W.C.
1892	Ward, Alexander Haustonne, Raneegunge, Bengal, India.
1897	Wells, Lionel B., M.Inst.C.E., 75, Haworth's Buildings 5, Cross Street, Manchester.
1880	Williams, Sir Edward Leader, M.Inst.C.E., Ship Canal Company, Spring Gardens, Manchester.

Members are requested to communicate to the Hon. Secretary all changes of address, also any omissions or corrections required in the list.

Lightning Source UK Ltd.
Milton Keynes UK
UKHW012235110219
337137UK00006B/1107/P